THE
GEOLOGIC STORY
OF THE
NATIONAL PARKS
AND MONUMENTS

Mt. Saint Helens erupting on May 18, 1980. (Photo by Keith and Dorothy Stoffel.)

THE
GEOLOGIC STORY
OF THE
NATIONAL PARKS
AND MONUMENTS

Fourth Edition

David V. Harris
Professor Emeritus of Geology
Colorado State University

Eugene P. Kiver
Professor of Geology
Eastern Washington University

Graphics by Gregory C. Nelson

John Wiley and Sons
New York   Chichester   Brisbane   Toronto   Singapore

COVER PHOTO:

Towering above Wrangell – St. Elias National Park,
18,008 feet above the Gulf of Alaska, Mt. St. Elias is
one of the world's most inspiring sights. Russian Explorer Vitus Bering named the mountain in 1741,
and because of its shape assumed it to be a volcanic
cone. It was so regarded for a century and a half,
until geologist I. C. Russell climbed its slopes and
dispelled the myth. Mt. St. Elias is composed of rocks
that were formed miles beneath the surface —
rocks that were later thrust upward to more than
three miles above sea level. Streams and then glaciers sculpted the pyramid that we see when looking northwest across the eastern edge of Malaspina
Glacier, the largest piedmont glacier on the continent. Wrangell – St. Elias National Park and Preserve,
set aside as a parkland in 1978, is now the largest
unit — about 12.4 million acres — in our National
Park System.

Photo by Charles G. Mull

All photographs not otherwise credited were taken
by David V. Harris

All color plates C1 – 32 can be found following text
page 272.

Library of Congress Cataloging in Publication Data:

Harris, David V.
  The geologic story of the national parks and monu-
ments.
  Bibliography: p.
  Includes index.
  1. Geology — United States.   2. National parks and
preserves — United States.   3. National monuments
— United States.   I. Kiver, Eugene P.   II. Title.
QE77.H368 1985   557.3   85-5286
ISBN 0-471-87224-5

10 9 8 7

Printed and bound by the Arcata Graphics Company.

PREFACE

It was in 1872 that our first — the world's first — national park was created, in the wilderness of Wyoming. Yellowstone was the first great geological museum to be set aside "as a pleasuring ground for all of the people." Later, in 1916, the National Park Service was established to preserve and protect an ever-increasing number of national parklands. The National Park System was vastly expanded in 1978 when President Carter designated ten new national parklands in Alaska. For wild scenery and fascinating geology, Alaska's parks are difficult to match; hence this revised edition is needed in light of the fact that with the addition of the new Alaskan parklands, the domain of the National Park Service was about doubled in size.

Presented in this edition are 92 areas administered by the National Park Service; in addition to the 49 national parks and 32 of the national monuments, eleven of the national seashores, lakeshores, recreation areas, and preserves are included. Although the Mt. St. Helens area is now under the jurisdiction of the U.S. Forest Service, it is discussed because of its unique geologic significance. (Refer to Appendix A for the addresses of these parklands.)

The plate tectonics theory of the development of the continents, ocean basins, and mountain ranges, proposed more than two decades ago, has revolutionized geological thinking. These new concepts, introduced in the earlier editions of this book, needed to be explored more fully. Partly — but by no means exclusively — for this reason, Dr. Eugene P. Kiver was invited to participate as co-author. It is our hope that we have presented complex concepts in an understandable fashion, whether or not the reader has had previous training in geology. For those geologists who would like more than our brief and simplified presentations, the bibliography serves as a guide to a bountiful supply of serious reading.

During this period of flux in geological thinking, it would be difficult to find any two geologists with identical ideas and interpretations. It seems appropriate to mention that, although we have labored over the problems together, the final wording throughout this edition is that of the senior author.

The co-author is mainly responsible for material on the plate tectonics of the Northwest and of the Appalachians, and for the addition of the following parklands: Channel Islands, Point Reyes, Lava Beds, John Day, Theodore Roosevelt, Cape Hatteras and Cape Cod, and Mt. St. Helens National Volcanic Monument.

Three brief trips into Alaska do not fully qualify a person to write at length on Alaskan geology; however, the senior author shouldered this responsibility with the capable assistance of Gil Mull, Jack Odum, and Greg Arehart. (These men are not held responsible for any misinformation that appears in this book; they did not have the opportunity to see the final draft!)

Several new maps, line-drawings and photographs, both color and black and white, have been added in this edition. The color photographs are now concentrated in a single section in the middle of the book.

Of particular interest to professors who are planning to offer a parks course are the more than 1000 color slides now available. For information, write to D. V. Harris, 107 Villa Lucia, LaLuz, New Mexico, 88337, or call (505) 437-3313.

We appreciate the assistance given by a large number of people in the Park Service, the U.S. Geological Survey, and our universities. In addition to those acknowledged in the third edition, we express special appreciation to Art Hathaway, Larry Hanson, Harvey Wickware, Tom Miller, E. B. Reed, E. Erslev, M. E. McCallum, Bob Johnson, Frank Ethridge, and Joe Weitz for their contributions to this edition.

Our sincere thanks go to Helen McClure, who deciphered our drafts and typed the manuscript.

Traditionally, the spouses of authors are responsible for books actually being completed. In addition to a multitude of other things, Cleo Harris disentangled the truth from many otherwise bewildering paragraphs. Barbara Kiver accompanied the co-author on visits to the national parks and provided strong encouragement and support during the revision of the book.

February 1985   **D. V. H.**
                                         E. P. K.

PREFACE
TO THE THIRD EDITION

The Geologic Story of the National Parks and Monuments was written for two purposes — to provide a textbook on National Parks geology and to make available a book for the general reader who is interested in what lies behind the scenery for which the Parks are world-famous. Behind this scenery are millions of years of geologic changes which have fashioned the landscape where plants grow and wild animals run. An appreciation of the total Park environment is based upon an understanding of the geologic story.

Several excellent geologic stories are now available — Lohman's on Arches and Canyonlands, Keefer's on Yellowstone, Love and Reed's on the Tetons, and others. But in this book, the 39 Parks, 30 of the most geologic Monuments, a National Lakeshore, a Seashore and a National Parkway — all 72 appear in a single volume. And it is written in simple language with as few technical terms as possible.

The Geologic Story is not intended to be an exhaustive treatise; such a book would be titled "Geology of the United States" and would be too heavy to lift and too expensive to buy. Instead, it is the drama of Earth history with scenes from the various Park areas where geology is at its best.

The Introduction contains a synopsis of geologic principles and processes for readers without prolonged exposure to geology. To assist the reader in becoming aware that geologic processes are related and interactive, I let the discussion flow logically from one topic to the one most closely allied, without subheadings shouting that we now have something entirely different, perhaps unrelated. Here, as in later chapters, much more could have been included. Therefore, instructors have many opportunities to improve the course by adding their own material.

Other successful innovations of the first edition were also retained. Glossaries are generally not overused, if used at all. As our readers refer to the index, they will find something more — definitions — in the Glossary Index.

From the beginning, special effort was made to keep the size of the book well within reason. Even so, I made a point of finding space for a brief introduction to the "new geology" — Plate Tectonics. Although still in its infancy, this new and earth-shaking theory has been almost universally accepted. Certainly it provides for explanations of many elusive problems in geology.

Experience in teaching the Geology of the National Parks and Monuments led me to organize the Park areas geographically, with the geomorphic province as the unit. Within each province, the geologic sequence is the same; therefore, the Park areas within a given province are related — like members of a family, each with its own distinctive features. The discussions of regional geology at the beginning of each chapter will enable readers to better appreciate the country they are traveling through as they go from one Park to another.

On occasion I broke with geological tradition in conveying my personal impressions of the Parks. And I even dared to gingerly salt and pepper the book with what Lorna Carter referred to as "sly humor" in her review of the second edition, in the *Journal of Geological Education,* November 1979.

Concern for the future of the Parks compelled me to mention the deterioration of certain Park environments in the hope that such awareness would increase public support of the Park Service in their almost impossible task of protecting these precious areas. With our backing, perhaps the pollution described here will become merely a dark interlude in National Park history.

In 1976, when the book was first published, there were comparatively few courses in Parks geology. By 1978 the number had increased significantly, due in part to the availability of color slides; therefore, expensive improvements were justified, including additional color photographs and drawings. The second edition is widely used in colleges and universities and in a few high schools. In addition, it is available in a number of libraries, Parks, and museums.

The third edition retains the same format and sequence, and most of the geological discussions remain essentially untouched. Significant changes were made, however — changes which fall into two main categories. First, the book was updated in several areas. For instance, Badlands is now a National Park, and there are several new Park areas in Alaska. In certain Parks, activities available to Park visitors have been altered. Also, several new references were added to the Bibliography. One of general interest is the July 1979 issue of *National Geographic* which is devoted entirely to the National Parks. Second, special attention was directed to making the illustrations more effective and to upgrading the quality and content of the photographs. Now, most of the truly colorful areas are in color. These improvements have resulted primarily from suggestions made by professors who have used the book, by naturalists and others in the Park Service, and by members of the U.S. Geological Survey. Their interest and assistance is recognized with sincere appreciation.

This book, which was written for the enjoyment of all, does very little for those blind persons who have never seen a canyon or a mountain. Perhaps you can suggest how it could be rewritten so that they, too, will be able to appreciate the beauty of the Park areas.

Ours is an ever-changing Earth. As this third edition goes to press, Mount St. Helens is showering southwestern Washington with volcanic ash and dust after 123 quiet years. When, in 1976, I suggested that we might see one of the Cascades volcanoes erupting (page 116), I did not anticipate that it would take place this soon! (For a view of the erupting volcano, see frontispiece.) Where and when will the next exciting geologic event occur? We can only be certain that this is not the final chapter.

April 1980   **David V. Harris**

ACKNOWLEDGMENTS

Many people, more than I can name, assisted me during the information gathering years and during the writing and revision of the manuscript. Their cooperation and encouragement made this book possible.

Special thanks are given to William D. Thornbury, who read the original manuscript and made valuable suggestions, particularly with respect to the regional geomorphology.

Robert L. Heller reviewed a later version and many of his constructive criticisms are reflected in the final copy. Reviewers who remained nameless are also acknowledged for their helpful comments on certain chapters.

Sections of the manuscript were sent for review to various National Parks where the Superintendents and their staffs were invariably cooperative, helpful, and encouraging. Park Service personnel, including a few now retired, to whom appreciation is extended, include Hugh P. Beattie, John V. Bezy, Richard H. Boyer, Karen Brantley, David Butts, Stanley G. Canter, Roger Contor, J. V. Court, Donald H. DeFoe, John Good, Fred Goodsell, Dwight Hamilton, Robert C. Heyder, Marilyn Hof, James H. Holcomb, Robert Howe, David H. Huntzinger, Victor L. Jackson, Joe L. Kennedy, Steven R. Huber, David May, Ross A. Maxwell, Charles H. McCurdy, R. Alan Mebane, George E. Neussenger, John C. O'Brien, John J. Palmer, David A. Pugh, William Rabenstein, Larry Reed, Tom Ritter, Ralph Root, Robert H. Rose, Robert Rothe, Edwin L. Rothfuss, Richard H. Sims, Kent Smith, Jr., Steven Q. Smith, C. Tampow, Henry M. Tanski, Pat Tolle, Richard L. Vance, Philip F. VanCleave, W. E. Weaver, Harry W. Wills.

It is a particular source of pleasure to find former students on the staff of many Parks I visit, and for their special acts of helpfulness and kindness I want to express gratitude.

Several geologists of the United States Geological Survey provided significant information and advice on a number of problems and their help is recognized here. Particularly important are the contributions of Dwight R. Crandell, Wallace R. Hansen, N. King Huber, W. R. Keefer and Stanley W. Lohman. Lorna Carter, Technical Publications Editor, reviewed the second edition in the *Journal of Geological Education* and later made numerous suggestions which contributed substantially to the improvement of the third edition.

My colleagues, past and present, at Colorado State University, have been reliable sources of information and advice throughout the years, and their support and encouragement have been greatly appreciated. They are Lary Burns, John A. Campbell, the late Roy G. Coffin, R. Scott Creely, Donald Doehring, Frank Ethridge, Robert B. Johnson, Malcolm E. McCallum, Stanley A. Schumm, Tommy B. Thompson, James Waltz, and Joseph Weitz.

To J. Merle Harris, brother and colleague of long standing, go my thanks for encouragement and counsel throughout the undertaking.

Readers who find the book effectively illustrated are indebted, as I am, to my students, colleagues, and many friends who shared their choice photographs. The National Park Service and the U.S. Geological Survey also supplied photographs and maps. Credits are given in the captions, except for photographs of my own. My special thanks go to Gregory C. Nelson who transformed my crude sketches into meaningful works of art, some of which interpret accompanying photographs.

I wish to express my appreciation to all who provided criticisms and suggestions for the betterment of this edition, particularly to R. S. Creely, Frank Ethridge, R. E. Gernant, J. M. Harris, W. C. Hood, V. Matthews, R. J. Larson, J. D. Love, M. Lukert, and K. M. Pohopien.

I am deeply grateful to the Colorado State University Foundation, and particularly to Tom Grippen and Larry Tew, for arranging financial support for publication and distribution. Special thanks to Earl Howey for the initiation of the project and to Eric Mosher who supervised publication of the second and third editions. Kathy Hobson assumed the tedium of putting the first two editions together, with unusual efficiency and good will. Betty Jo McKinney was responsible for editing, revising and upgrading the third edition, and she merits special commendation. Special thanks also to the employees of B. Vader Phototypesetting, Colorado State University Photo Lab and Printing Services for the production of this edition.

I wish to express both appreciation and admiration to the following for handling burdensome chores such as checking the bibliography and the index, typing illegible manuscript, and preparing maps: T. Bouska, A. Blackstone. T. D. Harris, H. McClure, H. Oakleaf, L. Odum, P. Logan, and J. Tholen.

The last should be first. My wife, Cleo, was at my side in all of the Parks and Monuments. She assisted me in every phase of the writing and is largely responsible if the book is one that everyone can understand. Without her help and encouragement, it is doubtful if this book ever would have been finished. But she was always there.

April 1980   **David V. Harris**

CONTENTS

CHAPTER 2
ALASKA AND ITS PARKLANDS; THE ALEUTIAN RANGE PROVINCE 52

ALEUTIAN RANGE PROVINCE 56

CHAPTER 3
ALASKA RANGE PROVINCE 71

CHAPTER 4
PACIFIC BORDER PROVINCE 80

CHAPTER 5
CASCADES PROVINCE 114

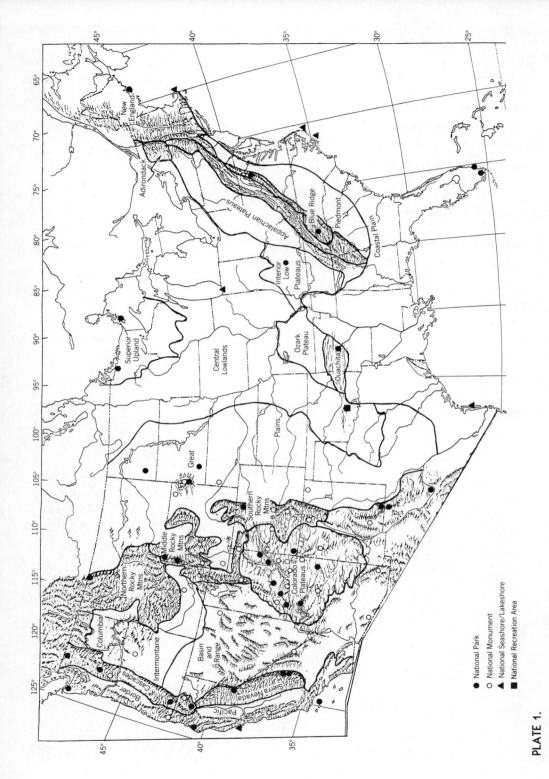

PLATE 1.

Geomorphic Provinces of conterminous United States. For names of the parklands shown here, refer to the province maps in the various chapters.

● National Park
○ National Monument
▲ National Seashore/Lakeshore
■ National Recreation Area

INTRODUCTION

Our national parks are a unique American heritage. Unlike many of our institutions that have their roots in European soil, the national park concept is our own. It was born in 1870 in the wilderness that is now Yellowstone National Park.

Prior to this time, little had been known of this remote area. Indians roamed the Yellowstone country but apparently they were tight-lipped about what they had seen. Historical records indicate that the first white man to see the wonders of Yellowstone was John Colter, one of Lewis and Clark's men, in 1807. His account of what he saw was too incredible for most people. Several years later, mountain-man Jim Bridger went into the Yellowstone country and came out with stories similar to Colter's, and similarly, they were ridiculed. According to Chittenden (1895), Bridger resented the fact that no one took him seriously and retaliated with fabrications so preposterous that they would give him some sort of renown. It is likely that his tales will stand tall for some time: very few people accept as fact his claim that the glass in Obsidian Cliff acts as

FIGURE I-1

Camp of F. V. Hayden's Geological Survey party on Yellowstone Lake in 1871. (Photo by W. H. Jackson, U. S. Geological Survey.)

a telescope lens to bring within shooting distance an elk 25 miles (42 km) away.

Eventually, curiosity as to whether there was any truth in Colter's and Bridger's stories led to the exploration of the Yellowstone country. In 1870 General Henry D. Washburn led a party of Montanans into the area to find out what was actually there. At the end of the summer they realized that here was something truly unique and that once it was publicized, people would flock in to see it. They could stake their claims and make a good thing out of it for themselves.

Sitting around a campfire, watching the flames lift into the darkness, is likely to bring out the best in a person, and finally Judge Cornelius Hedges startled his companions by saying that this area with all its awesome and beautiful features should belong to all the people to see and enjoy. In due time they agreed, and the idea of national parks took shape — parks to be used as "pleasuring grounds for all of the people."

Although Judge Hedges may not have been aware of it, almost 40 years earlier George Catlin had proposed to the government that a large tract of land around the Yellowstone geysers be preserved as the "Nation's Park" (Chittenden, 1895). Catlin is better known for his work of preserving records of the American Indian in his writings and in his hundreds of paintings, many of which are in the Smithsonian Institution in Washington.

Although Catlin must be given credit for the concept, it was Judge Hedges, N. P. Langford, and others who, by their tireless efforts, saw to it that this time the seed would grow and bring forth fruit. More evidence was needed to convince the skeptics in Washington, and in 1871 geologist F. V. Hayden and a party of scientists were sent to explore the area. Fortunately, their official report was convincing and on March 1, 1872, Congress established Yellowstone as a national park, the first in the world.

The national parks idea spread rapidly even into areas that were then remote. Australia was the first to follow our lead; then the New Zealanders indicated their approval by creating their first national park, later increasing the number to ten. Our neighbors the Canadians are, with good reason, proud of their Glacier, Banff, and Jasper national parks, along with others. Now, more than 100 nations have national parks or reserves; unquestionably the world's highest national park has recently been established on Mt. Everest. But the idea took shape as the flames rose above that campfire in Yellowstone.

THE NATIONAL PARK SYSTEM

The original national park concept has been expanded. Now certain park areas preserve sites of significant historic interest, like Valley Forge; others preserve archeological sites, such as Mesa Verde. There are also national seashores, national lakeshores, national parkways, national memorials, national preserves, and national recreation areas. Essentially all of the national parks and many of the monuments are noted primarily for their classic geologic features and their spectacular scenery.

In general, parks are larger and contain a wider variety of natural features than monuments. The main distinction between the two is the way they are established. An act of Congress is necessary for the creation of a national park, whereas the president can establish a monument. Thus a threatened area can be quickly set aside as a monument and, if there is justification, reclassified later as a national park. Recently Katmai and Glacier Bay in Alaska and Biscayne in Florida became national parks. Inasmuch as the difference is mainly administrative, the term *park*, as used in this book, refers also to monuments, except where otherwise specified.

By no means are all of the country's beautiful areas within the National Park System. Particularly significant are the vast areas that are administered by the U.S. Forest Service and the Bureau of Land Management. Then there are the state parks which serve much the same purpose as the national parks, namely, to preserve natural and historic features.

THE NATIONAL PARK SERVICE

When Yellowstone was established as a national park, much of the West was still untamed and so were many of the people who roamed these remote areas. They were rugged individuals who were in the habit of doing as they pleased and taking what was there. In general they were not at all impressed with the national park concept, and park boundaries were commonly ignored.

N. P. Langford, one of the Washburn party, was named as the first superintendent of Yellowstone National Park. Langford was given the title of Superintendent — and the nickname "National Park" Langford — but for five years, Congress neglected to appropriate any money for his salary or for the park. Consequently, he was powerless to carry out his duties. Poachers decimated the bison herd, and finally the U.S. Army was called in to prevent its total destruction. Serious damage was done to the formations in Yellowstone's thermal areas. Inroads were made in other parks as well, especially in Yosemite National Park where, despite John Muir's valiant efforts, the power interests prevailed and Hetch Hetchy Dam destroyed for all time one of the truly magnificent canyons in Yosemite.

Stephen T. Mather, a prominent Chicago businessman, was appalled at this sorry treatment of our national treasures and he wasted no words in expressing his feelings on the matter. The result was that he was given the job of managing all the parks and monuments. Mather was very persuasive. In less than two years, in 1916, Mather succeeded in getting legislation passed establishing the National Park Service within the Department of the Interior. As director, Mather devoted his life to

FIGURE I-2

Reflected in a quiet pool on Sugar Creek is the bluff of Mansfield Sandstone (locally called "millstone grits") in Turkey Run State Park, western Indiana. Created in 1916 to preserve hardwood trees, the park also protects relict plants hidden in the cool recesses of the deep ravines.

his job and when federal funds were lacking, he used his own money for the operation and improvement of the parks. At last, the way was paved for the preservation of the parks as prescribed by law more than 40 years earlier.

The National Park Service is headed by a director who is responsible to the Secretary of the Interior. In each of the parks and monuments a superintendent is in charge of operations; with a staff of assistants he or she sees to it that the natural, historic, and cultural features are properly interpreted for park visitors. Whether you go into the visitor center down at Oconaluftee or up at Ohanapecosh, the park people will do their best to make your visit both enjoyable and educational. They also do their best to protect the park from unthinking people who could unwittingly damage the environment.

Through the years, the National Park Service has evolved into an organization of well-trained men and women who are determined "to conserve the scenery and the natural and historic objects and the wildlife therein and to provide for the enjoyment of the same in such a manner and by such means as will leave them unimpaired for the enjoyment of future generations," as set forth in the original National Park Organic Act.

The administration of national parks is a challenging undertaking. Each park has its own particular problems, and all parks have a common problem — people with different ideas as to how the parks should be developed. Some believe that there should be *no* development; others are just as certain that there should be all sorts of accommodations and recreational facilities, with wide highways and parking areas in abundance. Yet, almost everyone agrees that our parklands should be preserved.

On the other hand, some people see land only in terms of the material things that it can produce, such as food, minerals, lumber, and energy. Now that the nation's appetite for energy appears insatiable, our parks are in greater jeopardy than at any time in their history. The

FIGURE I-3

Park Naturalist points out glacial features in the mountains above Nymph Lake in Rocky Mountain National Park. (Photo by National Park Service)

economic stakes involved in the exploitation of the nation's natural resources are high, and lobbying pressures on the administration and Congress to favor development over preservation are almost overwhelming. Given these circumstances, it is alarming to consider what can take place if people in high office have but little understanding of or concern for the environment, or for the future of our parks and wilderness areas. Without constant vigilance on the part of the general public, such officials can exercise their power to reinterpret the law. Although it is true that a substantial amount of geothermal energy could be developed in Yellowstone National Park, most Americans would not be willing to trade Old Faithful for a whole nest of hissing steam wells.

PARKS AND PEOPLE

National parks mean different things to different people. To one person a park may be an escape from a hectic life; to another it is a great place to ski. Some families plan their vacations around overnight stays in park campgrounds.

Scientists are particularly interested in the park areas. Biologists find in natural environments plants and animals that have been largely eradicated elsewhere. A geologist can see, on full display, classic examples of rocks and structures that reveal the history of the earth.

Most people visit the parks because of the beautiful scenery, still mostly unspoiled. They are likely to be more knowledgeable than visitors of earlier days. They are interested in the total environment, both physical and biological, and they want to know what is back of the scenery—what caused it to be as it is.

Behind the scenery and beneath the soil growing the plants that feed the animals are the rocks from which they were created. The story of the rocks and the landscapes and how they came to be is the geologic story of our national parks.

GEOLOGIC PRINCIPLES AND PROCESSES

This section is for the benefit of those who have had little or no previous training in geology. A number of geologic concepts are capsulized here and elaborated on later in the discussions of various parks. Although this treatment will probably be sufficient for most people, some may be interested in delving more deeply, and for them several geology textbooks are listed in the bibliography.

Geology is the study of the earth and its history, as revealed in the rocks. What is revealed in the rocks of the national parks and monuments, when pieced together, gives us a remarkably complete history of the earth.

The earth is constantly changing, and it has been changing throughout its 4.5-billion-year history. In some cases, as when Mexico's Paricutin Volcano was born in 1943 or when Iceland's Surtsey boiled up out of the sea, the changes are spectacular. In most cases, though, the changes are so gradual that many people assume that a mountain or a canyon "has always been there." But the mountain was not always there; Grand Canyon was no canyon at all 10 million years ago.

The concept of geologic time is one of the most difficult to comprehend. Most people are in the habit of thinking of time in terms of human existence. In the early days people believed that mountain ranges were the results of "colossal convulsions" within the earth and that they were formed almost overnight. Later, when the true meaning of fossils became apparent and biological evolution was finally accepted, the immensity of geologic time was gradually recognized.

Much later, when the radiometric method of age determination was developed, we had for the first time the means for assigning approximate ages in terms of years or millions of years. Although the method is highly technical, the principles are given here in simplified form. Certain forms of several chemical elements are radioactive; that is, they decay spontaneously and at a known rate. For example, one form of the element uranium decays at an extremely slow rate, so that after approximately 4.5 billion years, one-half of the uranium is left. After another 4.5 billion years, one-half of that half, or one-fourth of the original amount, is left. Therefore, this form of uranium has a *half-life* of 4.5 billion years. Radioactive carbon has a much shorter half-life, only about 5700 years. Radiocarbon dating is therefore used for age determinations of extremely young rocks, those that are less than 50,000 years old. Other radioactive elements have half-lives that are between these two extremes.

Unfortunately, many rocks do not contain minerals that have radioactive elements suitable for dating purposes. And even though a mineral in a sedimentary rock contains a radioactive element, the age determined is likely not the age of the sedimentary rock but, rather, the age of the original igneous rock in which the radioactive mineral crystallized. (An exception is a very young rock, less than 50,000 years old, in which radioactive carbon was formed.) For these reasons, and others, many formations are assigned only approximate ages based on extrapolations of data. Nevertheless, the radiometric method represents significant progress, a breakthrough in geologic dating.

How does a geologist go about deciphering the history of the earth, the complex structures exposed in the mountains, and the surface features we see everywhere? Prior to 1785, there was no systematic approach; instead, there was mere speculation. In 1785, a Scotsman by the name of James Hutton published a paper in which he stated that the geologic processes now operating are the same processes that have, throughout geologic time, changed the earth and formed it as we see it today. Thus, "the present is the key to the past." This simple concept, which became known as the Uniformitarian Principle, is fundamental in geologic thinking. In fact, geology became a science when the Uniformitarian Principle was

found to be sound and acceptable. For the first time, geologists had a rational, logical approach to use in interpreting the structures and features of the earth.

Although Hutton was the first to publish, others also deserve much credit; John Playfair undertook the task of restating in more understandable form many of Hutton's ideas. And later on, in 1830, Charles Lyell published his three-volume *Principles of Geology* in which the Uniformitarian Principle was firmly interwoven. It was a monumental work and was largely responsible for the acceptance of the Uniformitarian Principle.

The earth is made up of rocks that differ in composition, structure, and color; with a few exceptions each of these rocks is composed of several minerals. For example, granite is made up of the *essential minerals* quartz and orthoclase feldspar, and one or more of the following *accessory minerals:* muscovite mica, biotite (black mica), hornblende, or garnet. A few rocks, however, are composed of one mineral; for example, pure limestone contains millions of tiny crystals of the mineral calcite.

Concentrations of certain minerals can be found in fractures and other openings in the rocks. When such minerals have economic value and the quantity is sufficient, it is called an *ore deposit*. Frequently, two or more minerals containing lead, zinc, copper, iron, tungsten, or other valuable metals occur in these concentrations. During the mining boom in the Leadville District in Colorado, large quantities of gold, silver, lead, and zinc were taken out. In the United States, natural resources of all kinds were once abundant — inexhaustible in the minds of many people. But now an alarming number of our nonrenewable resources have been exhausted, or nearly so. Clearly, conservative use of what we have left is indicated, in order to postpone the day when we become one of the "have-not" nations — when we will be forced to mine our landfills.

The term *mineral* is used in both a general and a specific sense. When we speak of mineral resources we include oil, gas, and coal, none of which is a true mineral. To mineralogists, minerals are naturally occurring, inorganic, solid substances, each with its own specific physical properties (hardness, color, crystal form, cleavage, etc.) and its own chemical composition. Consequently, by making certain simple tests and observations we can identify most of the common minerals that we find in the field. For the less common ones, the mineralogist uses the petrographic microscope, which measures optical properties, or X-ray analysis.

A pocketknife is a good tool with which to determine hardness. Minerals that are too hard to be scratched with the knife are "hard" minerals; those that can be scratched with a fingernail are "soft." With a bit of practice, you can determine the hardness fairly accurately, so that you can make use of mineral identification tables (see Table I). *Cleavage* is another diagnostic physical property. When a mineral is broken, it may split along one or more flat, planar surfaces which are cleavage planes. A few minerals, such as quartz and pyrite, do not have cleavage; instead, they break (or fracture) along irregular surfaces. By checking the cleavage in the tables, you will likely find the mineral that has the same cleavage and hardness as your specimen. For less common minerals, refer to tables in a good mineralogy textbook, such as Hurlbut's (see Bibliography).

A mineral that develops without being interfered with by other minerals forms a distinct crystal. Quartz forms a six-sided (hexagonal) elongate crystal, whereas feldspars are usually blocky, as shown in Figure I-4 (see Color Section).[1]

Crystal form is related to, among other factors, chemical composition. In halite (NaCl), the two elements sodium (Na) and chlorine (Cl) are stacked one on top of the other and held together by a chemical bond, thus form-

ing a rectangular pattern that results in a cube. Several minerals with relatively simple chemical compositions, such as galena (PbS) and pyrite (FeS_2), typically form cubes. Many of the more complex minerals have more complicated crystal forms that reflect intricate internal arrangements of atoms. Most of the silicates — minerals that have one or more elements combined with silica — have complex internal structures. For example, orthoclase has one atom of potassium (K) and one of aluminum (Al), in combination with three of silicon (Si) and eight of oxygen (O); its chemical formula is $KAlSi_3O_8$. With but few exceptions, each mineral has its own distinctive crystal form; once you learn to recognize it, you will be able to identify the mineral without touching it. Crystallography, the study of crystals, is one of the most beautiful facets of mineralogy.

Color is usually the first physical property observed; unfortunately, it is not diagnostic in most cases. Most light-colored minerals occur in two or more colors. For instance, quartz may be colorless (clear like glass), or white, or rose-colored, or amethyst, or smoky (black). There are several minerals, however, that can be identified solely on the basis of color: azurite is a beautiful, deep azure blue, pyrite has its distinctive brassy yellow color, and magnetite appears only in black. But when considering the mineral kingdom as a whole, color is not as reliable a physical property as hardness, cleavage, or crystal form.

In time, you can learn to take a few shortcuts. If you see a black mineral that might be magnetite, you don't need to check hardness or cleavage; rather, find out whether it responds strongly to a magnet, because magnetite is the only common mineral that is highly magnetic. Or, if you pick up a light-colored mineral and feel that it is *too* heavy, you can be almost certain that it is barite, which has a specific gravity that is more than 50 percent higher than any other common light-colored mineral.

Minerals are formed by different processes, the simplest of which is by crystallization from a water solution, as a result of evaporation. You

---

[1] Readers will find it of interest to refer to the color photographs before reading each chapter. Figures 1-4, 1-6, etc., are color pictures of geologic features in Chapter 1.

TABLE 1

Minerals

More than 2000 minerals are found in the rocks of the earth's crust; the 20 given in this table are the most common. The physical properties of hardness[a] (H) and cleavage[b] (Cl) are most helpful in the identification of minerals. A few minerals have a characteristic color, but for most minerals the color varies greatly due to impurities.

CHEMICAL SYMBOLS: Al-aluminum; Ca-calcium; Fe-iron; K-potassium; Mg-magnesium; Na-sodium; Si-silicon.

Mineral	Properties	Miscellaneous
AMPHIBOLE A group of complex Ca, Mg, Fe, Al silicates	H: 5−6 Cl: 2 planes at about 60°	Generally dark-colored to black. Hornblende is a common variety.
BIOTITE Complex silicate of K, Fe, Al, and Mg	H: 2.5−3.0 Cl: 1 plane, perfect cleavage	Black mica.
CALCITE Ca carbonate	H: 3.0 Cl: 3 planes, not at right angles	Generally white. Transparent variety—Iceland Spar.
CHALCEDONY (Cryptocrystalline quartz) Si dioxide	H: 7 Cl: none; conchoidal fracture	Commonly white to gray; translucent.
CHLORITE Complex hydrous Mg, Fe, Al silicates	H: 1−2.5 Cl: 1 plane (like mica)	Color: grass-green. Common in metamorphic rocks.
EPIDOTE Complex Ca, Fe, Al silicates	H: 6−7 Cl: 1 plane	Pistachio green.
GARNET A complex silicate mainly of Ca, Fe, and Al	H: 6.5−7.5	Commonly brown to red; some varieties are semiprecious stone.
GYPSUM Hydrous Ca sulfate	H: 2.0 Cl: Perfect in 1 plane; imperfect in 2 other directions	Colorless to white. Fibrous variety—satinspar.
HALITE Na chloride	H: 2−2.5 Cl: 3 perfect at right angles	Rock salt; salty taste.
HEMATITE Fe oxide	H: 5−6.5 (earthy varieties softer)	Generally red to reddish-brown; may be black. Most important iron ore.

TABLE I

Continued

KAOLINITE Hydrous Al silicate	H: 1–2	White when pure. One of the important clay minerals.
LIMONITE Hydrous Fe oxide	H: 3–5.5	Yellow, brown to black.
MAGNETITE Magnetic Fe oxide	H: 5.5–6.5	Strongly attracted by a magnet.
MUSCOVITE K, Al silicate	H: 2–3 Cl: perfect in 1 plane	White mica; may be transparent.
OLIVINE Mg, Fe silicate	H: 6.5–7	Yellowish-green to brownish-green; olive green.
ORTHOCLASE FELDSPAR K, Al silicate	H: 6 Cl: 2 planes at 90°	Typically pink; important constituent of granite.
PLAGIOCLASE FELDSPAR Na, Ca, Al silicate	H: 6 Cl: 2 planes not at 90°	Generally white to gray.
PYRITE Fe sulfide	H: 6–6.5 Cl: none	Fool's Gold; brassy-yellow color. Source of sulfuric acid.
PYROXENE Ca, Mg, Fe silicate	H: 5–6 Cl: 2 planes nearly 90° apart	Augite is a common variety; greenish black.
QUARTZ Si dioxide	H: 7 Cl: none; uneven to conchoidal fracture	Hexagonal (6-sided) crystals common. Important constituent of granite and rhyolite.

[a] *Hardness* is the resistance to scratching or abrasion. A copper penny has a hardness of 3; a fingernail about 2.5; the steel in a good pocketknife about 5.5.

[b] *Cleavage:* When broken, most minerals will split along one or more planes of weakness. Quartz (and a few other minerals) has no planes of weakness in the crystal structure and therefore has no cleavage.

can observe this process by dissolving a teaspoon of table salt (NaCl) in a half-cup of hot water; then, using a flat pie pan, evaporate it slowly over a low flame. With evaporation, the solution becomes saturated with sodium (Na) and chlorine (Cl) which will combine to form crystals of artificial halite (NaCl). When crystallization begins, allow the solution to cool very slowly and observe the tiny cubic crystals, using a magnifying glass if necessary. In a few hours, you observe what has taken place many times during geologic history, where over thousands of years sea water evaporated in enclosed basins, thus forming layers or beds of salt.

Minerals also form when *magmas* and *lavas* cool. Should you time your trip to Hawaii wisely, you could observe minerals being formed as lava cools and crystallizes. Lavas cool rapidly, too rapidly for the crystals to grow to appreciable size; therefore, most or all of the minerals in lava rocks are too small to see with the naked eye. A few glassy, green crystals of *olivine* — crystals that grew in the slowly cooling magma before it reached the surface — might be visible. When large masses of magma crystallize far beneath the surface, all of the crystals grow to considerable size, large enough to recognize and identify. Granite, the most common coarse-grained rock, is formed under such conditions.

Another way minerals form is by crystallization directly from the vapor state. Hematite (iron oxide) and other minerals have been observed in the process of crystallization around tiny volcanic vents where extremely hot gases are pouring forth. You could have observed this process in the Valley of Ten Thousand Smokes in Alaska, if you had been up there when the vents (fumaroles) were active. Possibly two or three vents are still emitting gases.

The world of minerals is large; more than 2000 minerals have been described and cataloged. Of this vast array, however, almost all of the minerals that you will likely find can be identified by searching through the 20 described in Table I.

Most minerals are relatively stable in the geologic environments where they were formed. Even so, very gradual changes are brought about on the surface by such processes as weathering and erosion. And if the rocks containing the minerals are moved into a distinctly different geologic environment — where temperatures and pressures are extremely high — they will either melt or will be profoundly changed by the process known as *metamorphism*. By these and other geologic processes rocks are broken down, thus destroying their identities; but other rocks are constructed from their remains, in a never-ending cycle — the Rock Cycle.

In visualizing the Rock Cycle, *magma* is a good starting point. Magma is molten rock material that is beneath the surface; the term *lava* is applied to the same material that has reached the surface. When magma (or lava) cools sufficiently, minerals begin to crystallize out, and when all of the liquid has become solid, it is called *igneous* rock. If such rock is exposed, perhaps by erosion, it is *weathered;* that is, it is broken down, both physically and chemically, and is ready to be eroded away. The loosened materials that are transported away by some eroding agent (water, wind, or ice) and then deposited are sedimentary rocks-in-the-making. The loose materials must be cemented together or compacted sufficiently to form a solid rock. If these *sedimentary* rocks are subjected to enough pressure and heat, they become unstable and are transformed into *metamorphic* rocks. Then, if the metamorphic rocks are subjected to sufficiently high temperatures, they melt, thus forming magma. In this way the Rock Cycle is completed (indicated by the heavy lines in Figure I-5).

There are also other possible changes: Sedimentary rocks may be weathered and reconstructed into younger sedimentary rocks, or they may be melted into a magma. Igneous rocks may be metamorphosed or they may be melted. And metamorphic rocks may be weathered and reformed into sedimentary rocks (indicated by the light lines).

The Rock Cycle can be used as a skeletal outline in the study of geology. Here, a synthesis of some of the more important aspects of geology is presented. It is not complete, but for those without geologic training it may provide sufficient background for the understanding of this book, with occasional reference to a geology textbook (see Bibliography).

Before proceeding further, perhaps we

FIGURE I-5

The Rock Cycle, showing the main classes of rocks and the geologic processes involved in the changes from one to another.

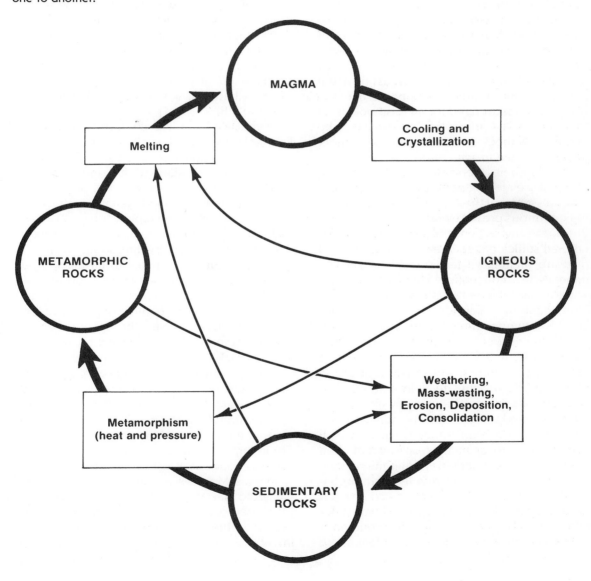

should consider the problem of the origin of magma, lest someone assume (as the ancients did) that magma is present everywhere, beneath a thin "crust." Actually, almost all of the earth is solid, except for a liquid envelope around the small inner core at the earth's center, and for masses of magma that underlie specific limited areas such as part of the Hawaiian Islands and perhaps Yellowstone.

We know that temperatures increase with depth below the surface, but we must also remember that pressures increase in the same manner and that pressure tends to prevent rocks from expanding and melting. Consequently, we have an extremely complex system on our hands. Exhaustive discussion of the problem would require more pages than we have space for, or need for, in this book. Brief treatment of the principles, however, may be helpful—without inflicting any permanent psychological damage.

At a depth of several miles the temperature is above the melting point of rocks under surface pressure, but they remain solid because of the enormous pressure of the overlying rocks. But if, for any reason, the temperature is increased sufficiently to overcome the effect of pressure, melting will take place. If radioactive minerals are concentrated in certain deep-seated rocks, the heat generated by their disintegration may provide the necessary increase in temperature. Or, in an active fault zone where rocks are being thrust over other rocks, the heat produced by friction might be sufficient. Heat from this mechanism would be particularly significant deep in a *subduction zone,* a topic discussed later in this section. On the other hand, should widespread, rapid erosion remove huge quantities of rock material from a mountain range, the resulting reduction of pressure would perhaps permit melting, directly or in combination with temperature increases.

Because of the difficulties involved in determining with accuracy just what conditions exist where magmas form, the problem of the origin of magma, like certain other geologic problems, may not be completely solved for some time. But with new additions to our knowledge and with new techniques, we are now somewhat closer to the solution than we were a century ago.

Magmas differ in chemical composition, mainly because of the differences in the mineral composition of the particular rocks that are melted. Because silicon (Si) and oxygen (O) are the two most abundant chemical elements in rocks of the earth's "crust" (the outer layer averaging about 25 miles, or 40 km, in thickness), all magmas contain silica (SiO_2) in moderate to large quantities. For this reason, the silica content of igneous rocks serves as one of the two bases of igneous rock classification (see Table II).

High-silica rocks are composed mainly of light-colored minerals such as quartz (pure silica) and orthoclase feldspar (a silicate, $KAlSi_3O_8$). If the magma cooled slowly and the rock is therefore coarse-grained, it is called granite; if it is fine-grained, its name is rhyolite; and if it was chilled too rapidly for any crystals to form, it is obsidian (volcanic glass). Obsidian, like granite and rhyolite, should be light-colored; instead, it appears to be black! It is actually light-colored, as you will observe if you break off a very thin flake; you can read through it. There are enough tiny particles of black material dispersed throughout the essentially colorless glass to make it opaque and appear black.

Low-silica magmas are magmas in which the silica content is too low for the formation of quartz; instead, the resulting rocks are composed dominantly of *pyroxene* and other dark-colored minerals, in some cases *olivine*. *Gabbro* is formed from the slow cooling of such magmas, and basalt is the rock derived from such lavas. More will be said about basalt when we get to Hawaii and later when we look into the Craters of the Moon. Intermediate magmas produce either diorite or andesite, as indicated in Table II.

TABLE II

Igneous Rocks

Texture	Silica Content		
	High	Medium	Low
Coarse-grained	GRANITE	DIORITE	GABBRO
Fine-grained	RHYOLITE	ANDESITE	BASALT
Glassy	OBSIDIAN		BASALT GLASS (rare)
Vesicular	PUMICE		SCORIA

Volcanic tuff (fine-grained) and breccia (coarse, angular fragments) are common rocks formed in areas of explosive volcanic activity.

Some rocks are highly porous; pumice, rock froth in which volcanic gases were trapped, is light enough to float on water. Basalt scoria, with fewer gas-bubble holes, is heavier than pumice.

When *tephra* (solid, angular fragments) is blown out of a volcano, it forms *pyroclastic* rocks such as *tuff,* if fine-grained, or volcanic *breccia* (pronounced "brechia") if the fragments are mainly large.

Volcanoes build cones that differ in configuration, based on the building materials which, in turn, depend upon the explosiveness of eruption. Lying beneath the volcano in the control center is the magma, the critical factor in the system. In the magma is extremely hot and highly fluid, it will ordinarily be forced out of the vent without violent explosion. (The words *violent* and *quiet,* as used here, are only relative!) In these "quiet" eruptions, highly mobile basaltic lava pours from the vent and flows rapidly out away from the center, often for many miles. With successive eruptions, flow upon flow, a flattish, circular cone is eventually built up. These *shield cones,* so named because they resemble the round shields of ancient warriors, are the dominant type in the Hawaiian Islands.

In distinct contrast, when gases concentrate sufficiently in the volcanic vent, a violent explosion occurs, one that blasts through the solid rock, lifting large blocks and finer materials high into the air. The coarse materials fall back near the vent, building up a steep-sided *cinder cone* like the one in Figure I-7. Finer materials, mainly volcanic ash and dust, are carried for varying distances downwind. Explosive eruptions are common only where the magma is either silicic or intermediate in composition; however, when conditions are right, basalt cinder cones are built, as we will see when we visit Sunset Crater in Arizona.

Some volcanoes shift from violent explosions to quiet lava-producing eruptions; thus a layered cone is constructed, as shown in Figure I-8. The cone is built higher during violent

FIGURE I-6

Generalized vertical cross section of a shield cone composed of thick basaltic flows with thin ash layers between certain flows; also known as the "Hawaiian type" volcanic cone.

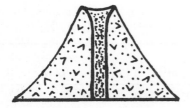

FIGURE I-7

Cross section of a cinder cone made up of large and small blocks and ash (cinders) which were blown out of the volcanic vent.

FIGURE I-8

Cross section of a stratovolcano or composite cone composed of alternating layers of lava and pyroclastic materials, forming high, steep-sided mountains.

FIGURE I-9

Generalized cross section showing intrusive igneous bodies, a volcano, and a lava flow.

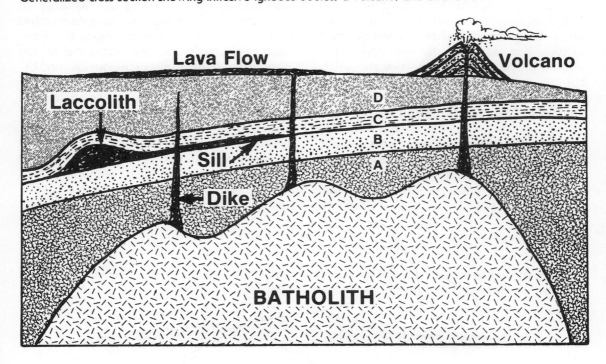

explosions and is extended outward during lava-flow eruptions. A large number of the world's majestic volcanoes are *composite cones* or *stratovolcanoes* — such as Fujiyama, Vesuvius, and many in the Andes and in our Cascade Mountains.

Igneous rocks that develop beneath the surface are found in igneous bodies of different sizes and shapes. By far the largest is the *batholith* in which most of the granite is found. Granitic magmas are formed by the melting of deep-seated rocks and are enlarged tremendously by continued melting of the roof rocks. Eventually, while the top of the magma chamber is still miles below the surface, melting ceases and later on crystallization begins. Because of the extremely large volume of the magma and the great thickness of overlying rocks, the heat escapes at a very slow rate, thus providing time for the growth of large mineral grains — hence the rocks are invariably coarse-grained.

As the magma cools, it shrinks, and eventually the reduction of volume of this huge mass is sufficient to produce severe stresses on the overlying rocks, which are broken, or "fractured," thus providing avenues for the escape of some of the still molten material. As shown in Figure I-9[2] this magma either solidifies in much smaller bodies such as dikes, sills, or laccoliths, or it escapes to the surface to form volcanoes or lava flows. Being much smaller and much closer to the surface, these intrusive bodies are composed of finer-grained rock, particularly those that are formed at shallow depths. Consequently, the latter are fine-grained and very similar to the rocks in the lava flows.

In your travels you are likely to observe in the wall of a canyon something similar to what is shown in Figure I-9. It is worth a stop to ponder what occurred and in what sequence. Which is the oldest rock and which is the youngest? The age of a rock refers to the time that it was formed, not where it was formed or where it is now. Where layers of sedimentary rock are exposed, the first of these rocks to be deposited is the bottom layer. In our cross-section we observe that the dikes cut through all of the sedimentary layers; therefore, the dikes are younger than the youngest layer, bed D. Since the rocks in the batholith and in the other igneous bodies are the same age, they are the youngest rocks. Formation A must have been there when the magma intruded; therefore, Formation A is the oldest rock exposed in the canyon.

When an intrusive igneous body is exposed, like the dike in Figure I-10, it means that erosion has been taking place over a long period of time. Imagine how long it must have taken to expose the huge Sierra Nevada batholith in California.

As indicated earlier, sedimentary rocks form from all kinds of pre-existing rocks, either by the consolidation of rock fragments or by chemical precipitation of material derived by chemical decomposition of pre-existing rocks. Ordinarily, those materials are transported a considerable distance from the source; in any case, before they can be transported they must first be loosened from the bedrock.

Weathering is the loosening process. When water penetrates into cracks or fractures and then freezes, the rocks are pried apart. Tree roots grow into the fractures and enlarge them. Expansion and contraction of the rocks as they are heated and cooled cause them to crumble. In such cases, physical forces are involved; the term *physical weathering* is applied to this process. When water combines with carbon dioxide they form a weak acid, carbonic acid, which attacks the minerals and decomposes them; this is *chemical weathering*. As an example, orthoclase feldspar is slowly decomposed

---

[2] Cross-sections are slices down through the earth's crust showing structures that are not exposed; plan-views show surface features as viewed from the air.

FIGURE I-10

Dike exposed by stream erosion near Willow Creek Pass in Middle Park, Colorado. (Photo by J. A. Campbell.)

and an entirely different mineral, namely *kaolinite clay,* is formed. *Calcite,* the main mineral in *limestone,* is taken into solution and carried away. Physical weathering and chemical weathering work hand-in-hand, each assisting the other, in the breakdown and separation of materials from the solid bedrock. The resulting soils vary widely, depending upon the *parent material* and the climate. Parent material derived from granite is distinctly different from basalt parent material; therefore the soils are different. A warm, humid climate is conducive to rapid weathering and soil development; therefore, *residual soils,* those formed in place, are generally deeper (thicker) than residual

soils formed in arid regions.

Plants are selective in their habitats. Climatic factors — temperature and precipitation — provide the basis for climatic zones, each with its own assemblage of plants; within each zone the soil determines the distribution of the plants. These ecological relationships are particularly well illustrated in park areas where the influence of humans is minimal.

The soil and other unconsolidated material do not accumulate indefinitely. They may move slowly downslope when freezing and thawing occur. On steeper slopes, a mass of the loose material, perhaps saturated with water, may slide down to the base of the hill, creating a *landslide.* These and other downslope movements are forms of *mass-wasting,* an important process in the leveling of the landscape. Note that there is no transporting agent involved; the force of gravity is sufficient.

Erosion is the removal and transportation of rock materials on the earth's surface. The face of the earth is gradually changed by erosion; high areas, even mountains, are eroded away, and the erosional debris is deposited in low areas, much of it in the sea.

The main eroding agents are water (streams and ocean waves), wind, and glacier ice. Although all play important roles, a significant part of the landscape bears the unmistakable imprint of stream erosion.

The evolution of landscapes, particularly those developed by stream erosion, is a fascinating subject. In mountainous areas where there are steep-walled canyons and fast-flowing streams with waterfalls and rapids abundant, we see a youthful landscape. One day this spectacular scenery will be gone: the sides of the canyons will be much less steep, stream gradients (slope of channel) will be flatter and more uniform, and most or all of the waterfalls will have vanished. It will become a *mature landscape.* But neither you nor I will be here to see it; it requires too long a period — millions of years. In order to see what the country will be like when it reaches the *mature stage* we must go to a place where the mountains have been eroded for a longer period, perhaps the Ozarks or the Catskills. In another several million years they too will be distinctly different; they will be worn down — in fact, they will be erased, and undulating plains will be there instead. In this *old age stage* of the *erosion cycle,* or *fluvial cycle,* the streams will be sluggish and unable to lower the land surface. Many mountain ranges have suffered this fate, but here or elsewhere, new mountain ranges have risen, thus preventing the land areas of the earth from becoming monotonous plains — *peneplains.*

What becomes of the vast amount of material eroded away during one of these erosion cycles? The streams, large and small, pick up the material loosened by weathering and carry it away. These materials, particularly sand and larger particles, act as abrasive tools to grind away at the bedrock. The impact of boulders against the sides and bottom of the channel breaks off chips or fragments of the bedrock. Abrasion and impact are particularly effective during floods when the stream is capable of moving coarse material along its bed, as *bedload. Potholes* are formed when, by the swirling action of the stream, boulders grind against the walls and floor and drill large wells in the bedrock floor, especially at the base of waterfalls. Here, the boulders are rounded and shaped, some of them into near-perfect spheres.

Other remarkable things happen to the materials as they are swept along, banging against each other in the violence of a turbulent stream. The feldspar minerals have good cleavage; that is, they break or split easily along planes of weakness. Consequently, they are readily broken down into smaller and smaller particles. Soft minerals such as the micas are ground down into tiny flakes. Small particles are formed by physical and chemical weathering along the channel and are thus made ready to be transported as *suspended load* by the next flood.

Quartz is less affected by weathering as it is carried to the sea. Quartz is the most durable of all the common minerals; it has no planes of weakness and can be broken only with difficulty. Consequently, there is an abundance of quartz sand along the seashore while many of the other minerals have been reduced to silt and clay.

At the edge of the sea, the coarsest materials are deposited first, near shore. Farther out, beyond the sand zone, silts and clays slowly settle out. In the right environment — warm, shallow water which is the habitat of lime-secreting marine animals — the material carried in chemical solution, mainly calcium bicarbonate, precipitates out in the form of the mineral calcite ($CaCO_3$), and from these limy muds *limestones* are formed.

Although most sedimentary rocks are deposited in the sea, as *marine* rocks, some are laid down in lakes or on land. Gypsum and rock salt are nonmarine rocks that are formed in large inland basins where, by evaporation, the solutions become sufficiently concentrated to cause chemical precipitation. Coal is another special case; it is formed in huge swamps where over a long period the plant materials that accumulate are buried and slowly converted to bituminous coal (soft coal).

Consolidation must take place to convert loose sediment into hard rocks. Pebbles and sand grains are cemented together by a cementing agent, usually silica or calcium carbonate, to form conglomerate and sandstone, respectively. Silts, clays, and calcareous muds are compacted by the weight of the materials deposited on top, forming shale and limestone.

In places, we find rounded structures, some spherical and others elongate, which because of their superior hardness have weathered out of the sedimentary rock in which they were imbedded. Some of these structures are *concretions* but others are *geodes*. Concretions may be formed on the sea bottom when masses of gluey material are rolled around by the waves; others form in loose sand, by continued cementation of the sand around a nu-

TABLE III

Sedimentary Rocks

Sediment	Sedimentary Rock
Gravel, cobbles, and boulders	CONGLOMERATE
Sand	SANDSTONE
Silt and Clay	SHALE
Calcium carbonate[a]	LIMESTONE[c]
Hydrous calcium sulfate[b]	GYPSUM
Sodium chloride[b]	ROCK SALT

[a] Chemical and biochemical precipitate

[b] Chemical precipitate

[c] DOLOMITE, calcium-magnesium carbonate.

cleus. As more and more of the sand is cemented, the concretion grows larger, sometimes to enormous size, as in Figure I-11. Geodes resemble concretions, but they are hollow and are therefore lighter in weight. They are formed by chemical solution that dissolves a rounded opening in limestone. Then, by chemical precipitation, layer after layer of material, usually silica, is deposited on the walls of the opening, partially filling the cavity. It is not uncommon to find geodes that contain excellent crystals of quartz or, in rare instances, beautiful amethyst. Some contain calcite crystals which, because they resemble canine teeth, are called dogtooth spar.

Fossils may be found in certain igneous and metamorphic rocks, but almost all fossils occur in sedimentary rocks. There are tree molds in the lavas of Hawaii and of Craters of the Moon National Monument; insects, leaves, and tree trunks are preserved in the volcanic ash of Florissant National Monument. Even fewer fossils are found in metamorphic rocks; however, here and there fossiliferous shales have been baked by the heat of intrusions and the fossils are not only recognizable but may be identifiable in the resulting *hornfels*.

Sandstones may contain fossils — the bones of the ''Terrible Reptiles'' in Dinosaur National

FIGURE I-11

Concretion weathered out of shales near Belen, New Mexico. (Photo by M. R. Isaacson.)

FIGURE I-12

Glimpses of evolution, beginning with primitive forms in the ancient Precambrian rocks. (For proper sequence, read from bottom to top.)

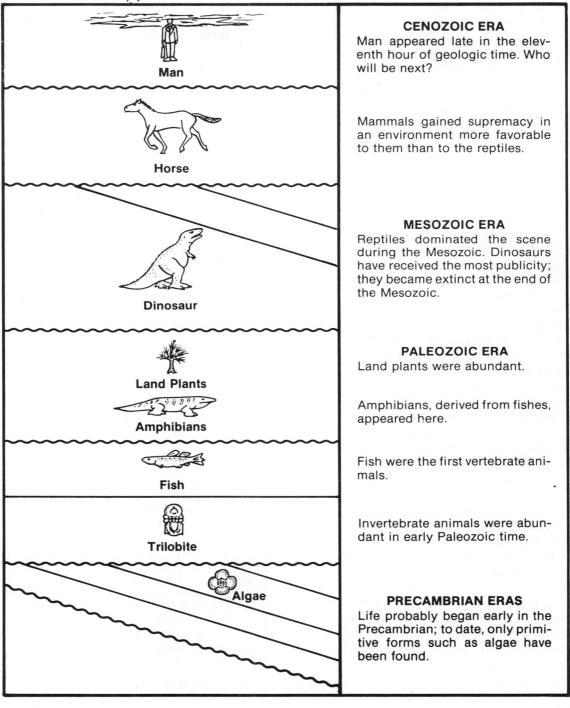

CENOZOIC ERA
Man appeared late in the eleventh hour of geologic time. Who will be next?

Mammals gained supremacy in an environment more favorable to them than to the reptiles.

MESOZOIC ERA
Reptiles dominated the scene during the Mesozoic. Dinosaurs have received the most publicity; they became extinct at the end of the Mesozoic.

PALEOZOIC ERA
Land plants were abundant.

Amphibians, derived from fishes, appeared here.

Fish were the first vertebrate animals.

Invertebrate animals were abundant in early Paleozoic time.

PRECAMBRIAN ERAS
Life probably began early in the Precambrian; to date, only primitive forms such as algae have been found.

Monument, for instance — but most fossils occur in limestones and shales. The limestones in Grand Canyon and in Big Bend contain excellent records of past life for *paleontologists* to decipher. By identifying the fossils and fitting them into the evolutionary scale, the geologic age of the formation is established.

Environments of the past can be reconstructed by using the evidence found in sedimentary rocks. Here again, fossils are invaluable keys to the past. Shales that contain marine fossils were obviously laid down in the sea, where the descendants of these animals now live. Dinosaur skeletons indicate a terrestrial (land) environment, in most cases low-lying swampy conditions. Sedimentary structures such as ripple marks, mud cracks, and raindrop impressions suggest deposition in areas that were covered by shallow water for part of the time.

Sandstone with eolian (wind) *cross-bedding* is indicative of a dry climate, perhaps a desert where there was little vegetation to prevent active wind erosion and deposition, as in the Sahara today. As our travels take us into park areas where sedimentary rocks occur, these bits of evidence will help us interpret environments that existed millions of years, even hundreds of millions of years, ago.

The development of metamorphic rocks is the next and final step in the Rock Cycle. Metamorphism means change in form. The changes are brought about mainly by heat and pressure, although on occasion, solutions derived from magmas play a significant role. Therefore, although weathering brings about change, weathering is not a type of metamorphism because it occurs on or near the surface where the rocks are at low temperatures and pressures.

Because metamorphism takes place at depth, no one has observed metamorphic rocks until, after a long period, they have been exposed by erosion. Their formation has been established by carefully and logically applying the principles of physics and geochemistry.

Those processes are too complex to be outlined here in more than the simplest way.

Most metamorphic rocks are formed by *regional* metamorphism far below the surface, in an environment where the rocks are subjected to enormous pressures, mainly shearing pressures. Picture a pack of cards flat on the table. Place your hand on top and push sideways; each card is *sheared* along the top of the card beneath it. This, in oversimplified form, is one phase of metamorphism. Moreover, at a depth of several miles, temperatures are also high. Minerals that were stable in a near-surface environment cannot withstand the pressures and heat of the metamorphic environment. Blocky minerals such as orthoclase are changed to platy minerals, mainly muscovite mica, when one part of the mineral is sheared over the adjacent parts. Most rocks formed in this way are *foliated;* that is, they are composed of *folia,* or thin leaves. *Slates, mica schists,* and *chlorite schists* are examples. There are also banded rocks called *gneisses* and massive (unlayered) *quartzite* and *marble* (see Table IV — Metamorphic Rocks).

On a lesser scale, thermal or *contact* metamorphism affects rocks that are in contact with igneous intrusions. The heat from the magma, in some cases aided by hot solutions that pen-

TABLE IV

Metamorphic Rocks

Original Rock	Metamorphic Rock
Sandstone	QUARTZITE
Limestone	MARBLE
Shale[a]	SLATE — PHYLLITE — SCHIST[b]
Granite	GRANITE GNEISS[c]

[a] Shale may be baked by the heat of intruding magma; the resulting rock is called HORNFELS.

[b] Slate is composed of microscopic plates of muscovite; phyllite and schist are composed of larger plates of mica, the product of prolonged metamorphism.

[c] Gneiss is made up of bands of light and dark minerals.

etrate into the adjacent rocks, bakes and hardens the *country rock.* A good place to observe the effects of contact metamorphism is in a road cut where a dike is exposed in contact with sedimentary rocks.

Although much remains to be learned about regional metamorphism, it probably occurs mainly in places where the earth's crust is being subjected to severe deformation by almost unbelievable forces. An area of active mountain-building would be such a place, where rocks near the surface are being folded and broken by faults. Below this so-called *zone of fracture,* the zone where shearing pressures dominate, is the *zone of plastic flow,* or the *zone of metamorphism.*

Mountain-building takes place in many ways, but most major mountain ranges are formed by lateral (horizontal) forces, the origin of which is not completely understood. At one time these forces were assumed to be the horizontal components of the downward force of gravity. It was believed that the earth was cooling and shrinking and that, as the interior continued to shrink, the *crust* was crumpled into mountains. But is the earth actually getting smaller? True, heat is flowing from the interior

to the surface, but heat is concurrently being generated by the decay of radioactive elements deep in the earth. Probably our planet is not changing materially in size. Then what causes the crust to be bent, broken, and thrust up in the form of mountains?

Until a few decades ago, geologic interpretations were based almost entirely on the rocks and structures exposed on the land masses; essentially no consideration was given to the evidence hidden in the vast ocean basins which cover more than 70 percent of the earth's surface. When geologists acquired a significant amount of information about the ocean floors, it became evident to a few imaginative minds that it was time to reexamine our time-honored geologic concepts. As is invariably the case when "sacred" ideas are challenged, the new concepts met with considerable opposition. Even so, a major upheaval in geologic thinking was in progress, and by the late 1960s the *plate tectonics theory* was widely accepted.

According to this theory the earth's crust is a mosaic of large plates that have been and are being moved about by forces generated in the plastic zone that underlies the crust.

By referring to the cross-section in Figure I-14, you will soon grasp the essentials of the plate tectonics concept. Note first that the ocean floor is being spread apart by the outward movement of the plastic material below the crust. Sea-floor spreading is taking place along the Mid-Atlantic Ridge and along similar ridges in the other oceans. Basaltic magmas which rise to fill the gap between the plates pour out onto the ocean floor, and the resulting rocks are added to the plates on both sides of the ridge. Where volcanoes develop and build their cones above the sea, islands are formed. Iceland was first a tiny island, like the Island of Surtsey a few miles to the south; now Iceland is about 70 miles (115 km) across. Can you picture an Atlantic Ocean so narrow that you could almost jump across it — a few hundred million years ago? Next, turn the

FIGURE I-13

Dike cutting through sandstone and shale, on Interstate 25 near the Colorado – New Mexico boundary. The shale in contact with the dike is now hornfels; the sandstone is now quartzite.

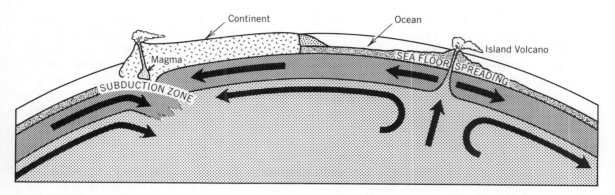

FIGURE I-14

Schematic cross section showing possible mechanisms involved in plate tectonics.

time-clock back one more tick and see that there was one gigantic continent, Pangaea, a composite of all of the continents in existence today.

As the continents moved apart, their leading edges collided with oceanic crust. In these collision zones, the rocks were crumpled, broken, and faulted, thus forming long ranges of coastal mountains. In places the heavier oceanic crust "dived" *slowly* under the mountains in what are called "subduction zones." Here, magmas that were generated at depth intruded upward toward the surface — to break through in places as volcanoes, as shown on the left in Figure I-14. The Andes Mountains, a classic example, result from the oceanic plate diving under the western coast of South America.

The earth is three-dimensional and thus plates may move in any direction. Along the west coast of North America the oceanic plate is moving generally northward and is carrying a slice of California with it — the part that is west of the San Andreas Fault. And parts of what is now Alaska were apparently rafted northward over long distances, perhaps from as far south as the equator!

The concept of moving continents is far from new. In 1912 Alfred Wegener, a German meteorologist, startled the world with his theory of "continental drift," but it was too

outrageous to be taken seriously. It was a "known fact" that the continents have always been where they are now! When the older author of this book was in school during the 1930s, some professors mentioned Wegener, suggesting that we shouldn't be wasting our valuable time on his "fantastic fantasy." After all, if we accepted his ideas we would have to throw away most of what we had learned for the past 100 years! Now, Wegener's concepts are essential to the plate tectonics theory — the "new geology."

Unfortunately, it was not until long after Wegener lost his life trying to prove his theory (he froze to death out on the Greenland ice cap) that his ideas were accepted. (For an account of the ridicule to which he was subjected, read Russell Miller's *Continents in Collision*, one of the Planet Earth series published by Time/Life in 1983.) It is a sad reminder of the tribulations of Galileo, minus the threat of physical torture, and of Darwin's harassment in more modern times. It is also a well-written chronicle of the evolution of geologic thinking — from the pronouncement of Archbishop Ussher up through the development of the plate tectonics concept. (As a warning, if you don't think much of plate tectonics and are dead set against being converted, by all means leave this book alone!)

Here we can only glimpse this earth-shaking

concept; later, in the geomorphic provinces where plate tectonics are active today, particularly in those bordering the Pacific, we will be in a better position to visualize what is taking place. Meantime, those interested in delving into the depths of the subduction zones will find an abundance of fairly heavy reading in the references given in the Bibliography.

Perhaps before accepting the new geology as being the final, the everlasting theory, however, we should pause for a moment and reflect on the fact that more than a century ago Charles Lyell was being congratulated for *his* "new geology." From this we may conclude that what we have now is merely the newest new geology and that revolutionary concepts will continue to be brought to light in the future. Nevertheless, the plate tectonics theory is without question the most important development in geology in our time.

In our discussion of the origin of sedimentary rocks, we left them as they were deposited, essentially flat-lying. Should you drive westward across the Great Plains, you will go through occasional road-cuts where horizontal sedimentary rocks are exposed. When you get to the Black Hills in South Dakota, however, you will note that the rocks are not flat-lying but tilted, and that on the east side they are tilted (*dipping*) to the east, away from the

FIGURE I-15

Cross section of folded sedimentary beds after prolonged erosion; restoration shown by dashed lines. Arrows indicate the direction of forces involved.

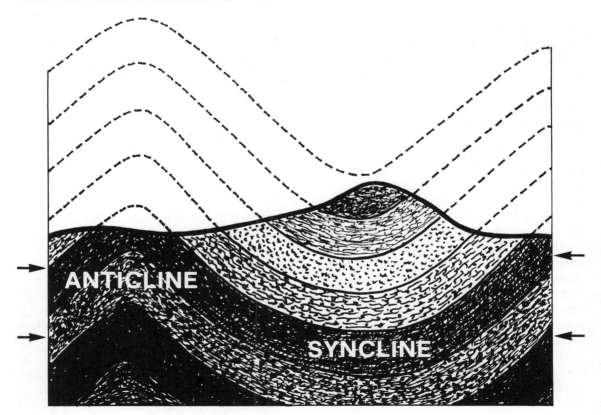

FIGURE I-16

Sand Creek Anticline in northern Larimer County, Colorado, looking north to the Gangplank in southern Wyoming. (Airview by J.A. Campbell)

mountains. Later, you will see much older metamorphic and igneous rocks that form the higher mountains. Still later, when you have crossed the Black Hills, you suddenly realize that here on the west side you are seeing the same sedimentary rocks, but that they are dipping in the opposite direction, toward the west. What happened? How did they get that way? That night, sitting around your campfire, suddenly the lights begin to flash and the answer is there. The Black Hills were domed up and the sedimentary rocks on top were eroded away, leaving the "stubs" dipping away from the core on all sides.

Then, if you swing down through Wyoming into Colorado and head west into the Front Range, again you will see sedimentary rocks dipping away from the mountains. Could the Rockies have been formed in about the same way as the Black Hills, only on a much larger scale?

By now you may have decided that after rocks are formed, they are deformed — at least in places. Many of the rocks have been bent, twisted, wrinkled (folded) and/or broken, depending on the type and intensity of the forces

applied. Two of the many *geologic structures* formed are *anticlines* (upfolds) and *synclines* (downfolds); they are discussed together because they frequently occur together.

You can simulate folding by using a stack of rugs to represent the different layers of rock. When you push against one side of the stack, the rugs buckle up and form an anticline; continue pushing and another anticline develops, with a downfold, a syncline, between the two. Of course, rugs are limber and fold easily; rocks are rigid and brittle, and they would surely break when pressure is applied. But under normal conditions, the pressures build up gradually, allowing time for the rocks to adjust internally; consequently, although some fracturing does occur, it is of minor importance. Also, except for those on the surface, the rocks are under great pressure, enough to prevent more than tiny cracks from forming. Thus, instead of breaking, the rocks are folded into broad, essentially symmetrical anticlines and synclines, as shown in Figure I-15. Note that anticlines and synclines are *structures* below the surface and that surface topography may not reflect those structures; here in our cross-section the synclinal structure is represented by a ridge.

FIGURE I-17

Cross section in which an asymmetrical anticline has been thrust faulted.

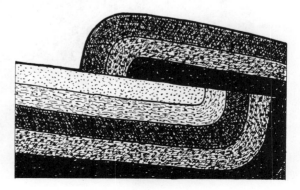

FIGURE I-18

Schematic cross section of a normal fault.

If compressional forces are long-continued, the anticline will become asymmetrical, as shown in Figure I-16. Structures such as Sand Creek Anticline occur in many places along the base of the Rockies and elsewhere. Some of these anticlines contain oil and gas — those in which a porous formation, usually sandstone, is capped by an essentially impervious shale which traps the oil.

In some cases, however, the pressures continue for a longer period than at Sand Creek; then the rocks are broken and *thrust faulted,* as shown in Figure I-17. Refer again to Sand Creek Anticline, noting the rocks shown in the distance. At the time folding occurred, the rocks now exposed in the anticline were buried deep beneath the rocks you see in the background. Sand Creek and its tributaries stripped off all of those rocks, exposing the hard rocks of the anticline; then Sand Creek cut a small canyon into the hard rocks, thus exposing the "innards" of the anticline. Such exposures afford a good opportunity for geologists to learn more of the details of the process of folding.

Thrust faults (see Figure I-17) may develop to enormous size. Later, when we travel to Glacier National Park, we will see a thrust fault that is more than 100 miles long, where the *displacement* (amount of offset along the fault) is more than 30 miles.

Folds and thrust faults shorten the earth's crust and are the results of compressional forces. But there are places, such as parts of the Basin and Range Province, where the earth's crust is being pulled apart. This causes faulting to occur, but the faults are distinctly different; one block simply slides down the sloping fault plane, responding to the force of gravity. There are many *gravity faults* and a few are extremely large. The Sierra Nevada is a giant *fault-block;* a system of gravity faults, also called *normal faults,* lies at the base of the precipitous east face of the Sierra. Figure I-18 shows in simplified form what the Sierra Nevada looked like before it was dissected by streams and later by glaciers.

FIGURE I-19a

Small normal faults offsetting beds of volcanic ash in a highway cut on U.S. 180 near Alma, southwestern New Mexico.

FIGURE I-19b

Cross-section sketch of I-19a.

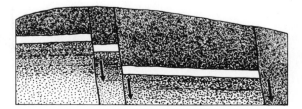

There are many more faults than most people realize, because in many places they are not well exposed. Many unusually straight mountain valleys are straight because the streams cut deeply into the broken, easily eroded material in fault zones; mountain passes are frequently at the head of two such valleys. Many faults are covered by overwash material, glacial deposits, and soil. Road-cuts, particularly the huge ones along some of the interstate highways, are likely places to observe faults. Note the small gravity faults in Figure I-19, where a 6-inch (15-cm) layer of white volcanic ash serves as a key bed in measuring the displacements. The displacement along the fault at the right cannot be measured because the ash bed is at an unknown depth below the roadbed.

Of far greater magnitude are the faults or fault systems along which huge plates of the earth's crust grind past each other. The San Andreas fault system in California is our best known example; here, the Pacific plate is moving northward past the continental plate. The displacements are almost entirely in a horizontal direction in contrast to both gravity and thrust faults which have important vertical components of movement. The San Andreas and similar faults are generally called *transcurrent faults*.

Faulting consists of sudden adjustments interspersed with long periods of quiet, during which stresses gradually build up—to the breaking point. Then, without warning or accurate prediction, displacement occurs that can be measured in inches or in tens of feet. Regardless, these abrupt displacements generate earthquake waves that radiate out from the center. The major earthquake in 1959 that rocked the Yellowstone area was caused by gravity faulting; maximum displacement was 19 feet (6 m), almost all vertical movement. There are a number of possible causes of earthquakes and earth tremors, including volcanic explosions, but most of the devastating quakes are generated by faulting.

Folds and faults generally document single events in the geologic history of an area, perhaps of a mountain range; *unconformities* record a series of events, in some cases involving many millions of years. An unconformity is a buried erosion surface, one that represents a gap in the geologic record. Figure I-20a shows a continuous sequence, a *conformable sequence*, in which the shale (sh) bed B was laid down on top of the sandstone (ss). Then, the limestone (ls) bed C was deposited on the shale, without any interruption. In I-20b, perhaps a hundred miles away, we find the same sandstone and the same limestone, but the

FIGURE I-20a

Conformable sequence of sedimentary rocks A, B, C, laid down in sequence.

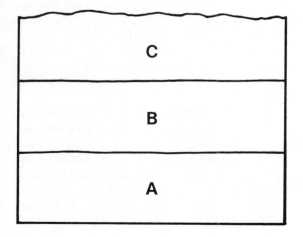

FIGURE I-20b

Parallel unconformity or disconformity; bed B was eroded away before C was laid down.

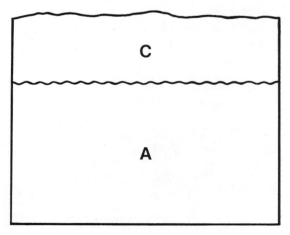

shale bed B is missing. Assuming that all of these rocks were deposited beneath the sea, the following sequence is indicated for the second area: the sandstone was deposited and then part of the shale, as in the first area; then the area was uplifted above the sea and the shale was eroded away; later, with lowering beneath the sea again, limestone was deposited on the erosion surface, converting it into an unconformity. Because the beds above and below the unconformity are parallel, this type of unconformity is called a *disconformity,* the simplest of the three types.

Suppose that several beds were deposited and then folding and uplifting began. As the folded beds appeared above the sea, erosion by streams would begin to plane them down to sea level. Eventually they would be truncated, as shown in the lower part of Figure I-21. Then with the submergence of the area, deposition of beds Y and then Z would take place. This unconformity is called an *angular unconform-ity* for obvious reasons. Although exceptions are possible, generally much more time is involved here than in the case of the disconformity, inasmuch as folding is an extremely slow

FIGURE I-21

Angular unconformity; the rocks were tilted after bed *D* was deposited. Then, after the area was leveled by erosion, beds *Y* and *Z* were laid down.

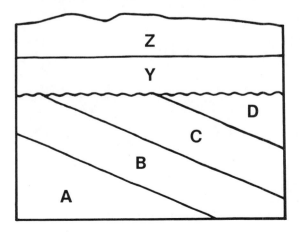

process. And frequently the folding is of widespread significance, as in the building of mountains — folded mountains. Actually, Figure I-21 is a simplified cross-section of the rocks exposed in northernmost Colorado where the Front Range meets the Great Plains. The tilted rocks were deposited in the geosyncline before the Rockies were born and were subsequently folded and uplifted above the sea; during and after the uplifting, these rocks were eroded by streams, and finally an erosion surface was developed across the tilted beds. Still later, streams deposited huge quantities of material on the erosion surface, thus forming the angular unconformity. To complete the sequence, the Rockies had to be uplifted to cause the streams to cut down in order to expose the unconformity and to expose Sand Creek Anticline as we see it today. By determining the geologic age of the rocks above the unconformity and of the youngest tilted formation, geologists were able to establish the time when the Rocky Mountains were born.

The third type of unconformity is called the *nonconformity.* Here, the rocks below the unconformity are deep-seated rocks, either metamorphics or granitic igneous rocks or a combination of the two. These rocks, formed miles beneath the surface, had to be unroofed and then deeply eroded during the formation of the erosion surface; thus the length of time is simply mind-boggling. Then, as with the other two types, sedimentary rocks or lavas were deposited on the old erosion surface. Thus we see that unconformities contribute much in the deciphering of ancient history — geologic history. They provide essential information not only in localized areas but in entire mountain ranges as well.

The geologist looks at a mountain or mountain range in terms of the dominant geologic processes involved. Volcanic activity was the process that formed the Hawaiian Archipelago, a gigantic mountain chain. Folding was responsible for the Black Hills, as discussed, and the Uinta Mountains of Utah, a huge anticline. Block-faulting has formed many small ranges and the majestic Sierra Nevada. Thrust faulting was the main process involved in the creation of the mountains in the Glacier National Park area. In some cases, as in the Southern Rockies, two processes are of about the same importance; closely related, both the results of lateral compressive forces, folding and thrust faulting were the primary processes in the Southern Rockies and in many other large mountain systems.

Mountains generally are products of internal geologic processes; regardless, all mountains are subjected to the external forces of erosion. Stream erosion is at work on all mountains; those that are high enough to receive sufficient snowfall, where more snow falls than melts each year, are being eroded by glaciers — those mighty rivers of ice that have produced our most spectacular mountain scenery. The sharp spines, or *horns,* were formed by three or more glaciers enlarging their *cirques,* grinding away the mountain on all sides; Matterhorn in the Alps is a beautiful example.

To some people, glaciers are mysterious creatures of the past, now to be seen only in distant lands. Actually, there are thousands of glaciers on the earth, and hundreds of them are close at hand, many of them in our national parks. They are smaller than at certain times in the past but will expand again if history repeats itself. In fact, a significant number have been expanding since about 1945.

Glaciers are some of man's best friends.

FIGURE I-22

Unconformity (nonconformity) in the Grand Coulee area, Washington. Cretaceous granite batholith was unroofed by erosion before layers of Miocene basalt were laid down on top. (Photo by E. Kiver.)

They store water and release it gradually during the summer. Streams that would otherwise "go dry" are maintained throughout the summer, thus supplying critical irrigation water in dry regions. And they provide cities with reliable supplies of water and hydroelectric power.

Someone said that they are our "frozen assets" and so they are, up to a point. But contemplate what our situation would be if all of our assets were frozen and much of the United States, including the Corn Belt, were beneath glacier ice. Food and energy would be in short supply; even living room would be near-critical.

Because glaciers have come and gone several times in the past, many scientists have anticipated that another advance would be upon us "sometime in the future." Until recently, the assumption was made that climatic changes that bring on glacial advances are invariably slow and that we would have more than ample time to adjust to the change. Recently, however, evidence has been found that indicates that in at least one instance the change was extremely rapid. If true, this opens up a Pandora's Box of large dimensions — for the experts who plan for food, energy, and population.

The ongoing cooling trend which began about 1945 has included both positive and negative extremes in temperature and precipitation, and not all areas have been affected alike. Nevertheless, a significant number of glaciers in most of the glacier areas of the world are now advancing — glaciers that had been retreating for about 100 years. There is no reason to assume that this latest climatic change will be extremely rapid and long-continued, thereby causing a rapid expansion of the glaciers; there is also no reason to assume that it will not be!

Here we are mainly concerned with glaciers, present and past, and how they have changed the face of the earth, especially in our national parks. What causes large masses of ice to move? Glaciers that form along the high divides in the mountains move down the valleys, apparently under the influence of gravity. Without question, gravity is one of the forces, but when we find places where glacier ice has moved *uphill*, it becomes necessary to envision a more complex mechanism than mere gravity movement.

Exhaustive treatment of the mechanics of glacier movement is beyond the scope of this book; however, to avoid the subject entirely would be sheer negligence. First of all, ice is a peculiar rock — different from all others in one respect: when sufficient pressure is applied, ice will melt, even at temperatures considerably colder than its normal melting point. Therefore, when snow that is converted to ice continues to accumulate year after year, an icefield is formed.[3] An icefield is stationary; movement within the ice mass is necessary for it to become a glacier. When an icefield becomes sufficiently thick, the weight of the overlying ice provides the pressure to melt the ice at the base. The meltwater flows into openings and refreezes — a process called *regelation*. Expansion during refreezing creates local pressures and causes additional melting and refreezing. Thus, the lower layer of the glacier is the active layer — the layer that is responsible for the movement of the entire glacier. Although such factors as temperature and slope affect the thickness required, ice that is 200 feet thick provides enough pressure to activate the glacier — to convert an icefield into a glacier.

Glacier ice moves in the direction of least pressure — downvalley for valley or mountain glaciers, and out away from the accumulation centers of continental glaciers, or ice sheets. Melting is more rapid at the ice front than elsewhere, and in order for the glacier to advance,

---

[3] The term *icefield* is sometimes used incorrectly. For example, the Columbia Icefield in the Canadian Rockies should be called the Columbia Ice Cap Glacier.

FIGURE I-23

Alpine or valley glacier in Glacier Bay, Alaska. Note medial moraine (black ridge). McBride Glacier and Muir Inlet (fjord).

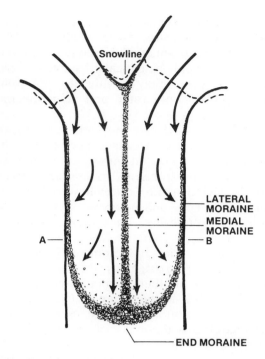

FIGURE I-24a

Schematic plan view of lower end of a valley glacier; arrows show direction of ice movement.

FIGURE I-24b

Cross section of A – B, Figure 24a.

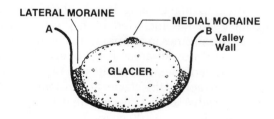

the amount of ice moved forward must exceed that which is lost by melting. Somewhere on the glacier the loss by melting just equals that gained from snowfall. Above this line — called the *snowline* — is the *accumulation zone;* below the snowline, where melting exceeds the accumulation, is the *zone of wastage*. It is here that boulders and smaller materials carried in the ice are deposited by the melting of the ice along the sides and at the front of the glacier. Thus, along the sides, side or *lateral moraines* are formed. *End moraines* — ridged deposits at the end or front — are formed where melting equals the forward movement of the ice front.

As the glaciers grind their way down the valleys, they deepen and widen them, and the V-shaped valleys are remodeled so they are U-shaped. As you drive up a canyon that has been only partially glaciated, you will immediately recognize the glaciated section when you reach it, even though the end moraine may be absent, having been eroded away. In the U-shaped glaciated portion you will see glacial polish on the rocks along the sides, and close up you will find *glacial grooves* and scratches, or *striations*. Thus, the glaciers have left their own special imprints on the landscape, features both bold and delicate.

Not all glaciers originate in mountains. *Continental glaciers,* or *ice sheets,* develop on relatively flat terrain, generally in the high latitudes. Those that covered northern United States originated in central Canada; they extended southward at least four times during the *Pleistocene Epoch* and overran essentially all areas north of the Missouri and the Ohio rivers. Unlike the mountain or *alpine glaciers* which sharpen topographic features, ice sheets tend to smooth the topography by shaving the hilltops and filling the valleys. Thus the areas covered by these glaciers were also profoundly changed, but in a different way.

The third type of glacier is found at the base of mountains, on the flat *piedmont;* therefore the term *piedmont glacier* is a logical designation. Valley glaciers which are now confined to the higher mountains were much longer 20,000 years ago, and some of them pushed out beyond their canyon mouths and spread out over the piedmont. With but few exceptions, the piedmont glaciers of today are confined to the high latitudes. Malaspina Piedmont Glacier in Alaska is a classic example.

Live glaciers, some of them highly active, are abundant in the park system. There is no better place to see them in action; by observing these, we can better understand glacial features elsewhere that were fashioned by the glaciers of the past.

With this brief introduction, the reader is invited to enjoy a beautiful book, *Glacier Ice,* in which Post and La Chapelle (1971) present outstanding photographs, many from the air, of glaciers in all parts of the world. Complete with text material, this book is an excellent reference on glaciers and glaciology.

We have seen that the topography in certain areas is clearly the work of streams, in others the work of glaciers. In still other places, the landscapes were unmistakably fashioned by the wind, particularly in dry regions where the sparse vegetation is unable to hold the soil in place. Large dune areas are common in true deserts. But sand dunes are also common in other areas — wherever loose sand is abundant — such as seashores and lakeshores. Dunes are common features in a number of the park areas: Great Sand Dunes, White Sands, and Death Valley national monuments, and in Padre Island National Seashore and In-

diana Dunes National Lakeshore. However, the dunes of ancient days were even more widespread; the Navajo Sandstone in Zion and other Colorado Plateau parks show the same eolian cross-bedding that is found in modern dunes. By observing how the modern dunes are being formed, we can re-enact the *Jurassic* scene.[4]

Wind erosion occurs over large areas, but in certain places large depressions — *deflation basins* — were formed where the wind concentrated its fury. Such depressions are likely to be filled with water for short periods following rainstorms.

Along seashores, and to a lesser extent along lakeshores, the work of the waves is clearly marked. In the bold wave-cut cliffs in Acadia National Park and on the opposite side of the continent in Olympic National Park, the force of the endless waves is evident. *Sea caves, sea arches,* and *stacks* are other phases of the ever-changing oceanscape.

Thus far, we have concerned ourselves mainly with surface features. In some places, what has been developing in the underworld for centuries of centuries is for some people even more fascinating. Caverns, some small and some mammoth, are found in most of the states; in the park system there are several, including Carlsbad Caverns (New Mexico), Wind Cave (South Dakota), Lehman Cave (Nevada), and Mammoth Cave (Kentucky), the largest cavern system in the world.

What do these caverns have in common? The climate varies from very wet to very dry; some are in flat country while others are beneath mountains. But almost all are in limestone, the most common rock that is susceptible to chemical solution. Rainwater charged with carbon dioxide percolates down into fractures in the limestone, dissolving it and enlarging the fractures. Certain limestone beds are more soluble than others, and the caverns

---

[4] The Jurassic is a geologic period during the Mesozoic Era; refer to the geologic time scale on page 34.

are developed laterally along these layers. Then, if many caverns are connected and there is an outlet to a surface stream, an underground river develops that erodes and enlarges the cavern system. Eventually, with enlargement, the cavern roofs collapse, forming depressions called *sinkholes.*

Most caverns are adorned with *stalactites, stalagmites,* and other features formed by chemical precipitation. (Remember that the "tites" hang down and the "mites" stand up!) When they join, *columns* are formed. Usually, the stalactites develop along fractures in the roof; eventually, with enlargement, a continuous curtain is developed. There appears to be no end to the fascinating features that surface creatures seldom see.

Most solution caverns are in limestones, but in a number of places they are found in marble, the metamorphic equivalent of limestone. Calcite, the main mineral in both marble and limestone, has essentially the same solubility wherever it occurs. The "Marble Halls of Ore-

gon" in Oregon Caves National Monument and Crystal Cave and others in Sequoia National Park are in marble.

GEOLOGIC TIME

People are in the habit of thinking of time in terms of years, months, and days; consequently, when geologists talk casually about millions, even billions, of years, some people become ill at ease. But when we stand on the rim of Grand Canyon we see many layers of sedimentary rock, each of which required thousands or even millions of years to form. And when we look deeper, down into the inner gorge, and see metamorphic rocks which are older still, we suddenly realize that we must develop a "feel" for the immensity of geologic time if we are to appreciate the ancient history so vividly written in the rocks.

Our present conception of geologic time was developed slowly and with great difficulty. It was a long road from Archbishop Ussher's 1650 decree that the earth was created in 4004 B.C. to the generally accepted view that the earth is about 4.5 *billion* years old. Through the years, many different methods were used in attempting to determine the age of the earth; however, without exception, each method involved assumptions that ultimately caused it to be discredited. Finally, only a few decades ago, the radiometric method was developed and is now accepted as reliable. The oldest rock that has been dated is about 4 billion years old, and it is surrounded by rocks that are somewhat older. Perhaps still older rocks will yet be found, thus extending the record back even farther.

Long ago, geologists saw the need for divisions and subdivisions of geologic time, and a standard geologic time scale was agreed upon. The *eras,* of longest time span, were set up to represent the time of one worldwide *orogeny* (mountain-building) to the next. We know now that orogenies occur more frequently but

FIGURE I-25

Grand Canyon, John Wesley Powell's "Book of Geology," where some of the continent's oldest rocks are exposed in the Inner Gorge.

in restricted areas; nevertheless, the eras appear to be here to stay. Rocks formed during the Paleozoic Era contain fossils, remains of former life; rocks older than Paleozoic do not, or so it was believed at that time. Therefore, Paleozoic — which means "ancient life" — was an appropriate name. Mesozoic ("middle life") and Cenozoic ("recent life") are successively more recent eras. The time before the Paleozoic was essentially an unknown. It was rather arbitrarily divided into two eras, the older Archeozoic ("beginning life") and the younger Proterozoic ("fore life"), without the realization that each of these two eras was of longer duration than all three of the later eras. Now we know from radiometric dating that geologic time prior to the Paleozoic amounts to almost seven-eighths of all geologic time; in addition we know that each successive era is significantly shorter than the one before it. Although we now have a different concept of geologic time, the original system, imperfect though it is, serves its intended purpose reasonably well.

The Paleozoic and younger eras were divided into shorter time-intervals called *periods;* the periods of the Cenozoic Era were divided into *epochs,* as indicated in the Geologic Time Scale. Note that the oldest era, the Archeozoic, is at the bottom and that the youngest era, the Cenozoic, is at the top.

One commonly used designation is "Precambrian." The Cambrian Period is the oldest period of the Paleozoic Era; Precambrian is a term used for the time and the rocks that are older than Cambrian. In many areas, the ancient rocks are so complexly intermixed that precise ages have not yet been determined.

Geologic time scales appearing in textbooks differ in certain details. For example, in some cases the Quaternary Period is divided into the Pleistocene and Recent epochs. In most later books, "Recent" has been replaced by the designation "Holocene." Here, as you see in Figure I-26, the Pleistocene is the only epoch of the Quaternary Period.

The Pleistocene or Glacial Epoch, which began about 3 million years ago, was the period during which the glaciers advanced over the land several times, only to melt away (retreat) during warm intervals called "interglacials." The retreat of the last great ice sheets began about 11,000 years ago. The warming of the climate continued until about 5000 years ago when it was considerably warmer than now; the period from about 6000 years ago to 4000 years ago is called the Thermal Maximum.

Again the climate cooled, beginning about 4000 years ago, and as existing glaciers expanded, many cirque glaciers were reborn and later extended far downvalley. The climate fluctuated, as did the glaciers. During the 1600s the glaciers moved down the valleys in the Alps, and one village, St. Jean de Perthuis, was overwhelmed. The helpless villagers could do nothing but stand and watch the monster demolish and overrun one after another of their homes. (There goes the neighborhood!)

If we look into the early history of geology we find that in 1839 Charles Lyell defined the Pleistocene as the period since the Pliocene, as reported by Flint (1971). Later, the interpretation somehow developed that the Pleistocene ended "about 10,000 years ago," at which time the "post-glacial" or Recent (now Holocene) Epoch began. Although this view is now widely accepted, a number of geologists object to it, pointing out that in many places in the world glaciers are large and active today. Would anyone have the temerity to argue that the ice age is over in Greenland and Antarctica, or even on Mt. McKinley?

Matthes (1942) introduced the term *Little Ice Age* for the period beginning about 4000 years ago, during which time there were at least three significant glacial advances, the latest beginning during the 1940s. When we get to Glacier Bay, Alaska, we will examine the records of this and earlier advances. The term *Neoglacial* suggested by Porter and Denton

FIGURE I-26

Geologic Time Scale.

ERAS	PERIODS	EPOCHS	DOMINANT LIFE
CENOZOIC	Quaternary	Pleistocene 3 m.y.*	"Age of Mammals"
	Tertiary	Pliocene	
		Miocene	
		Oligocene	
		Eocene	
		Paleocene 70 m.y.*	
MESOZOIC	Cretaceous		"Age of Reptiles"
	Jurassic		
	Triassic	225 m.y.*	
PALEOZOIC	Permian		"Age of Invertebrates"
	Pennsylvanian		
	Mississippian		
	Devonian		
	Silurian		
	Ordovician		
	Cambrian	600 m.y.*	

PRECAMBRIAN ERAS   Approx. 7/8 of   Geologic Time
    PROTEROZOIC
    ARCHEOZOIC
(Other divisions of the Precambrian are used in certain areas)
                        4500 m.y.*
			(Fossil Record
			Very Incomplete)

*m.y.—million years since the beginning of era, period, or epoch.

(1967), is used instead of Little Ice Age by most workers today.

To summarize, the Pleistocene continues. Many scientists strongly believe that we are living in an interglacial and that huge ice sheets will again overrun large sections of the continents. If this proves to be true, the Little Ice Age will give way to a colossal icing that will provide one of the greatest challenges that civilization has ever faced.

Before leaving the subject of glaciation, we should take note that although the Pleistocene is frequently referred to as *the* glacial period, actually it is merely the latest of several; glaciers were widespread during a number of geologic periods, such as the Permian and several times during the Precambrian. The latest glaciation is uppermost in our minds because essentially all of the glacial features we see today were formed during the Pleistocene.

THE GEOMORPHIC PROVINCE

The regional concept involving geomorphic provinces is used in our treatment of the national parks. A *geomorphic province* is an area in which the rocks, geologic structure, geologic and geomorphic history, and landforms are similar. Within a province, a certain area may have detailed features that distinguish it from surrounding areas. Such an area is referred to as a "section"; the Colorado Piedmont and the High Plains are two sections of the Great Plains Geomorphic Province.

The parks within a geomorphic province are related geologically, although each has its unique features. For example, the four national parks in the Cascades Geomorphic Province are all different, but much of their geological development is the same. The geology of the park areas will unfold with, first, a summary of the geomorphic province and then a discussion of the distinguishing features of its national parks and monuments. By following this sequence, hopefully the relationships between the parks within each province will be apparent.

We will begin with the Hawaiian parks, go by way of Alaska to the Pacific Northwest, and then cross the country to the East Coast.

CHAPTER 1

HAWAIIAN ARCHIPELAGO

The Hawaiian Archipelago is composed of a linear group of islands and *atolls* that extend from the Island of Hawaii northwestward beyond Midway to Kure, a distance of about 1600 miles (2666 km). The Leeward (Westerly) Islands northwest of the Hawaiian Islands are mainly atolls (rings of coral islands), but volcanic rocks are exposed among the organic reefs toward the southeastern end of the chain. A common origin — submarine volcanism — is envisioned for the entire archipelago. According to Emery (1955), a fissure developed first at the northwestern end and, with attendant submarine volcanism, rippled gradually southeastward. Thus, by the time the volcanoes built the Hawaiian Islands, the older islands to the northwest had been eroded down to sea level and converted into atolls. Kauai, at the western end of the Hawaiian islands, has been deeply eroded during the long period since the last eruption; the same is true of Oahu but to a lesser extent. The only land-based volcanoes currently active, Mauna Loa, Kilauea, and Mauna Ulu, are on the big island of Hawaii. Perhaps we can anticipate that one or more of the submarine volcanoes southeast of Hawaii will one day build new islands. This interpretation, involving the migration of volcanism, was until recently generally accepted.

Now, a new theory of origin of the archipelago has gained favor with many geologists. Instead of having volcanic activity begin at Kure and migrate gradually southeastward to where it is active now on Hawaii, the new theory requires the "fire" to remain stationary and the earth's crust to migrate over the fire, northwestward from Hawaii to Kure. This "assembly line system" is pictured in S. W. Matthews' article, "This Changing Earth," in *National Geographic* (1973). This is part of the so-called New Geology as outlined in the Introduction.

Although we accept the general concepts of the plate tectonics theory, we are reluctant to abandon the theory of migrating volcanoes. The evidence presently at hand does not appear to favor one theory to the exclusion of the other. Conceivably, while the oceanic plate was moving northwestward, volcanism was migrating in the opposite direction, as believed originally. This dual interpretation is suggested in part to advise against jumping from one idea to another without first considering the possibility that the two may be mutually compatible.

Many of the islands and atolls in the archipelago are only slightly above sea level; in contrast, Mauna Kea (13,784 feet; 4202 m) and Mauna Loa (13,680 feet; 4140 m), both on Hawaii, rise high above the Pacific. Haleakala on Maui is 10,023 feet (3056 m) high, and three of the other Hawaiian Islands rise more than 4000 feet (1220 m) above the sea, high enough to produce marked climatic differences. The northeast trade winds drop the bulk of their moisture on the windward side; as a result, much drier conditions obtain on the leeward, southwest side. A prime example is Kauai, where on Mt. Waialeale (5170 feet; 1576 m), the annual rainfall exceeds 450 inches (11.4 m), while less than 20 inches (0.5 m) falls on the leeward side. Although close to the equator, Mauna Kea and Mauna Loa on Hawaii and Haleakala on Maui rise high enough above the sea to receive snowfall. Consequently, geomorphic processes vary in type and intensity, and the resulting landforms differ significantly from one place to another. Therefore, the Hawaiian Archipelago is actually more than a single geomorphic province, although it is so regarded in this general discussion.

FIGURE 1-1

Map of the Hawaiian islands.

Hawaiian Islands

Eight main islands together with small offshore islets make up the Hawaiian Islands. All are south of the Tropic of Cancer, and part of southernmost Hawaii is less than 19° north of the equator. Rising to heights as great as 13,784 feet (4202 m) above the sea, and more than 32,000 feet (9756 m) above the seafloor, here may be the greatest continuous relief on the globe.

Shield volcanoes and *volcanic shield clusters* (Wentworth and Macdonald, 1953) make up the Hawaiian Islands. The Island of Maui was originally two volcanic islands that have been joined together. A cluster of five main volcanoes have merged to form the Island of Hawaii.

Eruptions in the Hawaiian islands are mainly of the so-called "quiet type" in which highly fluid basaltic lavas pour out of the vent, flow rapidly downslope, and spread out over large areas. Flank eruptions, also common, occur when lava breaks through along the flanks of

FIGURE 1-2

Looking north over Diamond Head, a tuff cone on the island of Oahu, southeast of Honolulu.

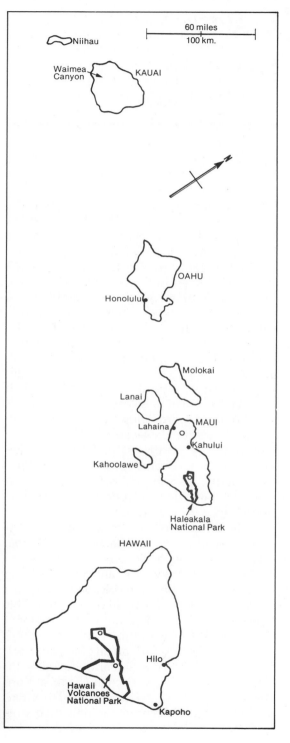

the cone. The resulting shield cones are broad, and the slopes are much less steep than those of the pyroclastic (cinder) cones formed by explosive eruptions. (Refer to sketches of volcanic cones in the Introduction.) The gas concentration in the throat of the volcano is the main factor that determines the violence of the eruption. When steam, the main gas, concentrates sufficiently to break through a well-sealed vent, a violent eruption occurs. Occasionally, when sea water enters the magma chamber or comes in contact with hot rocks, the steam thus generated breaks through violently, resulting in the buildup of steep-sided *tuff cones* such as Diamond Head on the Island of Oahu.

Because Hawaiian volcanic eruptions are relatively quiet and safe, they have been studied in great detail, and many articles and books about them have been published. One particularly interesting and comprehensive book was written by Stearns (1966), after more than 40 years of intensive field work. In his *Geology of the State of Hawaii,* Stearns outlines a sequence of development and destruction of Hawaiian volcanic mountains. Beginning with highly explosive eruptions on the sea floor, he traces the volcanoes through several stages after they have raised their heads above the sea — the shield-building stage, followed by the collapse of the tops to form the huge calderas such as Mokuaweoweo on Mauna Loa. Mauna Kea has reached a more advanced stage in which, after a period of quiescence, andesitic materials are poured out and blown out of the vent. The "bumps" on the flanks are actually cinder cones of considerable size. Later, after the volcanoes are dead, streams and ocean waves continue without interruption their work of planing off the islands and reducing them to sea level. Thus, the cycle is completed except for the work of the corals and other life forms that build the atolls. All stages are represented within the Hawaiian Archipelago.

When did the Hawaiian Islands appear as islands above the sea? Most workers, including Wentworth (1927) and Macdonald (1960), favor a Pliocene-Pleistocene period for their development. Many radiometric analyses have been made, but the samples analyzed are from surface rocks rather than from the deeply buried rocks of the embryonic islands. Until more complete data are obtained, a Plio-Pleistocene age is assigned, recognizing the probability that much of the youngest island, Hawaii, was built during the Pleistocene.

Basalts are the most abundant rocks in the Hawaiian Islands; however, as already indicated, andesites are found in localized areas, and *trachytes* (rhyolite without quartz) are present but rare.

Some of the basalts contain the mineral *olivine* as an important constituent. Olivine is a high-temperature mineral and, as such, is likely to be the first mineral to crystallize, thus forming large crystals, or *phenocrysts.* In places the phenocrysts are large and flawless and are used as semiprecious stones generally marketed under the name "peridot." Other main minerals are calcium-rich *plagioclase feldspar* and *augite,* or some similar *pyroxene. Magnetite* is an accessory mineral that varies considerably in amount from place to place.

Basaltic lava flows fall into two main types. If the lava is hot (1050° to 1200°C), it is highly fluid and flows rapidly to the place where it cools and solidifies; the resulting surface is relatively smooth (see Figure 1-3). This is a pahoehoe (pronounced paahoyhoy) flow. In contrast, if the surface layer of a highly viscous flow solidifies and then additional movement of the flow takes place repeatedly, the crust is broken into angular fragments, small and large. When such a flow, called an aa (ah-ah) flow, comes to rest, its surface is extremely rough and jagged. Refer to Figure 1-4 in the color section.

Wave Erosion

There is a never-ending battle between the waves that seek to destroy the islands and the volcanoes that periodically "roll back the

FIGURE 1-3

Kipuka standing above and surrounded by 1935 pahoehoe lava flow, on north flank of Mauna Loa. (Photo by E. Kiver.)

waves" and push the shorelines farther seaward. When the volcanoes become inactive, only the destructive processes continue; consequently, one by one the Hawaiian Islands will be planed off at sea level and the fringing reefs of today will become offshore barrier reefs and, with subsidence, atolls. Midway Island and others apparently developed in this way.

Wave-cut platforms and wave-cut cliffs are common features around the islands, particularly on their northeast (windward) sides. Waterfalls have developed where hanging valleys empty their waters over the cliffs. The streams, many of them relatively short, have been unable to cut down fast enough to keep pace with cliff recession.

But not all of the products of wave erosion can be seen on the surface. Below the surface at various depths are flat-topped *seamounts;* they were once islands that were planed off and then submerged when the sea floor sank. Further evidence of subsidence is found in the submarine canyons which, according to Stearns (1966), were cut into hard rocks by streams. Thus, crustal movements in the mid-Pacific are well documented.

Mass-Wasting

In the islands, where earthquakes are common, where in places steep slopes expose unstable rocks, and where there is an abundance of moisture, landslides and other downslope movements are geologic hazards. The problem is particularly serious where volcanic ash has been altered to slippery clay. When these materials accumulate in quantity and are then saturated during a rainstorm, the resulting mudflows race downvalley for long distances. Stearns cites a large prehistoric mudflow that moved down Kaupo Valley on the island of Maui, one that is more than 350 feet (107 m) thick on land and of unknown thickness beneath the ocean.

Stream Erosion

The extent to which streams have dissected the islands depends mainly on the age of the particular island. Kauai, the oldest of the main islands, is deeply dissected and is in a more advanced stage of the fluvial cycle (cycle of stream erosion) than any of the others; in parts of Kauai the erosion cycle has reached the mature stage. On the youngest island, Hawaii, early youth is represented in many places, particularly on Mauna Loa and Kilauea where the

FIGURE 1-5

Anomalous drainage pattern on the Island of Kauai.

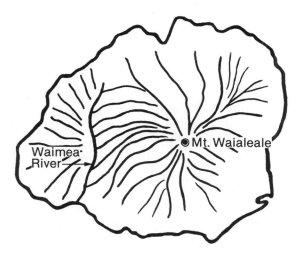

valleys formed by erosion are soon filled with lava. On Kohala, which has long been inactive, deep canyons have been cut into the rainy northeastern flank.

Drainage patterns on volcanic cones are typically radial. The water concentrates in ready-made "valleys" between flows and begins its downcutting and valley development there. These are *consequent* streams, streams whose routes are the consequences of the uneven surface. Valley development continues only until the next eruption fills the valley with lava or other debris. The water seeks out another route of the same kind, and this stream is also consequent. On Kauai, an anomalous situation exists. Although a radial drainage pattern is well developed on most of the island, Waimea River flows southward across the west flank. Structure has subsequently modified the original simple radial pattern. A north-south fault zone provided a zone of weakness in which erosion proceeded more rapidly than elsewhere. Waimea River, beginning as a short stream at the south end of the fault, cut headward in the easily eroded material in the fault zone and diverted one by one the headwaters of southwest-, west-, and northwest-flowing streams. Waimea River, a *subsequent* stream, has cut a deep canyon across the west flank of the volcanic cone, a canyon known as the Grand Canyon of the Pacific. Waimea Canyon not only has an unusual origin but also affords an excellent opportunity to study the anatomy of a shield volcano. The suggestion that it be set aside as a part of a new national park deserves enthusiastic support.

Glaciation in the Hawaiian Islands

Glaciation, the geologic process least expected in the Hawaiian islands, is clearly recorded on Mauna Kea, north of Hawaii Volcanoes National Park. Glacial features were first recognized by Daly (1910); later studies by Stearns (1945) indicate that a 26- to 28-square-mile ice cap occupied the peak during the late Pleistocene Wisconsin glaciation. It has been suggested that Mauna Loa, only 104 feet lower, may also have been glaciated. If Mauna Loa was glaciated, the glacial features are now buried beneath recent lavas. Mauna Loa is much younger than Mauna Kea and reached its present height much later. Although we cannot positively rule out the possibility of glaciation, it appears probable that Mauna Loa was not high enough, even during Wisconsin time, to support glaciers.

Should glaciers again become as widespread as they were 20,000 years ago, both Mauna Kea and Mauna Loa would likely be covered with ice caps.

Volcanoes and People

A short walk through the Devastated Area in Hawaii Volcanoes National Park brings into sharp focus the destructiveness of volcanoes. In 1881 lava flows from Mauna Loa menaced Hilo, 13,680 feet (4140 m) below and 40 miles (65 km) away. In 1942, when a lava flow was only 12 miles (19 km) from Hilo, army bombers were called in to try to divert the flow. They failed, but fortunately the eruption soon ceased and the town was safe — temporarily. Picture an aa flow moving slowly but relentlessly over papaya groves and then houses, as the people of Kapoho experienced in 1960. In September of 1977 lava spewed forth from a new vent on Kilauea's East Rift immediately above the coastal village of Kalapana. As the flow approached the village, army scientists attempted to divert the flow to one side. The consensus is, however, that Kalapana was saved because the eruptions ceased at precisely the right moment; the flow slowed, cooled, and stopped. Pele, the Volcano Goddess, was compassionate one more time! In 1982, 1983 and 1984 there were more eruptions, and still more can be expected.

HAWAII VOLCANOES NATIONAL PARK

Hawaii had a national park long before attaining statehood. In 1916 the volcanoes Mauna Loa and Kilauea on Hawaii and Haleakala on Maui were designated Hawaii National Park. This was only a few years after Dr. T. A. Jagger, volcanologist from Massachusetts Institute of Technology, established his Volcano Observatory on the north rim of Kilauea Caldera. Much of what is now known about shield volcanoes was learned at this observatory which, for a number of years, has been operated by the U.S. Geological Survey. In 1961 each of the two sections gained park status, Haleakala National Park on Maui and Hawaii Volcanoes National Park on Hawaii.

Hawaii Volcanoes National Park extends from Mauna Loa eastward to include Kilauea and the area to the south, down to the Pacific. The Kilauea area, the Kau Desert to the south, and part of the coastal area are accessible by road; an 18-mile (28-km) hiking trail leads up to Mokuaweoweo Caldera atop Mauna Loa.

Climatic Contrasts

The Crater Rim Drive takes the park visitor through lush tree-fern jungles; the Hilina Pali Road skirts the edge of the Kau Desert, southwest of Kilauea. Here in the "rain-shadow," classic examples of pahoehoe lava and other features are well exposed in an area essentially devoid of vegetative cover. Due to the lack of moisture, weathering and soil formation take place at a very slow rate as compared to the adjacent well-watered jungle areas. The contrast between these areas is the result of climatic differences.

Volcanoes in Action

Both Mauna Loa and Kilauea are active and are still in the shield-building phase. Mauna Loa

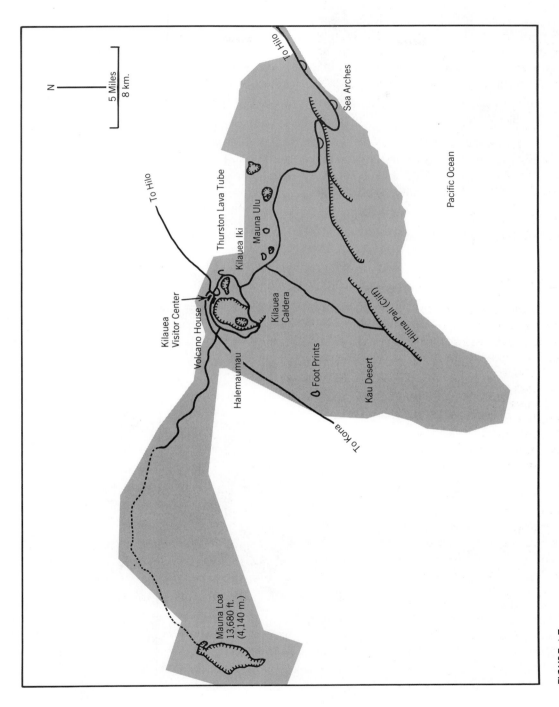

FIGURE 1-7
Map of Hawaiian Volcanoes National Park.

FIGURE 1-8

Tropical vegetation along Kilauea Iki Trail.

had been erupting on an average of about every three and a half years, prior to the latest major eruption in 1950. Volcanic activity appeared to shift eastward with Kilauea's eruptions in 1952, 1954, 1955, 1959–60, and 1967–68, and in 1969 to 1973 with those of the new "Growing Mountain," Mauna Ulu on Kilauea's east flank. But when, in 1975, lava started pouring out over the floor of Mauna Loa's summit caldera, this interpretation was questioned; now Mauna Loa is being monitored, in view of the possibility of a major eruption.

Although Hawaiian eruptions have been relatively tame, there have been exceptions: When Kilauea erupted in 1924, dust clouds

FIGURE 1-9

"Fires" smoldering in Mokuaweoweo Caldera, October 1977. (Photo by G. Arehart.)

FIGURE 1-10

Kilauea Iki Caldera in 1965. A lava lake during 1959 eruption.

were blasted more than 20,000 feet (6098 m) into the air and large blocks of rock were thrown a half-mile (0.8 km) from the vent (Macdonald and Hubbard, 1961). There was one fatality, the only one in the park's history caused by volcanism. Definitely recorded, although details are lacking, is another explosive eruption which in 1790 killed a group of Hawaiian soldiers. Their footprints are there, bringing into stark reality the tragedy which befell these men. Swanson and Christiansen (1973) have attempted to reconstruct the sequence of events of that tragic day. From the window of Volcano House, however, park visitors have been able to observe, with reasonable safety, many aspects of the geologic process that built the Hawaiian Islands.

Craters and Calderas

The terms *crater* and *caldera* have unfortunately been used interchangeably without regard for mode of origin. (The basin of Crater Lake in Oregon is a caldera, as is Katmai Crater in Alaska.) Both craters and calderas are topographic depressions at or near the summit of volcanoes. A volcanic crater has a simple origin; it is the funnel-shaped depression, the upper part of the vent through which lava and solid, fragmental material have been ejected. (Craters may also be formed by impact; examples are Meteor Crater in Arizona and some of the craters on the moon). Craters vary greatly in size but are generally small as compared to most calderas. The evolution of a caldera is much more complex, and the sequence may vary from one caldera to another. Mokuaweoweo Caldera on Mauna Loa and Kilauea's caldera are subsidence features, formed by the caving in of roof rocks and enlarged by infall of wall rocks. Picture a dome-shaped shield volcano that is being intruded by magma. Heat from the magma melts the roof rocks, and in this phase-change from solid to liquid there is an increase in volume. In this way the dome expands, upward and outward, until eventually cracks form in the roof and a large block of rock founders and sinks in the magma. Immediately adjacent blocks fall in, enlarging the depression and thus forming a caldera. It is not unlikely that some explosive activity may precede the foundering of the first block; in fact, there are cases, such as Crater Lake, where ejected material in the surrounding area indicates that violent explosions preceded collapse. On Hawaii, however, it appears that it was primarily subsidence that formed the calderas. In either case, flank eruptions that lower the level of the magma may contribute to the enlargement of calderas.

Mokuaweoweo Caldera atop Mauna Loa is about three miles (4.8 km) long, 1.5 miles (2.4 km) wide, and as much as 600 feet (183 m) deep;[1] almost vertical walls rise above its relatively flat floor. East of Mauna Loa, Kilauea's summit caldera is about 2.5 miles (4 km) across and about 450 feet (137 m) deep. There is a "hole" in the floor on the southwestern side of the caldera; this is Halemaumau (the Fire Pit), the active vent in Kilauea Caldera and the alleged home of Pele, the "Goddess of Volcanoes." Actually Halemaumau is a caldera within a caldera. With the one exception already mentioned, recent eruptions have been non-violent; the 1924 eruption served as a reminder that Pele still has her moments!

---

[1] The depth of an active caldera changes with each eruption as lava is poured out over the floor or when the lava lake solidifies at a new level.

The "Pit Craters" along the Chain of Craters Road southeast of Kilauea are in reality calderas, although several are quite small. Kilauea Iki or "Little Kilauea" is larger and is located immediately adjacent to the main caldera. It was here, in 1959, that a major eruption began. Lava fountains rose to heights of more than 1900 feet (580 m). A large area, now known as the Devastated Area, was buried beneath ash and cinders, and part of the Crater Rim Road was covered. Early in 1960 the scene changed; local faulting took place several miles to the east, and a flank eruption developed above the coastal town of Kapoho east of the park. Lava flowed down the slope and overwhelmed all in its path including most of Kapoho. The lava that reached the sea cooled rapidly, and *pillow lava*, composed largely of basalt glass, extended the shoreline seaward, adding about 500 acres of new land to the island.

Early in 1969, as during periods preceding earlier eruptions, tremors were recorded on the observatory *seismographs*, instruments that record earthquakes. However, the eruption took place, not at Mauna Loa or at Kilauea, but about six miles southeast of Kilauea near the Chain of Craters rift zone. Pit craters Aloi and Alae are only a mile apart, but the breakthrough took place between them, and a new volcano, Mauna Ulu, was born on May 24, 1969. The name, which means "growing mountain," is appropriate because a new shield volcano has taken shape. (Possibly, as some suggest, Mauna Ulu is merely the latest of Kilauea's vents.)

Once again, the maps of Hawaii were obsolete; some of the contours were changed, certain streams were translocated, and the shoreline of the park was shifted seaward as lava flows pushed into the Pacific. These submarine flows were observed and photographed by Swanson and others (1973) while the lava was still in motion. The activities of Mauna Ulu are further described in a general article by Parsons (1973), in the 1973 Time-Life book *Hawaii*, and in the U.S. Geological Survey's movie, *Fire Under the Sea,* which is as exciting as it is instructive.

This is geology in action; the earth is ever changing. What is happening in Hawaii now is what took place in many areas many times in the past. By observing and recording the activities of Kilauea and Mauna Ulu, we can reconstruct the general sequence of events during earlier periods when similar volcanoes were active in other places. The present is, indeed, the key to the past.

Volcanic Gases

Water vapor, or steam, is the most abundant gas involved in volcanic eruptions. According to Macdonald and Hubbard (1961), water vapor in Kilauea gases averages about 70 percent, with carbon dioxide, the other main constituent, about 14 percent. The gas recognized immediately is sulfur dioxide; it is not only unpleasant but hazardous to all living things. Sulfur Banks, a short distance west of park headquarters, are the result of *fumarolic action;* here, malodorous hydrogen sulfide is an obvious constituent. Chlorine gas, abundant in some places, is found only in traces in Hawaiian eruptions.

Features Related to Volcanism

Lava tubes are formed when lava breaks through and drains out from under a solid crust that is sufficiently thick to remain as the roof. Thurston Tube, near Kilauea Iki, was formed in this way. When lava oozes out in front of the foot of the flow and solidifies in an elongate bulbous mass, it is called a lava toe. Lava tubes and lava toes have been forming downslope from Mauna Ulu during recent eruptions.

Where the end of a lava tube is below sea level, the pounding of the waves within the tube may develop a vertical chimney along fractures. These *blowholes* erupt as "cold-water geysers" with each major wave. (Blowholes may also develop from sea caves by wave erosion in many different rock types.)

Areas covered by recent flows are essentially barren of vegetation. There are "islands," however, where there is a lush growth of trees and bushes. These are "kipukas" (pronounced kipookas), areas that have not been covered by recent lava flows. Elsewhere, soils have not had time to develop; hence there is little or no vegetative cover. Soils form at a rapid rate in the islands, except for the dry rainshadow areas. The warm and humid climate is conducive to chemical weathering and soil formation; also, basalt is very susceptible to chemical weathering, particularly the ash and cinders.

Fossils are uncommon in igneous rocks, but when trees are overwhelmed and encased by

FIGURE 1-11

Footprints left by Hawaiians crossing ash that was erupted from Kilauea in 1790. A number of warriors were killed by the ash cloud. (Photo by E. Kiver.)

FIGURE 1-13

Map of Haleakala National Park.

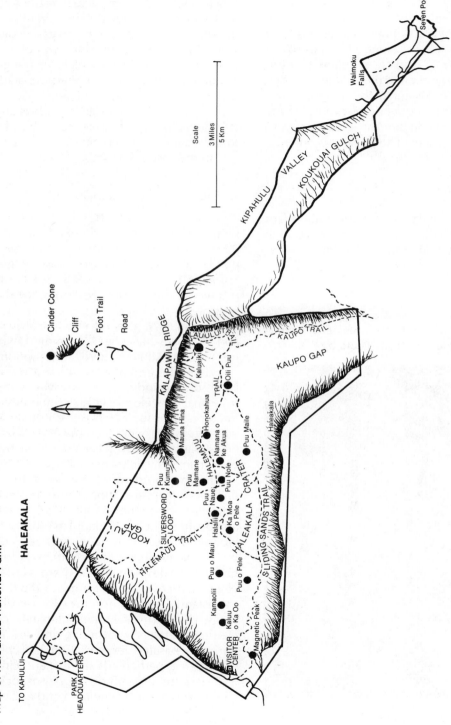

lava, their impressions — fossils — are left in the rock. Tree Molds can be seen near the road to Kipuka Puaulu Bird Park, northwest of Kilauea.

Volcanic eruptions in Hawaii are frequent, and thus the earthquakes attract less attention than in other tectonic areas. However, when on November 29, 1975, a 7.2-magnitude quake hit the southeastern section of Hawaii, it was publicized as the heaviest shock for more than a century. Moderately severe earthquakes are fairly common; the *Atlas of Hawaii* records several shocks of 5.3 and greater magnitude between 1925 and 1973, the date of publication. Damage in the 1975 earthquake was serious, and there was one death; thus earthquakes are another of Hawaii's geologic hazards.

Erosional Features

Clearly, volcanic activity dominates the scene, although shore features such as wave-cut cliffs and wave-cut platforms record the action of the Pacific Ocean in restricting the growth of the island. Sea caves are formed by wave action along fractures and other zones of weakness in the rocks. When sea caves form back to back along the sides of *headlands* (rock promontories) and are then enlarged, they merge to form sea arches. Excellent examples are readily accessible near the east entrance of the park.

There are numerous cliffs a few miles inland; the term *pali* has been applied to these features. The origin of palis is uncertain; perhaps they are modified fault scarps. Or they may be sea cliffs that were formed when the sea was higher or the islands were lower. One process is probably dominant in certain cases and another process, or combination of processes, is responsible elsewhere.

Stream-cut valleys are shallow and poorly developed in this volcanic park. Eruptions are too frequent to allow for appreciable valley development; instead, lava flows periodically fill the valleys, forcing the stream to shift to a new route.

Hawaii Volcanoes National Park is famous for its active volcanoes. But here, as in most other parks, other geologic processes are also at their work of relandscaping the earth.

HALEAKALA NATIONAL PARK

The name Haleakala (pronounced hah-lay-ah-kah-lah), which means "house of the sun," is related, according to legend, to a project by the Polynesian god Maui to capture the sun and persuade it to move at a slower pace across the sky — so his mother could get all of her work done! Although this early attempt at daylight-saving failed, the name Haleakala prevailed. A sunrise picture across the top of the mountain is convincing evidence that the name is appropriate.

Established first in 1916 as a section of Hawaii National Park, Haleakala gained full status as a national park in 1961. Originally the park included only the upper part of the 10,023-foot-high volcanic cone; recently it was extended down the east side of the mountain to the sea, adding a beach environment, the Seven Pools area. Spectacular Waimoku Falls, which are at the head of one of the valleys, can be reached by hiking up through the tropical forests.

In order to reach the Seven Pools area, you drive a slow road from the town of Kahului around to the southeastern part of the island; to reach the crater rim you follow a tortuous road up the west side of the cone. First view is from Kalahaku Overlook, one you will not forget.

Haleakala is distinctly different in appearance from Mauna Loa and Kilauea. Haleakala is highly colored whereas the others are mainly black. It is considerably older and many of the iron-bearing minerals in the rocks have been altered to form the yellow and red iron-oxides so much in evidence.

The "crater," or caldera as it has also been

called, is about 7 miles (11 km) long and 3 miles (5 km) wide, and more than half a mile deep. Actually, it is neither a crater nor a caldera in the true sense because the present topography has resulted mainly from stream erosion. At its peak, the mountain was about 12,000 feet (3659 m) high; it was lowered to an extent when the caldera was formed but much more by the streams that dissected the caldera. Now, the highest point is 10,023 feet (3056 m) above the sea.

FIGURE 1-15

A mature Silversword plant on the flank of Haleakala.

Geologic Sequence

After the volcano built its cone above the sea, it constructed a basaltic shield volcano. Then, subsidence of the summit area formed an elongate caldera with two low saddles or gaps in the rim, one on the north and the other on the southeast. In time, the streams that flowed out through Koolau Gap and Kaupo Gap changed the caldera floor into a stream-carved landscape. Much later, volcanic activity began again. This time andesitic materials were ejected partly as lava flows but mainly as pyroclastics which formed cinder cones such as 600-foot-high Puu O Maui. Evidently some of the eruptions were highly explosive, because cinders and ash were distributed widely, particularly on the leeward flank of the cone. The latest major eruption probably occurred about 1750 (Macdonald and Hubbard, 1961); however, two small flank eruptions sent lava flows down into the sea around 1790.

By no means should Haleakala be regarded as an extinct volcano; it is a quiescent or dormant volcano that might erupt again at any time. Periodic earth tremors serve to remind us that although calm prevails at the surface, there is still some activity below.

Many of the rocks which you see are andesitic pyroclastics, but the bulk of the cone is made up of basalt flows. Some of the flows are porphyritic, and large olivine phenocrysts are common. Blanketing the basalts in many places are the andesitic materials, mainly cinders and ash.

Haleakala's Other Features

Within the caldera, there are many interesting features for energetic hikers or for those on horseback to enjoy. Cabins, available by reservation, are situated at strategic locations for those who wish to stay two or three days. Lava tubes, Bubble Cave, and Pele's Paintpot are among the attractions that can be visited.

Vegetation is sparse or lacking in much of the caldera due to the low precipitation and the high permeability of the cinder floor. Growing in some of the high, dry areas within the caldera and around the rim is a remarkable silvery, yucca-like plant, the silversword. When it matures, perhaps in 20 years, its tall stalk produces purple flowers once; then it dies. In the 1920s it was nearing extinction, having been eaten by wild goats and dug up for sale by dealers in ornamental plants. Under the protection of the National Park Service, it is now making a comeback. A few plants are found in similarly high places on the Island of Hawaii. Also unique is the nene, or Hawaiian goose, which has renounced its aquatic heritage and lives only on land, even within the caldera.

The eastern part of the park is an oasis where trees, grasses, and ferns are abundant. Rare plants grow here, including the apeape with leaves up to 5 feet (1.6 m) across. Birds of many kinds decorate the jungle, among them two endangered species — the Maui parrotbill and the nukupuu, formerly believed extinct. Before about 1900, birds were far more numerous in the area than now, and ornithologists had reported seeing about 25 kinds of birds that are no longer found. Scientists found that they had succumbed to disease carried by mosquitoes that were introduced to the island when sailors emptied the dregs of their near-empty water barrels containing mosquito larvae (brought from Mexico) into the stream where the barrels were being refilled near the port of Lahaina.

Haleakala is small compared to Hawaii Volcanoes National Park. Although still possible, it is unlikely that it will stage a spectacular eruption for you. However, it is different; it represents an older stage in the volcanic cycle than either Mauna Loa or Kilauea. Shield-building was followed by caldera development; then there was a long period of erosion, after which there was the final (?) spasm of volcanism — the cinder cones. Perhaps we see here the future of Mauna Loa and Kilauea, and even the embryonic Mauna Ulu.

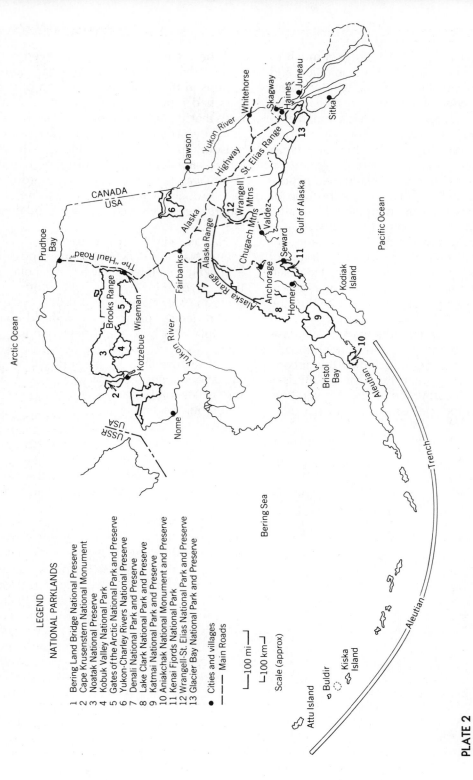

LEGEND

NATIONAL PARKLANDS

1 Bering Land Bridge National Preserve
2 Cape Krusenstern National Monument
3 Noatak National Preserve
4 Kobuk Valley National Park
5 Gates of the Arctic National Park and Preserve
6 Yukon-Charley Rivers National Preserve
7 Denali National Park and Preserve
8 Lake Clark National Park and Preserve
9 Katmai National Park and Preserve
10 Aniakchak National Monument and Preserve
11 Kenai Fjords National Park
12 Wrangell-St. Elias National Park and Preserve
13 Glacier Bay National Park and Preserve

● Cities and villages
——— Main Roads

├──────┤ 100 mi
├──────┤ 100 km
Scale (approx)

PLATE 2

Map of Alaska showing national parklands.

CHAPTER 2

ALASKA
AND ITS
PARKLANDS;
THE
ALEUTIAN
RANGE
PROVINCE

Alaska reaches from below sea level in its *fjords* up to 20,320 feet (6194 m) atop Mt. McKinley. With its endless mountains, volcanoes, glaciers, and fjords, Alaska has dozens of areas that are "naturals" as national parklands. Until recently, however, only four — Mt. McKinley, Glacier Bay, Katmai, and Sitka — had been designated as national parklands.

"Alaska" is derived from the Aleut word *Alakshak,* meaning "mainland" (Worldmark Encyclopedia of the States, 1981).[1] Probably Vitus Bering, or possibly Georg Wilhelm Steller, his "science adviser," was the first white man to learn this word while exploring the Alaskan coast in 1741 under commission of the Czar of Russia. This voyage was one of storms and shipwreck; many of the men died of scurvy. Captain Bering, in failing health, died and was buried on the lonely island that bears his name (Ford, 1966). Hardly could these men have dreamed that what they had done would lead to the extension of the Rus-

sian empire onto a "new" continent. Nor could they have known that Alaska would eventually attract people with violently opposing views as to how the land should be used.

Some of its people, old and new, find Alaska's oil and minerals more beautiful than its scenery. Other Alaskans want nothing more than the opportunity to live their own lives, free from crowds — without being fenced in. And of course we must not forget the native Alaskans, who had prior rights to the land! Could Alaska accommodate all of these diverse interests, even without the outsiders who wanted to set aside large areas as parklands — so that future generations would have something beautiful and unspoiled to enjoy?

Congress tussled at length with the land-use problem in Alaska, but understandably did not speak with one voice; instead it was hopelessly deadlocked. Finally, in 1978 President Carter exercised his executive powers and set aside 10 new areas as national monuments. Then in 1980 Congress passed the Alaska National Interest Lands Conservation Act, which upgraded both Katmai and Glacier Bay to park status and redesignated most of the new national monuments as national parks and/or preserves.[2]

In the vastness of Alaska, an area of more than 586,000 square miles, we now have 15 national parklands. Of these, Aniakchak, Katmai, Lake Clark, Kenai Fjords, Wrangell-St. Elias, Glacier Bay, Denali (formerly Mt. McKinley), Gates of the Arctic, and Kobuk Valley are clearly geologic parks and monuments. The remaining park areas were set aside mainly for their archeological and/or historical values; however, geology has played a part, directly or

---

[1] Another possible derivation is given by Ford (1966): the Aleut word was *agunalaksh,* which means "the shores where the sea breaks its back."

[2] A national preserve is a park area in which trapping and sport-hunting are permitted under supervision that will prevent any of the animals from becoming endangered species. Portions of certain parks and monuments are designated as preserves.

indirectly, in all of them, particularly in Klondike Gold Rush National Historical Park. We discuss Bering Land Bridge National Preserve first because it was here that humans first set foot on the North American continent.

BERING LAND BRIDGE NATIONAL PRESERVE

Access to many of Alaska's parklands is difficult. Some progress can be claimed, however, since the "first American" hiked across the Bering Land Bridge from Siberia, perhaps 28,000 years ago. The Bering Land Bridge National Preserve, on Seward Peninsula in northwestern Alaska, may be the place to find the first human habitation on the continent. It is probable that there were crossings at different times; land bridges were there during several periods when sea level was low—when the glaciers were most extensive (Hopkins, 1967).

Nome, park headquarters for the preserve, is on the south shore of Seward Peninsula. No roads lead from Nome into the preserve, but charter flights travel north from Nome or southwest from Kotzebue. There are no local guides, but the bush pilots are generally able to provide needed information. If you plan to do any backpacking, make certain that you are prepared for it, both physically and mentally. Write to the Park Service (see Appendix A) to obtain survival information well ahead of your arrival.

You will have an opportunity to see the Eskimos in their home territory, managing their reindeer herds and working at their arts and crafts. Plant life is abundant, and in late August the tundra is painted in many colors, some brilliant. Seals and whales can generally be observed in the waters of Bering Strait and Kotzebue Sound. Away from the coast, moose, wolves, and grizzly bears provide both interest and concern to "shutter-bugs." A camera with a zoom lens is highly recommended!

Little detailed geologic information on the preserve is now in print but the U.S. Geological Survey has studies in progress. The 1980 Alaska Geologic Map indicates that a broad area along the Chukchi Sea and Kotzebue Sound is largely covered with unconsolidated Pleistocene deposits of mixed origin. These deposits overlie Tertiary and Quaternary lava flows that crop out in several places. Some of the flows are apparently several million years old, whereas others poured out during the past few thousand years (Alison Till, personal communication, 1983). Precambrian rocks, some distinctly older than others, occupy extensive areas in the southern part of the preserve. The Serpentine Hot Springs, near the Continental Divide not far from the western boundary of the preserve, appear to be related to faults, some of them in granite. The Hot Springs are favorite haunts of the Eskimos of the region.

The preserve has been extensively glaciated, and some of the many lakes are the results of glacial action. Certain ones, like some of those shown here, may prove to be of unusual origin. Their basins may have been formed by gigantic explosions of gases that accumulated beneath the lavas. If this origin is definitely established, they will be properly called *maar craters,* a term that is already being used on a tentative basis.

A visit to Bering Land Bridge would be both difficult and expensive, but the experience of

FIGURE 2-1

"Maar craters." (Photo by National Park Service)

standing where the "First American" stood might make it well worthwhile. As a final word, the park service has made the suggestion that "local residents carry on their subsistence way of life within the preserve. Their camps, fish-nets and other equipment are critical to their well-being." Consideration and respect for their property and their privacy is likely to be mutually advantageous.

CAPE KRUSENSTERN NATIONAL MONUMENT

Across Kotzebue Sound from Bering Land Bridge (see Plate 2) is a treeless, windswept land — Cape Krusenstern National Monu-ment. The terrain is unusual, a series of essen-tially parallel ridges, more than 100 in all. Ac-cording to Park Service information, the ridges are old beach lines, each representing a differ-ent ocean level.

These beach ridges are treasure troves for archeologists. Artifacts and middens record occupation of the area during the past 5000 years; farther inland, obsidian projectile points were left behind by hunters who predate the Eskimos. Present-day Eskimos pitch their spring sealing camps along the foremost ridge.

Wildlife is abundant; in addition to the ani-mals that are commonly found in other north-ern parklands — grizzly bear, moose, and wolves — musk oxen are occasionally sighted. And, at various times of the year, walruses and polar bears are seen offshore, along with whales and several species of seals.

Access to the monument is by chartered aircraft or boat from Kotzebue, park head-quarters.

NOATAK NATIONAL PRESERVE

The Noatak River, one of Alaska's many Wild and Scenic Rivers, has its headwaters in the western part of Gates of the Arctic National Park. The part of the watershed that lies west of the Gates of the Arctic became Noatak Na-tional Preserve in 1980. The De Long Moun-tains extend along the north side, and the Baird Mountains are along the south side of the pre-serve. According to the Park Service, Noatak — an area of about 6.5 million acres — is "the largest undisturbed mountain-ringed water-shed in North America."

Large areas of the preserve are treeless tun-dra, which is crossed by thousands of caribou migrating to and from their calving grounds near the Arctic Ocean. Wolves are there to take care of any stragglers.

The valley and the mountains are carved out of sedimentary rocks of late Paleozoic and Mesozoic age. Devonian limestones and dolo-mites underlie much of the main valley; here the river has cut a deep gorge known as the Grand Canyon of the Noatak.

Access to the western part of the preserve is from Kotzebue and from Bettles for the eastern part, in both cases by chartered plane.

YUKON-CHARLEY RIVERS NATIONAL PRESERVE

From the earliest times, when the various peo-ples from Asia were exploring their new terri-tory, the Yukon was the main route of migra-tion and an important source of food and clothing. Huge mammals such as the wooly mammoth, the large-horned bison, and the short-faced bear — long since extinct — provided challenges for these early-day hunters (Guthrie, 1972).

Although the Yukon is the main thorough-fare for interior Alaska, traffic has generally not been heavy, with one notable exception. In 1896 the gold strike — not in Alaska but in Canada's Klondike region upriver from Alaska — changed the tempo on the Yukon for years to come. Although gold had been mined in Alaska as early as 1861, this was *the* Gold Rush, and the Yukon was the route to get there. It is quiet once again, and in Yukon-Charley Rivers National Preserve, peregrine falcons nest in the cliffs and bluffs. Here, on the Alaska side of the Alaska-Canada boundary, about 1.7 million acres of wilderness form this preserve. The preserve also contains archeological sites and here and there, relics left by the "stampeders," old tumble-down cabins and other trappings of man.

The Yukon is a truly fabulous river. Its headwaters lie on the Canadian side of Chilkoot Pass about 20 miles (33 km) up the Chilkoot Trail above Skagway, Alaska, north of Juneau. Should you decide to take the Chilkoot Trail, following the tracks the gold seekers made during the late 1890s, when you finally conquer the last steep pitch — the "Golden Stairs" — to the summit, you will be looking out over the upper drainage basin of the Yukon River, in Canada.

Rain falling on the north side of the Pass — only about 20 miles (33 km) from an arm of the Pacific Ocean — travels more than 2000 miles (3333 km) to reach the Bering Sea!

The trailhead is at Dyea, near Skagway, about 16 miles (27 km) from Chilkoot Pass, in Klondike Gold Rush National Historical Park. This trek should be undertaken only by those in excellent physical condition.

Keith Tryck and three other hardy outdoorsmen "did the Chilkoot Trail" and rafted down the Yukon, mainly to match the exploits of the old-timers. They ran the rapids through "waters that hissed of disaster" — as described by Robert Service, who put Alaska into verse. Tryck documents his travels in a fascinating *National Geographic* article (1975).

SITKA NATIONAL HISTORICAL PARK

Sitka National Historical Park, established in 1910, is the oldest of Alaska's parklands. Recorded here is the stormy history of Alaska — the struggles of the native peoples, the Eskimos and the Aleuts, and of the Indians, mainly the Tlingits and Athabascans. The Tlingits are skilled craftsmen, as their beautifully carved totem poles indicate.

Then the Russians came, first to explore and then to exploit. The site of the old fort, the last stronghold of the Tlingits, records the last battle — the last organized resistance of the Indians — in 1804.

ALEUTIAN RANGE PROVINCE

The Aleutian Range Province includes the Alaska Peninsula and the Aleutian Archipelago, an "island arc" that extends the United States to its westernmost point, the Island of Attu, west of the 180th meridian. The other end of the province is less well-defined; in fact, Wahrhaftig (1965) includes in his Alaska-Aleutian Range Province both of these long mountain ranges. Thornbury (1965) includes all of the volcanoes in the same province and draws the Aleutian Range boundary north of Mt. Spurr, west across Cook Inlet from Anchorage. When plate tectonics studies have been completed and the results fully analyzed, a number of province boundaries may be relocated, including this one; however, in this edition, Thornbury's designation is followed.

In the Aleutian Range, as in the Hawaiian Islands, volcanism has been the dominant geologic process during recent geologic time. But instead of quiet outpourings of lava and flat shield cones, violent eruptions of pyroclastic materials and steep-sided cones characterize Aleutian volcanoes.

There are probably about 80 major volcanoes in this 1600-mile-long arc; the exact number will not be known until all of the remote areas have been explored. Mt. Kialagvik,[3]

---

[3] Kialagvik and Shinningnellichahunga are two of the reasons why Alaska is on occasion referred to as the "land of mispronounceable names." Some are spelled so nearly alike that they may be confused; when in difficulty, consult *Alaska Place Names* (Orth, 1976).

FIGURE 2-2

Double Peak, one of many glacier-clad volcanoes in the Aleutians.

northeast of Aniakchak National Monument, was recently added to the list by Miller and Smith (1975).

At least 20 of the volcanoes have reached the caldera stage; Katmai Caldera, the youngest, was formed in 1912. The largest is Buldir, a submarine caldera that lies between the islands of Buldir and Kiska, far out on the Aleutian chain. According to Powers (1958), Buldir is 13 miles (22 km) wide and about 27 miles (44 km) long.

The Aleutian Range is an important segment in the Pacific "ring of fire" where volcanoes are erupting periodically — here and in Japan, the Philippines, New Zealand, and in the Andes of South America. Several eruptions occurred in Katmai National Park during this cen-

tury, and Mt. Spurr blew its top in 1953. Mt. Augustine, about 50 miles (83 km) north of Katmai, had been "smoking" periodically for years, and in 1976 it erupted, scattering ash and dust over its island domain and beyond.

The structure that is responsible for the volcanic activity in the Aleutians lies beneath the Pacific. It is the Aleutian Trench, the "surface" expression of the subduction zone that extends northward beneath the archipelago, as shown in Figure 2-3. Deep in this subduction zone, magmas are formed by the melting of deep-seated rocks; the magma bodies enlarge as more and more of the roof rocks are melted. Finally, gases concentrate in sufficient quantity to generate violent eruption and a volcano is born.

Early geologic work was largely reconnaissance, but recent studies by the U.S. Geological Survey have improved our understanding of Aleutian geology. Readers interested in detailed information will find it in such papers as those of Reed and Lanphere (1969, 1973) and those in their reference lists.

The rocks on the Alaska Peninsula are mainly Mesozoic and Cenozoic granitics and Paleozoic to Tertiary sedimentary rocks, discontinuously overlain by Tertiary and Quater-

FIGURE 2-3

Schematic cross section, south to north, across the Aleutians. Pacific Plate (left) plunging beneath the Aleutians.

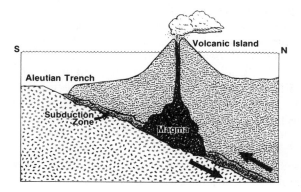

nary volcanics. The large masses of granite form an elongate batholith which, according to Reed and Lanphere (1973), intruded at different times from early Jurassic until the mid-Tertiary. Late in the Tertiary the peninsular mountains were formed by the upthrusting of an elongate block of the earth's crust, probably as the result of underthrusting by the oceanic plate.

Faults are difficult to locate in areas of recent volcanism, particularly where the bedrock is covered by glaciers and glacial deposits, if not by sea water. One major fault, the Bruin Bay, is mainly on land along the eastern border of the peninsula, extending for more than 180 miles (300 km) northward from Becharof Lake. Another, the Lake Clark Fault, extends northeastward from Lake Clark and apparently joins with Bruin Bay Fault west of Anchorage (Plafker et al., 1975). No recent movement has been observed; however, inasmuch as these faults are believed to be located in an active subduction zone, presumably they could be reactivated at any time.

Glaciers were widespread on the peninsula and on certain of the islands during the several advances of the Pleistocene. Determination of the glacial sequence has been complicated by the volcanic eruptions that interrupted the normal glacial regime. Indeed, many of the volcanoes were not there until after the last of the great glaciers vanished or were reduced to small remnants.

New glaciers were born during the Little Ice Age, and now there are too many to count, as all who have taken a flight down the peninsula are aware. They flank many of the volcanoes and other peaks and extend out in the form of valley glaciers. Although the mountains are not noted for their heights, the abundant snowfall and cool climate make this a favorable area for glaciers. For example, Mt. Veniaminof, a volcano about 75 miles (125 km) southwest of Aniakchak, is only 8225 feet (2492 m) high; even so, a large ice cap tops the mountain, and 15 outlet glaciers radiate in all directions. Early

in June of 1983, Mt. Veniaminof began a series of eruptions, thus ending a 39-year period of quiescence. The U.S. Geological Survey is monitoring the eruptions, and it is reported that the active cinder cone in the caldera spewed forth ash and volcanic bombs until October when lava flows began to appear (Betsy Yount, personal communication, October 18, 1983). At the time of writing, it seems inadvisable to try to predict future eruptions or their effect on the local glaciers. It can be said with assurance, however, that the Aleutians will provide considerable excitement for volcanologists for a long time.

The Aleutian Range Province contains three parklands, all on the peninsula. Lake Clark is located about 100 miles (160 km) southwest of Anchorage; Katmai is another 150 miles (250 km) farther out on the peninsula; and Aniakchak is about 400 miles (640 km) from Anchorage. Each has its own unique features; together the three park areas are geologically representative of the Alaska Peninsula.

ANIAKCHAK NATIONAL MONUMENT AND PRESERVE

Aniakchak National Monument is in a remote area on the Alaska Peninsula, about 400 miles (640 km) southwest of Anchorage. Apparently Aniakchak was known only to the Aleuts until 1922, but by 1929 field studies had been completed and a paper published by Knappen of the U.S. Geological Survey. The monument, about 580,000 acres of mountains, tundra, broad valleys, and coastal swamps, was established in 1978. Access is difficult and expensive. Float planes may be able to land on Surprise Lake; however, dense clouds ("cloud Niagaras," as they are called in the May 1982 *National Parks Magazine*) may flow over the rim and fill the caldera. At times the winds in the caldera are horrendous, as reported by those who have had their camps blown away by the *williwaws*. The 10-mile (16-km) hike east from the Port Heiden Airfield, through bogs, willows, and alder thickets, has been attempted by few people. It is suggested that you get specific advice directly from the Park Service in King Salmon, Alaska, and also that you share your plans with them so they will know where to search, just in case.

The Aniakchak Caldera is about 6 miles (9.6 km) across and is rimmed by cliffs as high as 2000 feet (606 m). Vent Mountain, a huge cinder cone, rises more than 2000 feet above the caldera floor, to about 3350 feet (1015 m) above sea level. Surprise Lake occupies a small area in the northeastern part of the caldera, leaving exposed most of the caldera's features — cinder cones, spatter cones, explosion pits, lava flows, and hot springs. (The calderas at Crater Lake in Oregon and at Katmai in the Aleutians are largely water-filled.) Surprise Lake is green to some and turquoise to others. Most of its waters are derived from runoff within the caldera rim, but highly mineralized waters from the hot springs contribute significantly, particularly to the temperature and quality of the water. Aniakchak River flows out of the lake, down through an awesome chasm in the east wall of the caldera. It rushes and tumbles down the mountain in torrents that challenge the most avid "whitewater bugs." Recently the Aniakchak was named as one of the National Wild Rivers; possibly no river deserves the designation more. The lower reaches are quieter, as the stream meanders

FIGURE 2-4

Airview east over Aniakchak Caldera, six miles wide from rim to rim, western end of Alaska Peninsula. (Photo by M. W. Williams, National Park Service.)

across miles of tundra and coastal swamps to Aniakchak Bay, on the Pacific side of the Alaska Peninsula.

The wildlife of Aniakchak, abundant and varied, includes brown bears, moose, caribou, foxes, seals, sea lions, sea otters, bald eagles, and sea birds. Sockeye salmon and trout thrive in the Aniakchak River but, according to Park Service information, these fish have their own distinctive flavor, due mainly to the soda and iron from the hot springs in the caldera.

Geology and Geologic Sequence

In the Aniakchak area, Tertiary and Quaternary volcanics overlie pre-Tertiary sedimentary rocks. Aniakchak is famous for its Quaternary volcanism, specifically its caldera formation. All of the features of the caldera are exposed, thus making it a geology textbook opened to the chapter on volcanic activity and caldera formation.

As shown and described on the geologic map by Detterman and others (1981), Jurassic, Cretaceous, and Tertiary sedimentary rocks are extensively exposed in the monument area. These rocks, along with Tertiary volcanics, formed the base on which the Quaternary volcanoes were superimposed. As elsewhere in the Aleutians, andesitic materials dominate, in the form of ashflows, breccias, lava flows, and intrusives. However, andesitic basalts and, as a late eruptive phase, dacites (intermediate between andesite and rhyolite) are also present. Dacitic materials are extensively exposed in the area of Aniakchak Peak, the high point on the south rim of the caldera. Within the caldera, volcanic mudflows cover certain large areas; alluvial fans cover other areas.

Marine terraces — flat benches that were planed off by the waves — are now about 100 feet (30 m) above sea level in the Aniakchak Bay area, on the Pacific side of the peninsula. On the opposite side, along Bristol Bay, these same terraces are only about 50 feet (15 m)

above the sea. Part of the uplift can be attributed to the removal of weight as the thick glaciers melted, but the difference in elevation of the terraces on the two sides can be explained only by tectonic activity that tilted the peninsula toward the west (Detterman et al., 1981).

Deposits related to the glaciers — moraines and outwash deposits — are known to be widespread in the lower areas, outside the caldera; higher up, around the caldera, we must visualize extensive deposits buried beneath post-glacial volcanics. Although early Pleistocene deposits are known, most are related to the late-Pleistocene Brooks Lake Glaciation, which began about 25,000 years ago and ended 10,000 to 8500 years ago. The height of Aniakchak Volcano during maximum glaciation may never be known, but it clearly was the main source of glacial ice within the national monument. Since the volcano was blown away, leaving only the caldera, the area has not been high enough to support glaciers.

The major geologic events in the formation of Aniakchak are as follows: The sedimentary rocks of Mesozoic and Tertiary age were folded and uplifted above the sea, forming the Aleutian Peninsula, probably in late Tertiary time. A subduction zone extending beneath the peninsula, as shown in Figure 2-3, produced heat and magma that moved upward, breaking through in places to form volcanoes. These magmas were andesitic rather than basaltic in composition; therefore, explosive eruptions were dominant. Eruption after eruption eventually built Aniakchak Volcano into a high mountain that supported large glaciers.

In the late Pleistocene, essentially one rock type — glacier ice — was dominant, with bedrock *nunataks* — some small, some large — protruding above the ice. With fluctuations in climate, the glaciers advanced and retreated during the 15,000 years of the Lake Wisconsin Brooks Lake Glaciation; then, with distinct warming of the climate about 11,000 years ago, a major retreat began. Presumably there were minor readvances, but by about 8500 years ago, the monument area was essentially free of ice.

The next major event probably took place about 3500 years ago when, by a series of colossal explosions, Aniakchak Caldera was formed. Immediately following each major explosion, the walls of the crater collapsed into the abyss, enlarging the diameter of the caldera so that it is now about 6 miles (9.6 km) from rim to rim.

Some of the explosions evidently shot fragmental materials quite high into the sky. At least, gigantic explosions are postulated by Miller and Smith (1977) to account for the momentum necessary to propel an ashflow over a high ridge far from the vent. The pyroclastic materials blown out were of all sizes, from dust to large blocks, and huge volcanic bombs have been reported, some as large as 2 feet (60 cm) in diameter (Detterman et al., 1981).

Although evidence for this next event has apparently not been reported, it is probable that the caldera was once full or nearly full of water, until it overtopped the rim or broke through the side at the weakest point. With tremendous volumes of water essentially exploding through weak, poorly consolidated rocks, the gorge through the east wall was probably cut in a relatively short time, leaving Surprise Lake in a depression below the floor of the caldera.

The caldera has been profoundly modified by successive eruptions during the past 3500 years. Vent Mountain, its cone well over a quarter of a mile high, had its latest blowout in 1931 when it spread ash and dust over a large area. In addition, many little spatter cones and explosion pits dot the lava flows that cover much of the caldera floor. The caldera walls have also been modified by landslides, rockfalls, and volcanic mudflows. The hot springs present today quietly remind us that heat energy is still down below — perhaps sufficient for the building of future cinder cones.

The plant cover of the Aniakchak area varies greatly, depending on the type and the timing of recent geologic events; thus, plant succession can be studied within the caldera. According to the Park Service brochure, "hardy pioneer plant communities are inching life into a silent moonscape."

Aniakchak National Monument will provide an unparalleled wilderness experience for the hardy souls who have the determination to get there and enjoy it.

KATMAI NATIONAL PARK AND PRESERVE

Katmai moved from obscurity into the limelight as a result of a volcanic eruption in 1912, one of the most violent ever. Explorations into this remote area, mainly by Griggs's National Geographic parties, led to the establishment of the national monument in 1918. In 1980 it was enlarged and renamed Katmai National Park and Preserve; now the park includes approximately 2.8 million acres.

Katmai is located about 250 miles (415 km) out on the Alaska Peninsula from Anchorage, with no roads leading into it. Katmai is an ideal spot for those who can get along without large crowds and loud noises.

Access is by commercial plane from Anchorage to King Salmon, and from there an experienced bush pilot can be hired for the short flight into the park. The one "road" in the park leads from Brooks Camp out to the Valley of Ten Thousand Smokes, 23 miles (37 km) away.

The Katmai area is populated mainly by wolves, foxes, moose, and bears, along with the few tourists who can be accommodated at Brooks and Grosvenor Camps. At Brooks Falls near the camp, brown bears can be seen fishing for salmon. If you meet one of these big fellows on the trail, it is well to show proper respect! He may weigh as much as 1500 pounds; no, he is not clumsy but very agile. Katmai is also inhabited by animals that are rare in other parks. The wily and sometimes

FIGURE 2-5

Map of Katmai National Park and Preserve.

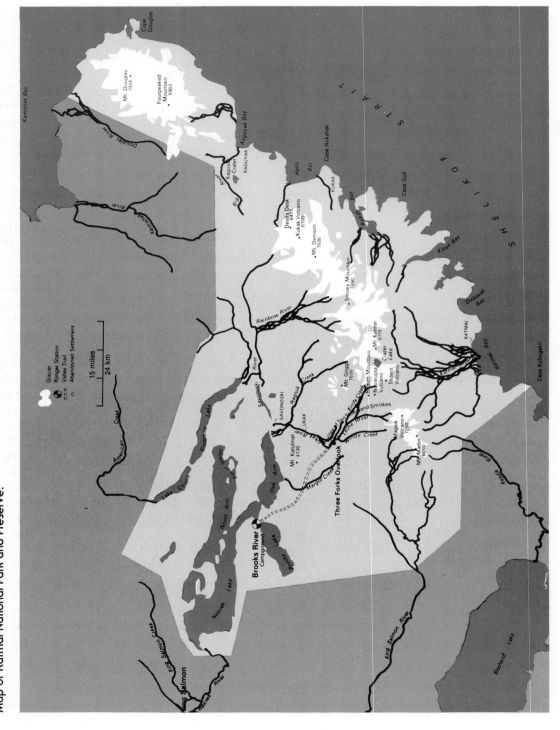

ferocious wolverine may make its presence known; and river otters, the slithering playboys of the animal world, are there seemingly for your entertainment. Perhaps, with the added space afforded by the enlarged park, the caribou and reindeer will return to their native haunts.

There are several good general references, one being Dale Brown's *Wild Alaska,* published in 1972 as one of the Time-Life Wilderness series. It is illustrated with color photographs that would be difficult to match. Much older, Griggs's *Valley of Ten Thousand Smokes* relates his experiences leading exploration parties into the Katmai wilderness, beginning in 1915. Although mainly a report on the 1912 eruptions, it includes a hair-raising account of their battles with the "williwaws," the violent winds that sweep into the area without warning. Griggs's book was published in 1922 by the National Geographic Society.

General Geologic Setting

Jurassic rocks — igneous intrusives, and *clastics* of the Naknek Formation — are extensively exposed within the park; however, the Tertiary-Quaternary volcanics are our main concern. Volcanic cones — Martin, Trident, Mageik, Denison, Griggs, and Katmai — rise prominently above the surrounding terrain.

Two large lakes, Naknek Lake and Grosvenor Lake, occupy much of the western third of the park. Brooks Lake, above Brooks Falls, empties into Naknek Lake. Savonoski River, in the northern part, drains westward into Lake Grosvenor; King Salmon River drains the southern part. In the central area of the monument is the famous Valley of Ten Thousand Smokes. The narrow strip east of the divide is drained into Shelikof Strait by Katmai River and other streams.

There are several glaciers along the divide, notably Hook, Hallo, and Serpent Tongue; many more are clustered around individual volcanic peaks. To date, however, it is not the glaciers but the volcanic features that attract most visitors to Katmai.

The Katmai-Novarupta Eruption

Prior to the 1912 eruption, the Valley of Ten Thousand Smokes was a broad valley covered with vegetation. In 1916, when R. F. Griggs finally succeeded in reaching the almost inaccessible area, he was startled to see that "the whole valley as far as the eye could reach was full of hundreds, no thousands — literally tens of thousands of smokes curling up from its fissured floor." Mt. Katmai was previously a volcanic cone about 7500 feet (2273 m) high; the eruption left a caldera about three miles across with a peak elevation of 6715 feet (2035 m). Six miles to the west at the head of the Valley of Ten Thousand Smokes was a new volcano which Griggs appropriately named Novarupta.

Early in June, 1912, when earthquakes became severe, the natives wisely withdrew from the area and, although there were narrow escapes, no casualties were reported. On June 6 the eruption began with a thunderous blast which sent more than two cubic miles of pumice, ash, and dust high into the air, and during the next few days another five cubic miles of *tephra* (solid fragmental material) were ejected. A huge cloud of white-hot ash, a *nuée ardente,* roared downslope and buried the valley under as much as 700 feet of pyroclastic materials, largely pumice and ash. The dust, which was blasted high into the air, even into the stratosphere, circled the globe for at least two years. Solar radiation in the Northern Hemisphere was significantly reduced, and temperatures were lowered accordingly. In the immediate area, sulfur gases killed all the vegetation; in a larger area, pumice ash covered the browse, and the moose that fed on it eventually starved as a result of having their teeth ground down by the abrasive pumice.

The sequence of events was not deciphered for many years. At first it was believed to be solely Katmai's explosion that had so ravaged

FIGURE 2-6

Katmai Caldera was formed in 1912, with post-1912 glacier on far wall (arrow). (Photo by National Park Service.)

FIGURE 2-7

Novarupta volcanic cone formed as the last phase of the 1912 eruption. (Photo by National Park Service.)

the landscape, but Griggs's discovery of Novarupta disproved this interpretation. The following sequence of events is now generally accepted (Curtis, 1958): The main explosion was at Novarupta, and the material in the valley was derived from this source. Magma was drained out from under Katmai, which then collapsed, thus forming the caldera. At first, this mechanism appears improbable, but when we consider the fact that Novarupta is more than 4000 feet lower in elevation than Katmai was at that time, the idea seems more plausible. Moreover, Griggs observed that the glaciers on the flanks of Katmai Caldera had not been melted, buried, or overrun by any nuée ardente. In addition, he noted that huge columns of steam were rising above both Mageik and Martin volcanoes, suggesting the possibility that a large body of magma underlay the entire area.

After Katmai Caldera was formed, a minor eruption built a small cone which rises as an island above the lake, like Wizard Island in the

FIGURE 2-8

Schematic cross section of Novarupta volcanic crater and volcanic dome (see Figure 2-7).

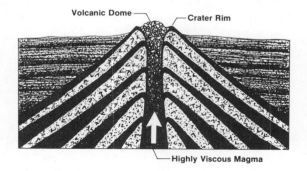

Volcanic Dome — — Crater Rim

Highly Viscous Magma

Crater Lake Caldera in Oregon. Katmai Caldera is similar to the one at Crater Lake except that it is somewhat smaller; it is about three miles long, two miles wide, and more than 4000 feet deep. Glaciers flanked Mt. Katmai at the time the caldera was formed; now, two glaciers have developed inside the caldera. One and perhaps by now both are more than a mile long. Well demonstrated here is the fact that under optimum conditions, glaciers can develop in a short period of time.

After Novarupta's explosive eruptions, highly viscous magma was pushed slowly up through its vent. It is postulated — although there has been no witness — that with only minor explosions, a steep-sided, dome-shaped mass of rhyolitic obsidian and rhyolite built up about 200 feet (60 m) high within Novarupta's crater. It is interesting to note that beginning in 1981 several geologists, including the co-author, had the rare opportunity to observe a similar dome as it gradually upswelled from the throat of the newly formed crater of Mt. St. Helens in the Washington Cascades. No significant modifications of the postulated sequence at Novarupta resulted from these day-to-day observations at Mt. St. Helens.

The Valley of Ten Thousand Smokes has changed greatly since Griggs saw it in 1916. By 1950 there were fewer than 100 live fumaroles or "smokes"; in 1964 the senior author found only five or six, and they appeared to be dying out. Around the fumaroles, hematite and other minerals had formed by sublimation of the volcanic gases.

The erosion of the 1912 deposits is almost unbelievable. The streams, particularly the Lethe and Knife rivers, have cut deep canyons through the pumice and tuff, and smaller tributaries have completely dissected parts of this 40-square-mile area, creating classic "badlands." Several physical factors contributed to this rapid erosion: Pumice is highly porous and floats readily; the extremely fine particles of tuff are also readily transported.

FIGURE 2-9

Stream-cut ravines in volcanic tuff, Valley of Ten Thousand Smokes, 50 years after the valley was buried by volcanic ash. (Photo by National Park Service.)

Moreover, these deposits are unconsolidated and therefore highly susceptible to erosion. Another equally important factor is the abundance of water — precipitation, snowmelt, and meltwater from glaciers. Here is an unparalleled opportunity to study the erosional process where all conditions are optimum.

How long will it be until this barren area is covered by vegetation? What has been the effect of sulfur fumes and volcanic dust on the surrounding area? Katmai offers an opportunity for ecologists to study the effects of volcanic activity on the plants and animals. It is also an idyllic retreat and an unexcelled fishing spot.

LAKE CLARK NATIONAL PARK AND PRESERVE

Lake Clark, a 50-mile-long glacial lake, is but one of many outstanding features in Lake Clark

FIGURE 2-10

The Tusks, glacially carved horns, near Merrill Pass in Lake Clark National Park. (Photo by P. Bradley, National Park Service.)

National Park, established in 1980 about 100 miles (160 km) southwest of Anchorage. Rugged, glaciered peaks, steaming volcanoes, fast-flowing rivers, and stormy-to-quiet inlets and bays make Lake Clark an exciting parkland.

Lake Clark National Park lies in two geomorphic provinces: The south end of the Alaska Range forms the western part, but most of the park is in the Chigmit Mountain section of the Aleutian Range. Three active volcanoes, Iliamna, Redoubt, and Mt. Spurr, rise prominently above the Chigmits. Iliamna erupted repeatedly during the three years ending in 1968, and Mt. Spurr, a short distance north of

present park boundaries, showered Anchorage with ash and dust in 1953.

Lake Clark occupies a considerable part of a broad glaciated valley and is fed by Tlikalila River, one of the National Wild and Scenic Rivers. The Tlikalila flows out of one of Alaska's Summit Lakes near Lake Clark Pass. Merrill River flows westward from Merrill Pass; and another river, called Another River, heads eastward a short distance north of The Tusks — actually horns.

For some, the shore features and the inhabitants of the coves and crags may be the best of the park. The shoreline, about 75 miles (125 km) long, extends from north of Tuxedni Bay southward into Chinitna Bay. Geologists investigating the oil seeps reported also on the wildlife: Kodiak and black bear, wolf, moose, lynx, otter, and beaver. There are also Dall sheep and caribou in the park, and with a little bit of luck you might see the bald eagle and catch a glimpse of the beluga whale.

No roads lead to the park, but motorists can get beautiful views of the volcanoes, and of some of the many glaciers, from across Cook Inlet on the Sterling Highway along the west side of Kenai Peninsula — providing the day is clear. Charter flights from Anchorage, Kenai, and Homer land at various points, often at Port Alsworth on Lake Clark. Provide some flexibility in your schedule; heavy clouds that may hover over the park for extended periods sometimes alter the best-laid plans.

Geologic Sequence

Geologists had previously explored areas of potential oil and mineral production, but the geologic story of Lake Clark began to be written in 1929, after Capps and his co-workers made their way into Lake Clark on foot, spending 40 of their 91-day field season en route, in and back out. In recent years, traveling to and from the area by plane or helicopter, geologists have been able to use their time more productively.

The Tuxedni Bay area has been studied in greatest detail because of its possibilities for oil and gas. Here, an exceptionally thick and complete sequence of bedded Jurassic rocks, about 26,000 feet (7880 m) in all, was mapped. Correlation was relatively simple because of an abundance of invertebrate fossils, including many ammonites and pelecypods (Detterman and Hartsock, 1966).

The oldest rocks in this part of the park are metamorphic — schist, marble, quartzite, argillite (dense, highly indurated shale), and greenstone (metamorphosed basalt) of Triassic

age. Resting on these rocks are the Jurassic graywacke (impure, "dirty" sandstone), conglomerate, siltstone, argillite, volcanic breccia, and tuff.

Few Cretaceous rocks have been found; evidently most that were deposited were eroded away late in the Cretaceous Period. Tilting of all of the rocks preceded the erosion, as indicated by the angular unconformity (see Figure I-21, page 27) between the Jurassic rocks and the overlying Tertiary sedimentary rocks and lava flows. Certain areas are capped by younger flows that poured out of the volcanoes during the Quaternary.

Viewed from the air or observed on topographic maps of Lake Clark, Kenai, and Tyonek

FIGURE 2-11

View across the upper end of 50-mile-long Lake Clark on a foggy day. (Photo by National Park Service.)

FIGURE 2-12

Twin Glaciers in Lake Clark National Park. (Photo by National Park Service)

quadrangles, the major structural feature of the park is clearly represented by the long, straight trough that angles southwestward across the area, with Lake Clark occupying a 50-mile (80-km) reach of the broad valley. The valley was carved out of the Lake Clark fault zone, in which the rocks are badly broken and easily eroded. Lake Clark Pass, at the head of the valley, and the valley in which Blockade Lake and Blockade Glacier are located, are all there because of the fault zone. According to Plafker and others (1975), the fault zone or *shear zone* is known to be as much as 900 feet (275 m) wide. It is highly probable that it was a zone of weakness long before the faulting began; at least it was along this line that granitic magmas began to intrude, from the Late Mesozoic into the Early Tertiary.

Precisely when faulting began is not known but apparently it was not until after a 38.6-million-year-old batholith was emplaced. Faulting offset this batholith about 3 miles (5 km), as

measured by Plafker et al. (1975). Another fault, the Bruin Bay, branches off from Lake Clark Fault and extends southward near the coast, entering the park in the vicinity of Tuxedni Bay. No evidence has been found for fault movement during the Quaternary; glacial deposits that blanket the faults have not been offset. Even so, from evidence outside the area, there is reason to regard both of these faults as potentially active.

Volcanic activity occurred during the Jurassic and again in mid-Tertiary time as reported by Reed and Lanphere (1973), but today's volcanoes did not begin their eruptive cycle until much later, at some time during the Quaternary. Like other Aleutian volcanoes, Iliamna and Redoubt are of the explosive type, as their steep-sided cones indicate. There is some variation in the composition of the volcanic materials but andesites and closely related rocks apparently dominate.

In areas such as this, where volcanic eruptions are frequent, streams have their courses blocked from time to time, thus interrupting the "normal" erosion regime. Eventually, however, the streams in the volcanic section of the park established their existing valleys.

Early in the Pleistocene, glaciers developed and presumably became widespread, although the evidence has been largely obliterated by successively younger glaciations. It is doubtful that Iliamna and Redoubt volcanoes began erupting before the first glaciers developed; at times, however, fire and ice have opposed one another.

Both volcanoes have glaciers streaming down their flanks; Iliamna supports 10, of which Tuxedni, the largest, is about 16 miles (26 km) long. Elsewhere in the park there are many other glaciers, some of them large, many without names. One little one with an intriguing name is Tired Pup Glacier. All are actively remodeling the landscape, deepening their valleys and sharpening the peaks. The glaciers have melted and readvanced several times. The exciting scenery is mainly the work of the Wisconsin glaciers that were much larger than those of today.

No two glaciers are identical; many factors, some very difficult to measure, control the behavior of glaciers. One glacier may be advancing while a nearby one is retreating.

Ice movement is distinctly different in the two glaciers shown in Figure 2-12. The one on the left is moving forward on a relatively smooth bed; the other is pushing out over a near-vertical cliff. Periodically, huge blocks of ice break off and fall down onto the ice below; this jumble of ice blocks is called an *ice-fall*. At the base, these blocks are all welded together into a solid mass called an *ogive* (pronounced o-jive). Note that the ogives near the top are essentially straight while those at the bottom are distinctly curved. From this it is clear that the middle part of the glacier is moving faster than the sides.

Essentially all of the many lakes were formed by glaciers, many by the gouging out of the bedrock or by the damming of the valleys with glacial debris — in the form of *end moraines*. Lake Clark and Lake Chakachamna — the two largest — are hemmed in by end moraines. Others are *kettle lakes,* formed when blocks of partially buried ice melted, leaving a depression. Still others are in depressions, small and large, on the uneven, irregular surface of moraines.

Lake Clark is a wilderness park. Certain parts are extremely difficult to get to; however, for those who are in good physical condition and know what survival means — in a country where the wildlife is truly wild — Lake Clark will surely be an unforgettable experience. Take warm clothes and don't forget rain gear!

CHAPTER 3

ALASKA RANGE PROVINCE

The Alaska Range is an arcuate chain of mountains, about 600 miles long, that extends northeastward from the Aleutian Range, thence eastward across south-central Alaska and southeastward into the western edge of Canada's Yukon Territory. In its southeastern part the range is comparatively narrow, but it becomes wider and higher in the Mt. McKinley section, where North America's highest mountain is located.

Geology and Geologic Sequence

The Alaska Range is one of the young mountain ranges of the world, and tectonic adjustments continue today. A major fault complex forms the framework for the range; many, if not all, of these faults are active at the present time, as frequent earthquakes indicate. Also, recent alluvial fans have been offset by the faults, and young valleys bend sharply at the faults.

The oldest rocks in the province are Precambrian metamorphics; Paleozoic and Mesozoic sedimentary rocks, some extensively metamorphosed, are prominently exposed in the higher mountains, whereas unmetamor-

FIGURE 3-1

Mt. McKinley and Mt. Foraker (far right). View from Eielson Visitor Center in Denali National Park.

FIGURE 3-2

Ice-capped Alaska Range mirrored in Summit Lake near Paxson. Gulkana Glacier (right) rises up into the low clouds.

phosed Tertiary sediments occupy many of the lower areas (Reed, 1961). Igneous rocks, mainly granites, form a number of the peaks, including parts of Mt. McKinley. In addition, Tertiary volcanic rocks — basalts, andesites, and rhyolites — are extensively exposed in a number of widely separated areas. To further

complicate the picture, there are Triassic pillow basalts that must have been dredged off the sea floor, probably from some distant location.

The dissimilarities between associated rock masses, both within and adjacent to the broad fault zone, formerly presented insurmountable difficulties in interpreting the geologic sequence; now, with plate tectonics concepts, the picture is gradually coming into focus.

Gilbert (1979) considers one of the faults, Hines Creek, to be the boundary between the North American Plate and the Pacific Plate, and believes that the collision began to affect the Alaska Range area late in the Mesozoic. When the northbound Pacific Plate made contact, it fractured and crushed the resident rocks; in places it ploughed into, under, or over those rocks, and the combined result was the upthrusting of the Alaska Range. Later, large masses of granitic magma intruded toward the surface, thus increasing the magnitude of uplift, especially in the Mt. McKinley section of the range. This general sequence, many geologists maintain, is the only one that logically explains this heterogeneous assemblage of rocks from widely separated localities.

As in many lofty ranges, faulting here was the main process involved in creating topographic relief; particularly striking is the precipitous north face of Mt. McKinley. From its summit at 20,320 feet (6194 m), the dropoff to the valley below is more than 16,000 feet (4878 m), in a horizontal distance of only about 15 miles.

Fault control of the direction of flow of certain streams is also striking; in places, streams have cut headward in the easily eroded fault-zone materials and developed valleys that are aligned counter to the regional slope. Prior to the coming of the glaciers, McKinley River flowed eastward in the Denali fault zone for about 15 miles; now Muldrow Glacier is using that same route, about at right angles to the north face of the mountain.

Glaciers and Glacial Streams

The glaciers of the Alaska Range are large and numerous because they are in the cold north and have an excellent source of moisture, the Pacific Ocean. Although in most mountains the larger glaciers are on the side protected from the sun, in the Alaska Range almost all of the large glaciers are on the southern flank where snowfall is greatest. However, Muldrow Glacier, 35 miles long, is a notable exception; because of its circuitous course it is by far the longest glacier on the north flank.

Rock glaciers are found in many glaciated areas but in the Alaska Range they are particularly well developed. At first glance, rock glaciers appear to be merely huge masses of *talus* material which have moved downvalley as rock streams. Inasmuch as many are fed by taluses, they were at one time believed to be forms of mass-wasting. What is not seen is more important than what is seen, namely an ice-core that is hidden inside all *active* rock glaciers. It is the ice mass that provides the motive power; therefore, a rock glacier is a special type of glacier, as Wahrhaftig and Cox established in 1959.

Glacial features such as horns, serrated ridges, and cirques are abundant. Cirque lakes, however, are uncommon because essentially all of the cirques are occupied by glaciers. In the lower areas, rock-basin lakes and morainal lakes are numerous.

Glacial and *fluvioglacial* deposits are also widespread in the lower areas. Moraines of all types, *eskers,* and outwash deposits cover large areas of the lower slopes and the flat *tundra* areas.

Streams flowing out from the glaciers, almost always loaded to capacity with coarse material, are *braided.* Their broad channels, together with constantly shifting streams, present some very difficult problems for the highway engineers.

The suspended load of the streams is *glacial flour,* rock material finely ground by the glacier. This material remains in suspension until it reaches the quiet water of a lake or the sea. Glacial streams often appear to have the consistency and color of an inexpensive chocolate milk shake.

In Chapter 2, we noted that the Alaska Range extends into the corner of Lake Clark National Park. Originally, Mt. McKinley was

FIGURE 3-4

Map of Denali National Park and Preserve.

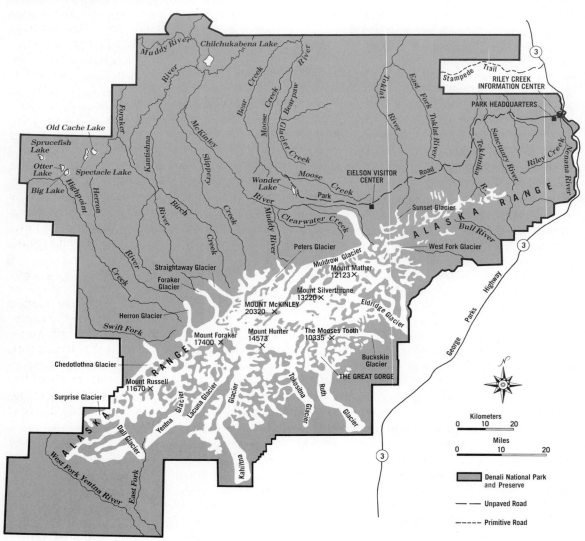

the only park in the Alaska Range Province; in 1980, it was enlarged and renamed Denali National Park and Preserve (see map, Plate 2).

DENALI NATIONAL PARK AND PRESERVE
(formerly Mt. McKinley National Park)

Denali, the Indian name for Mt. McKinley, is most appropriate because it means "The High One." Mt. McKinley National Park, established in 1917, was greatly enlarged in 1980 and renamed Denali National Park and Preserve. Mt. McKinley, however, is still the name of the 20,320-foot (6194 m) mountain that towers above all others on the continent.

The park was enlarged in order to include complete ecosystems and entire glacial systems within park boundaries. Denali now consists of approximately 6 million acres.

Access to the park, originally very difficult, was greatly improved in 1923 when the Alaska Railroad was completed. Now Alaska Highway 3 from Anchorage to Fairbanks essentially parallels the railroad, skirting around the eastern end of the park. For a time, cars were essential in order to get across the park to Wonder Lake; now, free bus transportation is available. In fact, private cars are allowed only as far as the Savage River Campground, about 12 miles (20 km) west of the park entrance. Inasmuch as cars are of little use in the park, you might consider riding the train from Anchorage or from Fairbanks, and travel leisurely through some of Alaska's most magnificent scenery.

Geology and Geologic Sequence

The Mt. McKinley segment is not only the highest but also the most complex part of the Alaska Range. Conceivably it is not by mere coincidence that this is true.

Early workers made substantial progress in mapping the Alaska Range, when consideration is given to the extent and remoteness of the area. Brooks (1911), Capps (1919), Reed (1933), and others did the early work. In 1961, J. C. Reed, Jr. published his U.S. Geological Survey Bulletin on the Mt. McKinley *quadrangle,* just before plate tectonics concepts were widely accepted or even well understood. The progress made during the past two decades is clearly reflected in the works of Gilbert (1976, 1979) who has concentrated most of his efforts on the Mt. McKinley area. His 1979 book, written for the benefit of park visitors, not only contains the modern interpretation of the geologic sequence but also a valuable guide to the geology along the 85-mile-long road across the park to Wonder Lake. For best results, carry his book for constant reference as you travel across this fascinating area. For the first time, Denali's visitors have available geologic information written especially for them.

The rocks of the Denali area are of many types, and they form essentially a collage or mosaic within the broad, intensely faulted zone — the collision zone between the northward-bound Pacific Plate and the North American Plate. Gilbert points out that the schists, quartzites and metavolcanics (metamorphosed volcanic rocks) that lie north of the Hines Creek Fault range in age from 350 million to about 1 billion years — distinctly older than the rocks south of the fault.

The scene is entirely different on the south side of the Hines Creek Fault; here, the rocks are "foreigners" — they were formed somewhere else and moved as plates and microplates to their present location. Although Gilbert does not attempt to explain precisely where the various rock masses came from, he has definite evidence, namely pillow lavas, that some of them were formed deep in an ocean basin.

The Pacific Plate moved generally northward, but offsets along many of the faults (Stout, 1973) indicate large-scale westward movement as well. As a result, there is a sharp bend in the structures in the vicinity of Mt. McKinley. The faults along the plate boundary bend southward at this point, creating a zone of maximum collision. Perhaps, as Gilbert suggests, this is the main reason why the Mt. McKinley segment has been thrust up much higher than any other part of the Alaska Range.

Certainly there are other factors. Large masses of magma, generated deep in the subduction zone, intruded upward to form large granite batholiths. Gilbert points out that these granites and associated metamorphics were somewhat lighter than the adjacent rocks and that, rather than being thrust downward into denser rocks, they were thrust upward. This upthrusting began during the Mesozoic and continued through the Cenozoic to the present. Recent faulting and periodic earthquakes indicate that it is still going on. Is it possible that North America's highest mountain could in the distant future become the world's highest mountain?

Throughout the Tertiary, the uplifted mountains were eroded and the debris was spread widely over the areas downstream. Twice during the Tertiary, volcanoes erupted and spread lavas and pyroclastic materials over extensive areas of the park. The light-colored rhyolites and the dark-colored basalts high on Polychrome Mountain are about 60 million years old. The multicolored materials along the park road in the Polychrome Pass area are highly altered volcanics of this series. The Mt. Galen volcanics, about 38 million years old, are widely exposed on Mt. Galen, about 4 miles (7 km) northwest of Eielson visitor center (Gilbert, 1979).

Glaciers: Past and Present

During the Pleistocene, glaciers developed in the mountains and extended down the valleys and far out across the lowlands. Vast amounts

FIGURE 3-5

Telephoto of Mt. McKinley from Eielson Visitor Center, in Denali National Park.

of material were deposited by the glaciers and by the streams flowing out from the glaciers. Lakes—some in kettles, the larger ones behind moranial dams—are widespread. Wonder Lake is where many park visitors get their prize pictures of Mt. McKinley—if the day is clear.

About 3 million years ago the climate cooled, and at high elevations glaciers developed in prompt response to the change. During their several advances, the glaciers completely relandscaped the mountains and deposited materials far out on the flats. How many major advances there were during the Pleistocene Epoch is not clear because much of the evidence here and elsewhere has been erased by more recent glaciers. The latest major advance, called the Wisconsin, reached its maximum 15,000 to 20,000 years ago. Today's glaciers are large, but even so, they are puny when compared to those of the Wisconsin advance. Lakes far out beyond present-day glaciers were formed in moraines deposited by earlier glaciers.

At present, glaciers blanket the tops of the Alaska Range and extend down as valley glaciers to the tundra areas below. Huge glaciers such as Eldridge and Ruth flow down the south slope. In fact, only one *large* glacier—35-mile-long Muldrow Glacier—is found on the north face; the smaller glaciers flow down steep valleys on the precipitous north face of the range and in a short distance reach elevations where melting balances ice movement. Muldrow follows the old McKinley River valley, which follows the Denali Fault; for this reason it travels 35 miles to get down to the same elevation reached by shorter glaciers such as the Straightaway.

In general, Mt. McKinley's glaciers had been slowly retreating since about 1850, as were almost all of the other glaciers of the world. But in 1956, Muldrow began to surge, and by 1957 its front had advanced about 4 miles. Such surges, some as early as the mid-1940s, were reported in other areas.

Denali's Attractions

"Mountain-watching"—or "watching for the Mountain"—is time-consuming at Denali. It is an elusive mountain, as many high mountains are, and sometimes park visitors look for it for days without success; clouds and fog protect it from the casual observer. Perseverance pays, because when it's there in all its glory, no other sight matches it. Nearby Mt. Foraker would also be famous if it had not been overshadowed; it is 17,390 feet (5802 m) high, and would overshadow every mountain in conterminous United States.

Distant views of glaciers present no problem, but to get close to or actually on a glacier requires advance planning and some degree of care. At present, Muldrow is the most accessible glacier in Denali. A short hike across the tundra south of the park road takes you to the

FIGURE 3-6

The Moose's Tooth is the highest of the Cathedral Spires in Denali National Park. (Photo by Dave Buchanan.)

FIGURE 3-7

Front of Muldrow Glacier (arrows) across McKinley River, a braided glacial stream.

edge of the bluff overlooking the braided McKinley River. Upstream, the river tumbles out of a tunnel in the toe of Muldrow Glacier. It may take a moment to realize that it is a glacier because the ice is completely buried beneath rock debris. There is a thick stand of brush growing on the glacier and if you are there at the right time, you may see a few caribou, or perhaps several hundred, feeding on the browse. Before attempting to negotiate the shifting streams of the braided McKinley River,

get some advice from a park ranger. The flow in a given channel fluctuates, and small streamlets may change rapidly into raging torrents, perhaps behind you! Also, people who do not venture into glacier tunnels are likely to live longer than those who do; tunnel roofs frequently collapse — and without warning.

Whether your interest is climbing icy peaks, observing glaciers, watching the grizzlies, caribou and Dall sheep, enjoying plant life, or admiring mountains, Denali National Park has it.

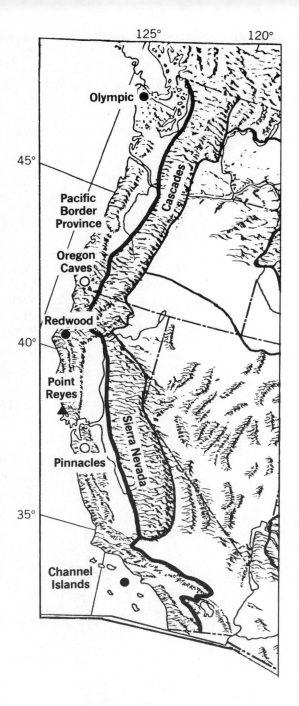

125°   120°

Olympic ●

45°

Pacific
Border
Province

Oregon
Caves
○

Redwood ●

40°

Point
Reyes
▲

Pinnacles
○

35°

Channel
Islands
●

Cascades

Sierra Nevada

● National Park
○ National Monument
▲ National Seashore/Lakeshore

PLATE 4

Pacific Border Province. (For Alaskan parklands, see
Plate 2.)

CHAPTER 4

PACIFIC BORDER PROVINCE

The Pacific Border Province is lean and long—almost 4000 miles (6666 km) long. Beginning with Kodiak Island, offshore from the Alaska Peninsula, the province includes the Kenai Peninsula and the coastal mountains around the Gulf of Alaska and extends southeastward through Alaska's panhandle and British Columbia, all the way south through Washington, Oregon, and California, to the tip of Baja California (see Index Maps, Plates 1 and 4).

In the northern part of the province, lofty mountains rise abruptly above the sea—mountains deeply indented by huge fjords. Here, countless glaciers actively relandscape the mountains—as silent reminders that this area is still in the Ice Age.

South of Olympic National Park in Washington, the coastal ranges are generally somewhat subdued. However, rugged, wave-cut cliffs, stacks, and sea arches present picturesque oceanscapes; elsewhere along the sea there are sandy beaches and coastal dunes. East of the coastal mountains a structural trough can be traced southward from Puget Sound in Washington through the Willamette Valley in Oregon, the Great Valley of California, and the Gulf of California.

FIGURE 4-1

Sea arch near Santa Cruz, California, 1971. See also Figures 4-1A and 4-1B in color section.

Tectonic activity characterizes the province. The gigantic fault zone bordering the continent has enabled the Pacific Plate to migrate northward for millions of years, carrying along microplates or slices of continental crust. But we must turn the clock back farther than that if we are to understand how the continent was pieced together.

Apparently mountain-building is, at least in many cases, directly related to the direction of plate movement. For example, during the Paleozoic when the Appalachians were being formed, the North American Plate was moving generally eastward, and it was the leading edge of the plate that was deformed. Later, during the Mesozoic, the direction of plate movement was reversed, and mountain-building began along what was then the western edge of the continent, in a subduction zone that was being shifted westward. The continental plate eventually collided with an oceanic plate that was eastward-northeastward bound. This stress situation was finally resolved in mid-Cenozoic time with the development of large transcurrent (horizontal) faults, such as San Andreas, Queen Charlotte in British Columbia, and Fairweather in southeastern Alaska. Since that time, the northward movement of the Pacific Plate toward Alaska has been the dominant tectonic influence in the Pacific Border Province. [Presented above are merely glimpses of what has been discussed in detail by many workers, among them Atwater (1970), Hamilton (1969), Maxwell (1974), and Dickinson (1979).]

To visualize how our earth has been broken into plates that have been shoved about and formed into mountains, the following analogy may be helpful: Picture a wide river that is flowing northward toward the Arctic Ocean several hundred miles away. The river is frozen over, but in its southerly reaches, water is flowing beneath the ice. When the spring thaw begins, the ice cracks into blocks, or "plates," some large, others small, some mere slivers. Soon the ice plates begin to move, carried

slowly northward by the flowing water under the ice. But in the cold North, the ice is still frozen solid, and here is the collision zone where the "plates" pile up in an ice jam. At first they create merely a ridge of ice blocks, but soon they become an elongate "mountain range," much like the mountains bordering the Gulf of Alaska. Some of the ice plates that have traveled hundreds of miles are now on top of, underneath, or beside blocks of "resident" ice — like some of the blocks of rock in the Chugach, Wrangell, and St. Elias mountains of southern Alaska. Thus the location of a plate of rock completely different from its neighbors in the Wrangell Mountains, yet identical to rocks in southern British Columbia, can now be reasonably explained using the concepts of plate tectonics.

Admittedly, ice blocks or ice plates on a river are distinctly different from 20- to 30-mile-thick plates of crustal rocks, and the plastic, slow-moving subcrust only remotely resembles water flowing beneath ice blocks. However, this analogy may help in visualizing what took place in the Pacific Border Province, particularly in the northern part.

Farther south, in Washington and Oregon, subduction zone deformation dominated the scene. Rocks that were carried along on the eastward-moving oceanic plate were crumpled up and thrust up into the coastal ranges, including the Olympic Mountains. Mountain-building began here in this general section early in the Tertiary, but the uplift that raised the mountains to their present height began much later, late in the Pliocene. (For an up-to-date interpretation of the tectonics of the Pacific Northwest, see Drake, 1982.)

Without question, tectonic movements continue in the Pacific Border Province today. The San Andreas fault system in California is of particular concern because of the millions of people living near it, and the recollection of devastating historic earthquakes: San Francisco in 1906, the Imperial Valley in 1940, and San Fernando in 1971. The earthquakes in

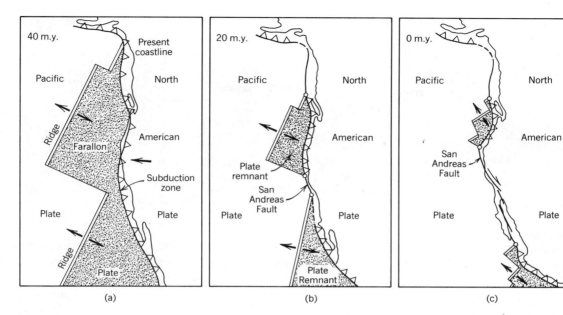

FIGURE 4-2

Diagrams depicting evolution of the Pacific Border during the last 40 million years. (Based on diagrams in Dickinson, 1979.) (*a*) Westward-moving North American Plate overrides Farallon Plate and begins to interact with the oceanic ridge about 40 million years ago. (*b*) Note the development of the San Andreas Fault and the change to right-lateral transverse faulting, about 20 m.y ago, where the oceanic ridge is overrun by the North American Plate. (*c*) Present-day plate elements along the Pacific Border. Small remnants of the original Farallon Plate remain, but right-lateral transverse faults have appeared where ridges are overrun.

Willamette Trough in Oregon and in Puget Sound in Washington are believed to be related to subduction activity. In southern Alaska it appears that all of the faults in the gigantic Border Ranges fault system are active or potentially active. In 1899, plate movements uplifted certain areas as much as 40 feet (13 m) and set off a major earthquake. Then in 1964, faulting of similar magnitude triggered the Good Friday earthquake that damaged large areas, particularly around Valdez and in downtown Anchorage. In summary, the Pacific Border Province is tectonically the most active province on the continent.

With its wide variety of fascinating geology and scenery, the Pacific Border Province contains a number of parklands in the United States and Canada. In a clockwise direction, they are: Kenai Fjords, Wrangell-St. Elias, Glacier Bay, Canada's Cape Scott, Strathcona and Pacific Rim, Olympic in Washington, Oregon Caves, and Redwood, Point Reyes, Pinnacles, and Channel Islands in California.

KENAI FJORDS NATIONAL PARK

Kenai Fjords National Park, consisting of about 587,000 acres of mountains, glaciers, and fjords, was established in 1980, under the Alaska National Interest Lands Conservation Act. The park is located in the southeastern

FIGURE 4-3

View along the side of Harding Icefield, with mountaintops rising as nunataks above the ice. (Photo by
D. Fellows, National Park Service.)

FIGURE 4-4

Sea lion and cub in Kenai Fjords National Park.
(Photo by M. W. Williams, National Park Service.)

part of the Kenai Peninsula, which extends for
about 150 miles (250 km) southwestward from
the Turnagain Arm of Cook Inlet, near Anchorage (see map, Plate 2).

The peninsula is split lengthwise by faults
that separate the Kenai Mountains on the
southeast from the Kenai Lowlands on the
northwest. Geologically, the Kenai Mountains
are a continuation of the Chugach Range,
which extends eastward from Anchorage into
Wrangell-St. Elias National Park.

The highest mountain rises only about 6800
feet (2060 m) above the sea but the Kenai
Mountains are deeply scored by countless glaciers. Much of the abundant precipitation falls
as snow, creating an ideal habitat for glaciers.
The Harding Icefield, an 800-square-mile ice
cap glacier, has a flattish surface about a mile
above the sea, with nunataks piercing through
(see Figure 4-3). The coast is extremely irregular; huge fjords extend far inland from the Gulf
of Alaska — fjords that were gouged out long
ago by Harding's outlet glaciers.

Between the fjords, long ridges or peninsulas extend out into the sea; here, stacks, sea
arches, cliffs, small islands, and jagged shorelines afford excellent habitats for seals, porpoises, whales, and sea otters. Then, there are
Steller's sea lions, named for Georg Wilhelm

Steller, the naturalist who, with explorer Vitus
Bering, was sent out by the Czar of Russia to
explore the Gulf of Alaska in 1741. Some 4000
raucous sea lions may be seen surging and lub-
bering over the rocks on the little islands in
Aialik and Nuka bays. The seaward ends of the
peninsulas, ice-free for centuries, support
rainforests of spruce and hemlock, ideal nesting places for bald eagles. Above treeline,
mountain goats frequent the craggy cliffs.

If you fly out over the park from Anchorage,
you can look down on the lakes, swamps, and
bogs in the vastness of the Kenai National
Moose Range, north of the park. Here, the
moose shares his range with goats, black and
grizzly bears, wolves, beaver, mink, trumpeter
swans, and loons. And, as you fly over Turnagain Arm, note the wide mud-flats, land only
since 1964 when the area was uplifted several
feet. The faulting that caused this uplift also
caused the Good Friday earthquake that damaged Anchorage so severely.

General Geology

More than half of the park is under ice but,
based on bedrock exposures at the ends and
on the seaward side of the icefield, most of the
rocks are Jurassic and Cretaceous argillites

FIGURE 4-5

Front of Exit Glacier in Kenai Fjords National Park. (Photo by National Park Service.)

(dense, highly indurated shales and siltstones), slate, graywacke (impure, "dirty" sandstone), and volcanic rocks of several types. However, certain of the peninsulas extending out into the Gulf of Alaska are in part composed of Tertiary intrusives, including sizable bodies of granite.

West of the park in most places, the Border Ranges fault system extends southwestward across Kenai Peninsula, ending in Kachemak Bay near Homer. Blocks and slices of many different rocks are found in this broad zone of intense deformation and dislocation. Most are metamorphics such as greenstone (formerly basalts and gabbros), schist and serpentinites; in addition, there are irregular masses of limestone, chert, gabbro, and granitic rocks.

Glacial Geology

The Harding Icefield is the crowning jewel of the national park. It is about 75 miles (125 km) long, extending from a few miles west of Seward southeastward through the park and into Kachemak Bay State Wilderness Park, near the southern tip of the peninsula. It was one continuous icefield 15,000 years ago when glaciers everywhere were much larger than at present; now there are two icefields, the main mass at the north being separated by a few miles from the small, irregularly-shaped mass to the south.

The term *icefield,* used in the strict sense, refers to a large mass of stationary ice that is too thin to flow under its own power. When the ice thickens sufficiently, generally to about 200 feet (60 m), the ice at the bottom becomes plastic due to the weight of the overlying ice. At this point the ice begins to flow out in all directions away from the center and the icefield becomes an ice cap glacier. The Harding Icefield is clearly an ice cap glacier because not only does it flow outward from the center, but it also feeds ice out into 34 outlet glaciers which extend out in all directions, like the arms of an overdeveloped octopus.

The outlet glaciers follow stream-cut valleys which they deepen and widen, steepening the valley walls. Those on the eastern side of the ice cap glacier pushed down steep canyons and out into the Gulf of Alaska, gouging out deep fjords. Here, the frontal ice is at the mercy of the waves during the fierce storms prevalent in the North Pacific. Therefore, tidewater glaciers here are subjected to greater destructive forces than are land-based glaciers, where melting is the only important destructive process. Under the present glacier regime, some of the glaciers still reach the sea but others, smaller and slower moving, now have their toes on land.

When the glaciers were at their maximum, those on the western side of the icefield ground down the canyons and out onto the Kenai Lowlands where some fanned out, forming piedmont glaciers. In other cases they followed well-defined valleys and extended themselves out across the flats toward Cook Inlet. Of these, Tustumena Glacier was probably the largest. It pushed out across the flats for almost 30 miles (50 km), where its terminal moraine forms a dam that holds back the water in Tustumena Lake, about 20 miles (33 km) long and as much as 7 miles (12 km) wide.

Farther north a similar glacier gouged out the basin occupied by Skilak Lake. South of

FIGURE 4-6

Tustamena Glacier (left) empties into Tustamena Lake; note icebergs on the lower right.

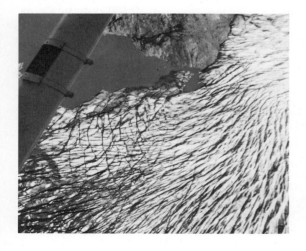

FIGURE 4-7

Looking down into crevasses in the frontal section of Tustamena Glacier.

Tustumena Glacier, several glaciers joined forces, turned southward and pushed out into Kachemak Bay, an arm of Cook Inlet.

The frontal section of Tustumena Glacier is clearly inactive; the crevasses are being widened by melting, and there is no evidence of forward movement. Meltwater is transporting debris out onto the delta that is being built out into Tustumena Lake.

Studies of the advances and retreats of the glaciers in and around Kenai Fjords National Park have not been made. In Glacier Bay, however, glaciologists have found that a general retreat, with minor advances, took place from about 1750 until about 1945; since then, several of the glaciers have made significant advances, some have slowed their retreats, and the fronts of others have remained essentially stationary. Probably the glaciers on Kenai Peninsula responded in much the same way to the fluctuations in climate, including the cooling trend that began during the 1940s. Interestingly enough, by coincidence the front of Tustumena is now at almost the same point that it was 20 years ago, as revealed in an old photograph. Undoubtedly, retreats and advances have occurred in the interim. Although Tustumena appears to be retreating at present, certain of the other glaciers may be advancing; each glacier behaves in response to a complex set of physical factors, and these factors may vary significantly between glaciers, even adjacent glaciers.

In summary, there is not sufficient information to warrant any generalizations regarding the present-day behavior of the glaciers on Kenai Peninsula, within or outside the park. Now that the park has been established we can anticipate that investigations will be made, and that we will then be in a position to do more than speculate on future glacier activity.

Access to the park is generally difficult, in part because of the weather that can change drastically in short periods of time; snow can fall on the Harding Ice Cap Glacier at any time of the year. There are no roads or camp-

grounds in the park. Seward, at the northeastern corner of the park, is park headquarters and the usual starting point for trips into or over the area. Charter aircraft are also available in Homer, at the opposite end of the park. The state ferry passes near the fjords on its way from Seward to Homer or Kodiak Island.

WRANGELL-ST. ELIAS NATIONAL PARK AND PRESERVE

Vitus Bering, exploring the Alaskan waters for the Czar of Russia, named St. Elias in honor of the saint of the day, in 1741. In 1978, about 12.4 million acres of the St. Elias, Wrangell, and Chugach mountains were set aside as a national monument; in 1980 it became the nation's largest park and preserve. Canada's Kluane National Park lies adjacent on the east, and together they became one of the World Heritage Sites.[1]

The St. Elias Range contains many lofty peaks. Canada's Mt. Logan (19,850 feet; 6015 m) is second only to Mt. McKinley, and Mt. St. Elias rises to 18,008 feet (5457 m) as the fourth highest on the continent. Mexico claims the third highest, Mt. Orizaba (Citlaltepec), 18,700 feet (5666 m).

Northwest of the St. Elias Range are the Wrangell Mountains, a series of volcanoes that rise above a volcanic plateau to as much as 16,000 feet (4848 m) above sea level. None of these volcanoes is known to have erupted during historic time, although there have been unsubstantiated reports of eruptions.

South of the Wrangells, across the broad Chitina River Valley, the Chugach Mountains rise abruptly above the coastal flats, merging

---

[1] World Heritage Sites are landmarks or natural areas of "outstanding universal value," designated by the United Nations Educational, Scientific and Cultural Organization (UNESCO), as discussed by David Douglas in the November/December 1982 issue of *National Parks Magazine.*

FIGURE 4-8

Mt. St. Elias, 18,008 feet high, dominates the St. Elias Range. (Photo by M. W. Williams, National Park Service.)

with the St. Elias Range at its eastern end. Glaciers are abundant in all three ranges, but the largest single ice mass is in the Chugach, the Bagley Icefield. This ice cap glacier feeds ice out into several outlet glaciers; one, the Malaspina, spreads out over the coastal flats to form the largest piedmont glacier in the world.

The park begins well below sea level in such fjords as Disenchantment Bay and extends to more than 3 miles above the sea; therefore, there is great variety in both flora and fauna. Forests of white and Sitka spruce are widespread, with aspen, birch, balsam, poplar, willow, and alder filling in the open areas. Moose, caribou, Dall sheep, mountain goats, black and grizzly bears, wolverines, wolves, lynx, coyotes, beavers, and otters struggle for sur-

vival on the land. Sea life, which is abundant particularly in Yakutat and Icy bays, includes humpback whales, porpoises, sea lions, harbor seals, and sea otters. Along the coast and on the rocky islands nearby, there are trumpeter and whistling swans, snow geese, sandhill cranes, gulls, guillemots, and kittiwakes, depending on the season. What most visitors least expect to see are the bison in the valleys of the Wrangells. They are not natives; instead, they were introduced about 1950 and have increased in number during the past few decades.

In addition to the above information, the first of the 1981 issues of the *Alaska Geographic* includes fascinating accounts of early settlement, including the various activities of

FIGURE 4-9

U-shaped glaciated valley. Chitistone River pours out of the hanging valley over Chitistone Falls, in the Wrangell Mountains. (Photo by National Park Service.)

Jack Dalton and of the boisterous era around Kennecott, the famous copper-mining camp in the Wrangells. Recognition is also given to the first inhabitants, the Athabascans, Tlingits, Tananas, and others who migrated here in the real "olden days." And the chronicles of assaults on Mt. St. Elias include those of Israel C. Russell in 1890 and 1891, and of the Italian Duke of Abruzzi, first to reach the summit, in 1897.

Microplate Tectonics

Three mountain ranges cannot but result in complexity of rocks and structures within the park, particularly when the rocks have been imported from distinctly different localities— some far distant. Because volcanic activity has dominated the scene in the Wrangells, they are distinctly different in character from the Chugach and the St. Elias where sedimentaries and metamorphics prevail, locally punctured by batholiths and other intrusives.

In recent geological papers, the term *terrane* is extensively used. A terrane, as distinct from *terrain*, is an area in which a particular group or sequence of related formations is prevalent. Thus, a microplate is regarded as one terrane and the adjacent microplate as another terrane.

The best-documented terrane in southern Alaska is called Wrangellia, because much of it now composes the Wrangell Mountains. Wrangellia has gained particular notoriety because a number of geologists believe that it traveled farther than any other microplate on its way up to Alaska—to be more specific, from down near the equator (Hillhouse, 1977). Slices of Wrangellia extend southeastward from the Wrangells. A splinter of the microplate was left behind in our next park, Glacier Bay National Park. Other slices fell off the northbound "barge" as far south as Vancouver Island, although Jones et al. (1978) described the sequence in more scientific terms.

Regardless of where Wrangellia originated, after the "great collision" took place, magmas that were generated in the subduction zone beneath the Wrangells slowly intruded toward the surface and eventually broke through and spread out as lava flows over much of the Wrangell area. Later, after many flows were piled one on top of another, the eruptions were confined to specific centers, and volcanic cones were built high above the lava plateau. Mounts Blackburn, Sanford, and Bona all top out at more than 16,000 feet (4848 m), and Mt. Wrangell and University Peak are more than 14,000 feet (4242 m) above the sea.

Which of the volcanoes will erupt first is not known, but Mt. Wrangell is being monitored closely. Mull (1981) reports that the University of Alaska's Carl Benson has recorded a marked increase in heat flux and solfatara activity since the 1964 earthquake. Sufficient heat has been generated to melt more than 500 million cubic yards of glacier ice from a single crater on the summit caldera's rim. Several miles to the west, mud volcanoes have built up low cones on the flank of Mt. Drum. Eruptions of these mud volcanoes occur when carbon dioxide and nitrogen gases concentrate in the warm waters below the surface (Mull, 1981).

Beneath the volcanoes and lava flows is a thick sequence of Triassic sedimentaries, including the Chitistone Limestone noted for the rich copper deposits at Kennecott, in the southern Wrangells. (The name Chitistone is appropriate because *chiti* is the Indian word for copper.) The mining boom ended in 1938, after more than a billion pounds of copper and almost 10 million ounces of silver had been produced. The Triassic age for these formations is well documented; invertebrate fossils are there in abundance (MacKevett, 1971).

The Nikolai Greenstone that underlies the Chitistone is also Triassic. (Here, the term *greenstone* refers to basaltic lavas that have been sufficiently altered to have a greenish color.) Pahoehoe and aa flows, similar to those in Hawaii, accumulated to a thickness of more than 10,000 feet (3030 m); pillow lavas have been found, thus indicating that the area was beneath the sea at times (Jones and Silberling, 1979). Certain of the flows contain amygdules — gas-bubble holes filled with secondary minerals. Some of the amygdules are calcite and zeolites, but native copper amygdules have been found. Where the base of the Nikolai is exposed, it is resting on limestone and argillite of Permian age.

The broad Chitina River Valley lies between the Wrangells and the Chugach Range to the south. Typical of streams fed by meltwater from glaciers, the Chitina is intricately braided in many of its reaches. It gathers the water from glaciers in all three of the ranges and carries it westward, emptying into the Copper River near the town of Chitina.

The collision zone, the Border Ranges Fault, forms the northern boundary of the Chugach Terrane which extends eastward into the St. Elias Range. The rocks are predominantly metamorphosed sedimentary and volcanic rocks that have been intruded by large granitic bodies (Mull, 1981). Along the south side of the Chugach Terrane, the St. Elias and Fairweather faults separate the highly metamorphosed rocks of the mountains from the essentially unaltered sedimentary rocks of the coastal plain.

The time when the upthrusting of the mountains began has not been established, although major plate movement is still actively taking place. Severe earthquakes have shaken the region repeatedly, and in 1899 and again in 1964, faulting uplifted certain areas as much as 40 feet. The latest major earthquake occurred in 1979. The sudden upfaulting and the resulting earthquakes are generally believed responsible for some of the phenomenal behavior of certain of the glaciers.

Glaciers in Action

Throughout the Pleistocene, glaciers have been relandscaping the park, advancing and retreating as the climate fluctuated. Although the early glaciers unquestionably did their part, much of the evidence was obliterated when the mighty Wisconsin (Late Pleistocene) glaciers plowed through the older deposits, reaching their maximum about 14,000 years ago (Denton, 1974). Then they retreated, in some cases rapidly, and during the Thermal Maximum, the valley glaciers were confined to the colder sites high in the mountains. A cooling of the climate about 4000 years ago initiated a period called the Little Ice Age, during which time there were a number of significant advances. Sharp (1958) states that the last major advance of Malaspina Piedmont Glacier took place about 200 years ago.

At present, there are about 200 active glaciers, about half of which have been named. A huge ice cap glacier covers the eastern part of the Chugach Mountains and extends eastward into the St. Elias. The part in Alaska is called the

FIGURE 4-10

Malaspina Glacier, one of the largest piedmont glaciers in the world, with Mt. St. Elias in the background. (Photo by Austin Post, U.S. Geological Survey.)

Bagley Icefield, that in Canada the Seward Glacier. Overall, this ice mass is more than 100 miles (160 km) long. It is certainly well-nourished; the winds that lift off the Pacific are heavily laden with moisture that produces extraordinarily heavy snowfall, perhaps as much as 600 inches (15 m) per year.

Malaspina Glacier, fed from the Bagley Icefield, has fascinated many people not only because of its enormous size, about 850 square miles, but because of the intricately folded and contorted moraines along its eastern margin, as seen in Figure 4-10. Sharp (1958) attributes this accordian-type folding to the fact that this marginal section of the ice was shoved uphill. The ice in the thicker middle section is as much as 2000 feet (606 m) thick, as seismic surveys indicate, and it lies in a basin that extends at least 800 feet (242 m) below sea level. Sharp reported that in 1954 the Malaspina was surging, clearly indicated by rapid enlargement of old crevasses and the noisy cracking of the ice in forming new ones. Probably an extraordinarily heavy snow accumulation several years earlier was responsible for the sudden increase in the rate of ice movement.

Farther west, outside the park boundary, Bering Piedmont Glacier has experienced similar surges (Post, 1969) and, according to Miller (1960), it was then very close to its maximum Wisconsin position. Martin River Glacier, a valley glacier west of the Bering, is also near its maximum position according to Reid (1970), who suggests that recent uplifts of the Bagley Icefield may have caused more ice to flow in the direction of the Bering and Martin River glaciers.

There are far too many glaciers to include in this book; however, Hubbard Glacier must receive our attention. It certainly would have if we had been there in September of 1899 when, in a five-minute period, it ran out a half-mile into Disenchantment Bay! This spectacular surge was triggered by one of the largest earthquakes ever to hit the North American continent. The sight of this mammoth glacier — five miles across its front — galloping along

at top speed must have been more than the prospectors had bargained for (Harrison, 1975). It surged again in 1973 and 1974 — without benefit of earthquake — albeit at a slower pace (Mull, 1981). This 80-mile-long valley glacier heads high on Canada's Mt. Logan, moves eastward and then southward across the border into the park. To be there standing in awe near the bold front of this tidewater glacier must be a rare experience indeed, judging from Figure 4-11 in color section.

Rock glaciers, discussed in the Alaska Range chapter, are also abundant in the park, particularly in the Wrangell Mountains. They form at the base of rapidly weathering cliffs and creep down the valleys, in some cases for several miles. As in all active rock glaciers, the ice hidden beneath the rock rubble provides the motive power. The thick blanket of rubble acts as insulation and prevents appreciable melting of the underlying ice; therefore, in some areas where the naked glaciers have melted away, rock glaciers are still active.

Where there are glaciers there are also waterfalls, and Wrangell-St. Elias has its full share. During periods of rapid melting when the rivers are roaring, the sight and sounds of the falls are particularly impressive. Chitistone Falls, in the eastern part of the Wrangells, is about 300 feet (91 m) high.

Most of the park is remote, miles from any road; even so, parts of Wrangell-St. Elias are more accessible than some of the other Alaskan parklands. If you are traveling the "long road," the Alaska Highway, you will skirt around Kluane National Park and into Alaska near the northeast corner of Wrangell-St. Elias. At Tok Junction, the Tok Cutoff leads south into the Copper River Valley along the western boundary of the park. From Anchorage, the Glenn Highway will take you east to Glennallen, a little town just outside the park boundary. If you want to get to McCarthy, in the Kennecott mining area, go south from Glennallen to Chitina and east on the McCarthy "road," preferably in dry weather. Don't miss

FIGURE 4-12

Rock Glacier creeping down toward the Chitistone River in the Wrangell Mountains. (Photo by G. Herben. Permission from *The Alaska Geographic*, reprinted from Wrangell-St. Elias: International Mountain Wilderness, Vol. 8, No. 1.)

FIGURE 4-13

Map of Glacier Bay National Park. Dotted lines show changes in glacier fronts since 1966.

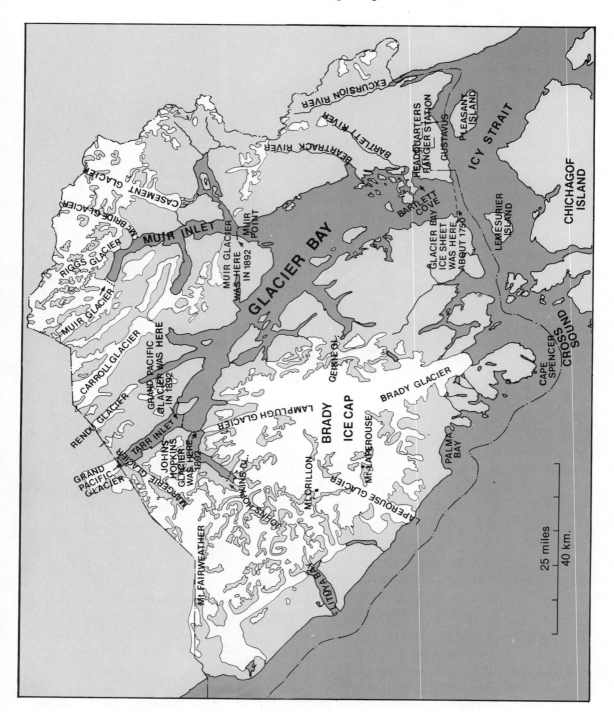

the old, narrow bridge that looks down 200 feet (61 m) to the Kuskulana River. If you stop at the end of the road and pull yourself across the Kennicott River on cables, you will be in McCarthy visiting with some of its eight residents (Mull, 1981).[2]

The Nabesna road leads into the north part of the park, ending at the old gold-mining town of Nabesna several miles north of the end of 75-mile-long Nabesna Glacier. The coastal area is more difficult; it is accessible by boat but squalls can develop suddenly and 60-foot waves are by no means uncommon. Charter flights are available at various points, including Anchorage, Fairbanks, Glennallen, Valdez, and Yakutat. Take sun glasses and a good supply of film.

GLACIER BAY NATIONAL PARK AND PRESERVE

Glacier Bay was established as a national monument in 1925 and enlarged and upgraded to park status in 1980. It now consists of about 3.3 million acres of mountains, glaciers, and fjords in southeastern Alaska.

The topography and scenery vary greatly, as there are parts of four mountain ranges within park boundaries. The high Fairweather Range lies along the west side, the tip of the St. Elias Range on the north, the Takhinsha Range on the northeast, and the Chilkat Range on the east. The highest peak, Mt. Fairweather, rises 15,300 feet (4670 m) above the Pacific Ocean. The mountains are heavily laden with glaciers, many of which funnel down into Glacier Bay, a mammoth fjord. To the west of the fjord is Brady Glacier, one of several large ice cap glaciers in the park system. Lituya, Fairweather, and other glaciers inch their way down the western face of the Fairweather Range toward

---

[2] The River and Kennicott Glacier were named for early explorer Robert Kennicott, spelled with an "i." Kennecott is the name of the town and the copper company.

the Gulf of Alaska. On the coast, La Perouse Glacier spreads out over the flats, thus forming a *piedmont glacier.*

Glacier Bay was one of John Muir's favorite haunts. He was curious as to how his favorite valley, Yosemite, had been formed and in 1879 he went up to Glacier Bay to find out. Once there, he became so involved with the glaciers that he had to go back — four more times — to explore more glaciers. In recognition, Muir Glacier and Muir Inlet, the fjord occupied in part by the glacier, were given his name. Muir's adventures in Alaska, as he defied all common sense by crossing glaciers and crevasses by himself, are recorded in a fascinating book, *Muir of the Mountains* (Douglas, 1961).

Two outstanding pictorial volumes on Glacier Bay are available. *Glacier Bay,* published as one of the 1975 volumes of the Alaska Geographic, and Dave Bohn's *Glacier Bay: The Land and the Silence* (1967) present all aspects of this fabulous national park.

General Geology

The most obvious rock in the park is ice — glacier ice. It blankets the highlands and extends as huge ice streams down into the fjords. When Henry Fielding Reid explored the area in 1892, a large part of Glacier Bay was buried beneath glaciers and snowfields. Even with most of the rocks covered it was obvious that it was a geologically complex area; all that he had to do was to examine a few of the glacial moraines and outwash deposits to see the wide variety of rocks of highly diverse origin. The glaciers were retreating at that time and, although some began to advance later, many areas are now free from the ice and available for geologic study. For example, a 10-foot-thick sill of gabbro was recently brought to light by the melting back of Plateau Glacier; thus the sequence of events in Wachusett Inlet is more nearly complete than before (Brew et al., 1976).

During the same period that additional areas of bedrock have come out from under the ice, more and more of the mysteries of plate tectonics have come out from the shadows. The combined result is that what was formerly a completely frustrating puzzle is now beginning to make sense—in the form of northwest trending terranes blocked out by faults and *suture zones.*

Wrangellia, discussed in the section on the Wrangell-St. Elias area, is represented in Glacier Bay by a narrow *suture zone* extending southeastward from Tarr Inlet beneath Brady Ice Cap Glacier (Brew and Morrell, 1979). Slate, schist, marble, and conglomerate are the main rocks in the suture zone. On both sides, there are many intrusive igneous bodies, mainly granitic in composition, probably all Tertiary in age. One large body of gabbro in the Fairweather Range is believed by Hudson and Plafker (1980) to have been emplaced about 40 million years ago. The rocks that were intruded include various metamorphics, argillite (dense, highly indurated shale), graywacke ("dirty" sandstone), and a few volcanics.

As more and more evidence is brought to light it appears that Glacier Bay is a collage of rocks—microplates, slices and slivers—most of which have been imported from different localities, some a considerable distance from the south. It is largely, if not entirely, within the Border Ranges Fault complex, the broad fault zone that surrounds the Gulf of Alaska. At least this seems to be the general interpretation favored by Jones et al. (1978).

Although movements along several of the faults were involved in bringing the rocks to their present locations, essentially all of the recent displacements have taken place along the Fairweather Fault, as reported by Plafker et al. (1978). Thus, Glacier Bay is apparently made up of broken old plates, all welded together except for the slice that is west of the Fairweather Fault.

No one who was in the Lituya Bay area on July 10, 1958 doubted that the Fairweather is an active fault. Displacements of as much as 21 feet (6.5 m) set off a 7.9-magnitude earthquake (Plafker et al., 1978). In turn, the cliff-face of the mountain crashed down into Lituya Bay, causing gigantic waves and much consternation among the fishermen who were there in their boats. For an exciting account, read the report by Miller (1960).

Streams that originally flowed straight across the fault now angle to the right within the fault zone. Therefore, the Pacific Plate is clearly moving northwestward, as it has been doing for many millions of years. The processes that built Alaska's mountains are still active today.

Glaciers: Past and Present

The glaciers of the past, abundantly nourished by snowfall derived from moisture-bearing winds directly off the Pacific, were both numerous and large. In fact, when the glaciers were at their maximum, practically all the area was under ice except for parts of the higher peaks and ranges that projected above as *nunataks,* large and small. At times, huge trunk glaciers developed and excavated gigantic fjords. Horns, aretes, and cirques were carved out of the mountains, reshaping the stream-cut topography completely.

Early Pleistocene events are virtually unknown because of the work of later glaciers, including those of the present. Wisconsin glaciers began their retreat from the lower areas about 10,000 years ago, and by about 7500 years ago they had receded to positions somewhat above their present locations (Goldthwait, 1963). After the Thermal Maximum these glaciers began to advance, perhaps about 3000 years ago. In the Glacier Bay area, the Neoglacial or Little Ice Age advance was a major one in which the trunk glacier moved forward about 60 miles (100 km), reaching its maximum position around 1750. In 1794, when Captain Vancouver sailed into Icy Strait, he saw a high wall of glacier ice at Bartlett

FIGURE 4-14

Air view of Fairweather Range, source of several glaciers in Glacier Bay.

Cove, near the lower end of Glacier Bay.

In reality this advance was not a continuous one; instead, there was advance, stillstand, then retreat, followed by advance. Here the record is unfolding as more and more trees that were buried by glacial debris are unearthed. Forests established themselves in the deglaciated areas during retreats, only to be buried during the next advance. Trees of different ages, based on carbon-14 determinations, record the different advances. Such trees have been found in various localities, particularly in the Muir Inlet; in 1969, tree stumps that were buried by the glacier about 1300 years ago were found in Geikie Inlet. As streams continue to expose more buried trees, the record of glacial advances and retreats will become more complete.

Just when the last major retreat began is uncertain, but in 1890 the glacier front was about 25 miles upfjord from its 1794 position. In 1890 John Muir could look across the front of Muir Glacier from his cabin; now the front is more than 15 miles (25 km) farther up the fjord. The front of the main glacier, the Grand Pacific, is almost 60 miles (100 km) above its 1794 position.

The most recent deglaciation provided excellent opportunities for ecological studies; not only have large areas been freed from the ice cover but records on when each area was deglaciated have been kept. In an area such as Glacier Bay, where moisture is abundant, plant succession is reasonably rapid and certain newly uncovered areas are already covered with dense thickets of alder and willows. W. S. Cooper (1937) was an early investigator in this area; more recently, Goldthwait et al. (1966) published a comprehensive report on the Muir Inlet.

Clearly the glaciers in Glacier Bay are extremely sensitive to climatic fluctuations. This was recognized immediately by William O. Field of the American Geographical Society on

FIGURE 4-15

Johns Hopkins Glacier has advanced several miles since W. O. Field first saw it in 1926.

his first visit to Glacier Bay in 1926; he established a field research project to record the changes from year to year. He and his assistants have made surveys and taken photographs from fixed camera points almost every year for more than 50 years. In 1969, the senior author participated in these field studies and gained a deep appreciation of the tremendous amount of work done by Dr. Field during this long period. Dr. Field's report, when published, will provide the most complete history of any glacial area in North America.

The retreat in progress in 1926 was catastrophic at times until about 1940; then deceleration of most retreats was followed by halting advances of certain glaciers. Grand Pacific, Johns Hopkins, and others on the east side of the high Fairweather Range have been advancing; in contrast, Muir and most of the other glaciers farther inland from the Pacific,

the source of their nourishment, have continued their retreat, but at much slower rates.

Early in this century, the front of Grand Pacific was in British Columbia, north of Glacier Bay. By 1948, it had advanced southward across the international boundary, and by 1969 it was approaching the terminus of Margerie, a former tributary glacier. It appeared then that Grand Pacific and Margerie would soon be reunited, perhaps within a dozen years. However, a visit in 1982 was a disappointment; there was still a gap of about 250 feet (75 m) separating them.

By referring to Figures 4-16 and 4-17, you can see how much narrower the gap was in 1980 than in 1969. Based on approximate figures obtained from Field (personal communication, 1982), there was an advance of about 2800 feet (850 m) from 1964 to 1979. But since then Grand Pacific has apparently slowed

FIGURE 4-16

Grand Pacific Glacier (lower right) and Margerie Glacier, a main tributary until Grand Pacific retreated upfjord; it had readvanced to this position by 1969. Note splashdown of iceberg from Margerie Glacier.

FIGURE 4-17

Grand Pacific and Margerie glaciers in 1980 (compare with Figure 4-16). Note the advance of Grand Pacific, almost overrunning the alluvial fan at the far end. (Photo by J. Luthy, National Park Service.)

down; in fact, the right-hand side has wasted away, leaving a pronounced embayment. But in 1982 it was obvious that Margerie's front had advanced significantly since 1969, with a pronounced bulge out into the deep water of the fjord. It is probable that the "breaking point" may soon be reached and that Tarr Inlet (fjord) may be afloat with icebergs of unprecedented number and size. If the cooling trend that began in the early 1940s is at an end, we will soon see the beginning of a recession. The climate still has the glaciers under its control.

The wildlife of Glacier Bay National Park is more varied than in many parks: humpback whales in Blue Mouse Cove, countless seals on and off the icebergs in Muir Inlet, the big glacier bear who had had his dinner of goat meat at one of our camera stations, and the timber wolf who made huge tracks past our little boat added excitement to our visit.

South from Glacier Bay the fjordland continues through southernmost Alaska and British Columbia. Then, in the northwestern corner of Washington is our next national park, Olympic.

OLYMPIC NATIONAL PARK

Within the park system there is no area of greater contrasts than Olympic National Park. Mt. Olympus (7965 feet; 2428 m) and other peaks are shrouded with glaciers. With the highest precipitation in conterminous United States, the glaciers are well nourished; so is the rain forest region on the Pacific slope. The situation is distinctly different on the eastern side. Moisture-bearing winds, which lose most of their moisture over the rain forest and glaciers, have but little to drop on this "rain shadow" area. In further contrast is the 50-mile-long section of the Pacific Coast where there are wave-cut cliffs, sea arches, and other features of the oceanscape.

Olympic National Park was established first as a national monument in 1909 and upgraded to park status in 1938. The park occupies a large part of the Olympic Mountains on the Olympic Peninsula in the northwestern part of Washington. The park headquarters and one of the visitor centers are at Port Angeles, about five miles north of the park. The glacier-clad

FIGURE 4-18

Sea stack and sea cliff along wilderness beach, Olympic National Park. (Photo by F. E. Mutschler.)

peaks can be viewed from the Hurricane Ridge Road south of Port Angeles. The Hoh River Road is one of the ways to get into the rain forest on the western slope, where the moss-draped giants of trees blot out the sun. On the east side, the dirt road up the Dosewallips River from the Hood Canal Highway leads into the dry eastern-slope section. Most of the park, the rugged and truly exciting parts, must be seen the hard way. About 600 miles (968 km) of trails penetrate into the back country along which there are numerous shelters.

The Geologic Story — Old and New

In the Olympics, Mother Nature relinquishes her secrets grudgingly. The terrain is extremely rugged and many areas are covered by dense vegetation or by glaciers.

In some mountain ranges — for example, the Front Range of the Southern Rockies — ancient rocks form the backbone, although the mountains themselves are relatively young. In the Olympics, however, the rocks are young and the mountains are younger, one of the youngest ranges on the continent.

Exposed in the large central section of the park is a series of rocks so intermixed and in places so metamorphosed as to essentially defy detailed interpretation. Apparently, marine sandstones, conglomerates, shales, and submarine pillow lavas and volcanic breccias have been highly contorted and metamorphosed.

A relatively simple structure was envisioned by Danner (1955) and other workers. According to this interpretation, the Olympics are a simple domed or anticlinal structure com-

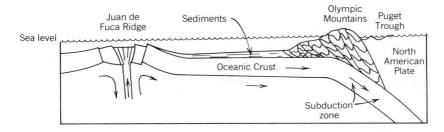

FIGURE 4-19

Collision of Juan de Fuca and North American plates "scrapes off" and deforms sedimentary rocks, forming the Olympic Mountains (refer to Figure 4-2). (Modified after Rau, 1980.)

posed mainly of the extremely thick Soleduck Formation, of Cretaceous age, with Tertiary rocks on the flanks of the structure. But the age of the Soleduck was subject to question; worm tubes, the only fossils, are less than diagnostic. Also, the thickness, believed to be many thousands of feet, was not acceptable to certain workers. Perhaps there was more information, more fossils to be dug up. The terrain was rough and many of the rocks were hidden beneath overburden of glacial deposits and glaciers. Even so, it was a puzzle that needed to be put together.

The combined efforts of a number of imaginative and determined geologists were required to solve this elusive problem. They had new tools to work with, and new avenues to explore. Perhaps these rocks had been completely deformed in a subduction zone and then thrust upward to be exposed by erosion. The fascinating story of the evolution of geologic thought is set forth by Tabor in his *Guide to the Geology of Olympic National Park* (1975).

Foraminifera — tiny fossils — were eventually found in the Soleduck Formation in the core of the range. They were identified as Eocene fossils; thus the Soleduck is much younger than it was formerly thought to be, and the concept of the simple anticlinal structure was no longer tenable. Instead, according to the new interpretation, Soleduck rocks of

the oceanic plate were thrust under the edge of the continent. In the subduction zone they were intensely folded, metamorphosed, and faulted, and the fault slices were thrust upward. Thus, instead of one extremely thick formation, the several slices gave the appearance of great thickness. With subsequent erosion these rocks were exposed in the core of the Olympics. At least, this is the interpretation that appears to explain the facts unearthed to date.

Glaciation

The Pleistocene glaciers that developed in the mountains were large, although none of the peaks is as high as 8000 feet (2440 m). Precipitation is clearly the dominant factor here. At present, more than 200 feet (60 m) of snow falls on Mt. Olympus during wetter years; how much more snow fell during the Pleistocene glacial advances can only be estimated. The glaciers advanced and retreated and formed the precipitous glacial features that we see today. Glacial deposits, particularly lateral moraines and end moraines, are numerous in many of the lower areas. Within these deposits all the rock types of the drainage basins can be found.

A serious problem developed, however, when granite boulders were found, because there is no granite bedrock in all of the Olympic Mountains. Some of these boulders are as much as 3000 feet above sea level; therefore they could not have floated in, ice-rafted from some distant point. As with most geologic problems, careful observations by persistent researchers produced a reasonable solution: The granite boulders had originally been a part of the Coast Range Batholith which is in British Columbia. Glaciers originating in the Canadian mountains pushed down into the lowland areas and coalesced to form a broad piedmont glacier which then headed southward and crossed the boundary into the Puget Sound area. Here it divided into lobes; one continued south into the Olympia area, and the other extended itself westward through the Strait of San Juan de Fuca north of the Olympics. The ice pushed down into the Puget Sound area at least six times (Tabor, 1975). At least once, this piedmont glacier was greatly thickened and extended up onto the flanks of the mountains as high as 3000 feet, to deposit the granite erratics.

The Canadian glacier was also responsible for a number of lake basins in lower areas. Near the northern boundary of the park, the ice scoured out the irregular, elongate basin now occupied by Lake Crescent and Lake Sutherland. After the ice withdrew, a large landslide rushed down from the north and separated the two lakes. (This explanation differs from the Indian legend in which Storm King Mountain hurled a big rock into the lake!)

FIGURE 4-20

Telephoto of Mt. Olympus, Olympic National Park, from Hurricane Ridge. (Photo by E. Kiver.)

Today's Glaciers

Perhaps as many as 60 glaciers, most of them small, are clustered around the higher peaks. The exact number is not known; in fact, the number has been changing. During the warming trend from about 1850 to 1945, the smaller glaciers became inactive icefields; after the cooling trend began, certain icefields thickened and became active glaciers again.

Six glaciers are draped around the summit of Mt. Olympus; of these, the Blue, White, and Hoh glaciers are more than 2 miles long. Blue Glacier has been studied in considerable detail by glaciologists from the University of Washington and the U.S. Geological Survey. Insight regarding the mechanics of glacier movement has been obtained from their observations and measurements in a tunnel excavated along the sole of the glacier.

The Oceanscape

Along the Pacific the waves tear away at the land relentlessly and endlessly. The continent is giving ground; the high sea cliffs provide evidence of that. As the sea cliff retreats, the wave-cut platform is enlarged and extended landward; at low tide, many intriguing creatures of the sea can be observed in the tide pools. The pounding of the waves, especially during violent storms, develops sea caves by dislodging blocks of bedrock. Where the rocks are locally more resistant to wave action, headlands—"points" or "heads"—extend out into the sea; Toleak Point south of Teahwhit Head is an example. When sea caves are developed back-to-back along fractures and faults in a headland, they are eventually joined to form a tunnel with a sea arch above it. Later, the arch collapses, leaving a *stack* isolated from the mainland. (Although the term *stack* is sometimes applied to other erosional features, geologically speaking the only stacks are sea stacks.)

Olympic National Park is a park of contrasts.

Take good hiking boots for the glaciers and sun-tan oil for the beaches, and by all means, take rain gear!

OREGON CAVES NATIONAL MONUMENT

In the Siskiyou Mountains in southwestern Oregon is a large cave that was discovered by a bear, a dog, and Elijah J. Davidson, in that order. Davidson was credited with the discovery, in 1874. Visitors came, most of them to see and to admire the "Marble Halls of Oregon," a few to break off the stalactites to take home as souvenirs. In order to protect the cavern formations from further vandalism, the U.S. Forest Service set the area aside as Oregon Caves National Monument in 1909. In 1933, the monument, consisting of 480 acres, was transferred to the National Park Service.

Most true caverns—those formed by solution by subsurface water—are in limestone. Oregon Caves are unusual because they were formed in marble.

The geologic story is lengthy, beginning early in the Mesozoic Era when during the Triassic period limy muds were deposited in the sea. When these limy muds were consolidated they became solid limestone on which thick sediments were deposited. Much later, during the Tertiary, there was mountain building, and the rocks far beneath the surface were profoundly changed—metamorphosed. Sandstones were converted to quartzites, shale to slate, and limestone to marble, the marble in which the caverns were eventually formed.

Mountain-building continued over a long period, and finally the marble and other rocks were thrust up to their present position and exposed by stream erosion. It was not until then that conditions were right for the formation of the caverns.

There are four main requisites for the development of cavern systems: (1) a soluble rock at or near the surface, (2) a well-developed fracture system, (3) subsurface water containing carbon dioxide, and (4) time. Here, marble is the soluble rock, severely fractured as the mountains were thrust up; and there has been an abundance of both water and time.

Carbon dioxide from the air combines with water to form carbonic acid which reacts with calcite, the main mineral in marble. When a fragment of calcite is put in hydrochloric acid, a violent reaction takes place and soon the calcite disappears. The material is still there, but as calcium bicarbonate in solution, not as solid calcium carbonate. With a weak acid such as carbonic acid, the reaction is extremely slow — much too slow for human observation. Thousands upon thousands of years must have been required for the development of Oregon Caves.

Although solution along fractures was the essential process initially, Oregon Caves were not enlarged to their present size by solution alone. Once the caverns were integrated into a continuous system and connected with a surface outlet, the water flowed through the caverns and, like a surface stream, River Styx eroded its bed and banks and greatly enlarged the passageways.

If only solution and underground stream erosion had taken place, the caverns with their bare walls would be relatively unexciting. But these caverns are adorned with many and varied formations that resulted from another process, chemical precipitation. When waters entering the caves from above were saturated with calcium bicarbonate, carbon dioxide was lost and calcium carbonate was precipitated — as stalactites hanging icicle-like from the ceiling and as stalagmites rising from the floor. A number of other *speleothems,* some weird and most irregular, were also formed. Evaporation also played a role in precipitation, but at least in the deeper parts where the relative humidity is extremely high, the loss of carbon dioxide is the main cause of precipitation.

These processes are active today. River Styx is still enlarging the caverns, the stalactites are growing longer and the stalagmites will be taller next year. Therefore, Oregon Caves are "living" caves. Eventually, however, they will be destroyed by collapse of the cavern roofs and by erosion of surface streams. But that will not take place for a long, long time.

Many other fascinating features — draperies in the Ghost Room, flowstone in Joaquin Miller's Chapel, the *soda-straw stalactites* and *helictites,* are well illustrated and discussed by Contor (1963) in his informative booklet, *The Underworld of Oregon Caves.* Also included are the animals and plants that grow in or frequent the caves. Bats are there but not in large numbers; most visitors do not see any of these elusive creatures.

REDWOOD NATIONAL PARK

Redwood National Park was created to preserve some of the earth's oldest living organisms. The older redwoods in the park were seedlings when Christ was born. (They are not the oldest trees, however, because the General Sherman Tree in Sequoia National Park is believed to be about 3500 years old.) These giant trees, some more than 350 feet (110 m) tall, were appropriately named Sequoias in honor of an early American who also stood tall; Sequoyah was chief of the Cherokee.

Established in 1968, Redwood National Park consists of about 58,000 acres of coastal areas of state and federal land in Del Norte and Humboldt Counties in northwestern California. Included within national park boundaries are three state parks — Jedediah Smith, Del Norte Coast, and Prairie Creek Redwoods state parks. Congress had been considering a proposal to add about 48,000 acres to the Redwood Creek Unit, near Orick, as protection from damaging logging operations; passage of this bill finally occurred on February 9, 1978.

The big trees are without question the main attraction. The mature trees have naked trunks up beach sand and piled it up as dunes in nearby areas.

The geologic story of the park does not unfold readily. The ridges and valleys of the Coast Range are abundantly watered, and the resulting thick soil and vegetation cover the bedrock almost everywhere, except in the coastal sections. Geologists had long been aware of the complexity of the area; now, with plate tectonics, some of the parts of the puzzle are beginning to fall into place. Although years will pass before the whole story is known, it now appears definite that most of the rocks in this

FIGURE 4-21

Columns and stalactites in Oregon Caves. (Photo by National Park Service.)

FIGURE 4-22

Looking up the trunk of the tallest tree in Tall Trees Grove, Redwood National Park. (Photo by W. E. Weaver, National Park Service)

extending up a hundred feet or more; their tops form a canopy that "creates the illusion of a great natural cathedral which visitors find both inspiring and humbling" (Sunset, 1965).

But Redwood has more for Park visitors than just trees. The beaches along 50 miles of ocean front are preserved in their natural state, free from the pilferings of beachcombers. From here, you see the results of the eternal pounding of the waves that have been shaping the shoreline. There are the sea cliffs and the stacks offshore, where sea cliffs stood at an earlier time. In places the winds have picked section of the Coast Range were deposited in a deep oceanic trench and were then faulted, metamorphosed, and thrust up to their present position.

The Jura-Cretaceous Franciscan Formation —in many places highly metamorphosed sediments and *greenstones (metavolcanics)*—and an older schist are the most abundant formations of the area. In some places, intrusives cut through these rocks; in others, the older rocks are buried beneath the Tertiary sediments or Quaternary sands and gravels.

The Coast Range Orogeny which built the mountains began in early Tertiary times. The area remains tectonically active today; adjustments, both horizontal and vertical, take place along the many branches of the complex fault system. After a long and involved history of uplifts and erosions, the mountains were shaped as we see them, unaffected by the glaciers that profoundly remodeled the Olympics and other mountains to the north. But the glaciers did have an indirect effect: As the level of the sea fluctuated during the glacial advances and retreats, marine terraces were formed at different levels. Uplifted terraces are now well exposed near Crescent City and at other points along the California coast.

The animal life of the park is highly varied because of the distinctly different environments—from the mountains to the ocean. In the redwood forests you may see beaver, raccoon, bobcat, deer, mountain lion, and black bear. Birds are numerous, but many species spend their time in the treetops well out of range. A herd of about 300 Roosevelt elk is maintained in one of the lower areas in the northern part of the park. The elk are sometimes unfriendly and can best be observed from a distance. Marine life is abundant and includes seals and sea lions, in addition to the multitude of smaller forms.

Why are the redwoods confined to such limited areas? Are they increasing or decreasing in numbers? When did they appear on the scene? A broad and evasive answer to the first

question is that only these few localities now provide a hospitable physical environment. Perhaps the rocks and the soils are responsible; however, the rocks are varied, and many of the same types occur outside the redwood forest as well. Thus, although the rocks and soils are suitable, the dominant factor must be the climate. High rainfall during the winters and dry summers with an abundance of fog— these within a moderate annual temperature range—must be the ideal climate for redwoods.

Today's redwoods are relics of the past. They evolved early in the Tertiary, and by Oligocene time they were abundant and widespread in parts of the western United States. Earlier, in Eocene time, they forested parts of the Yellowstone area. Since then, volcanoes have erupted many times and climates have changed repeatedly, particularly during the Pleistocene. Environments that were favorable became inhospitable, and now comparatively few of the redwoods are left.

Park visitors cannot but be inspired by the big trees that are links in the chain of life, past and future. But it is very difficult to meditate on their meaning when you are being subjected to the raucous wail of chain saws, the roar of logging trucks, and occasional dynamite blasts. The "great natural cathedrals" are no longer sacred!

While assured that the trees within the park boundaries will not be cut, we must nonetheless face the gloomy fact that the days of many of these giants are numbered. Logging in the surrounding areas, in some cases right up to park boundaries, has damaged the watersheds beyond belief; the huge wounds on the mountainsides are the worst forms of eye pollution. And with the removal of the trees and the burning of the slash, erosion has already been accelerated many-fold. Downstream, channels are in many places choked with erosional debris that forces the streams to undercut their banks. Now, the tallest tree in the world is only 25 feet (8 m) from the bank of Redwood Creek

FIGURE 4-23

Devastated area, the result of improper logging practices, is the source of the floods and erosion in the Tall Trees Grove downstream in Redwood National Park. (Photo by W. E. Weaver, National Park Service)

in the Tall Trees Grove. Of equal concern is the sudden rise in the water table caused by the filling of channels with erosional debris. According to park scientists, there is reason to doubt that the trees will be able to adjust their root systems to the waterlogged condition.

Moreover, the fogs — described by Sir Francis Drake as "the most vile, thicke, and stinking fogge's" — which provide essential moisture for the redwoods during the dry summers may be less abundant in the future. Graves (1974) reports that studies by Richard Janda of the U.S. Geological Survey indicate that when, because of clearcutting, only small islands of redwoods are left, the fog vanishes.

The tops of the trees slowly die and the trees are so weakened that they fall during heavy winds.

The redwood ecosystem has been severely damaged, but the total impact will probably not be known for years. Perhaps we can be optimistic enough to hope that most of the redwoods will somehow survive. At least for now, the tallest tree still stands.

POINT REYES NATIONAL SEASHORE

In our travels down the Pacific Coast, we first make contact at Point Reyes with the famous San Andreas Fault. Like the Fairweather in Glacier Bay, the San Andreas Fault is an important mover of the Pacific Plate, and it is the cause of many of California's major earthquakes.

Point Reyes National Seashore, established in 1962, is included with the national parks and monuments because of its particular significance geologically. It is now a peninsula about 40 miles (66 km) north of San Francisco; several million years ago it was far south of San Francisco, closer to Los Angeles; a few million years from now it will be a triangular island northwest of its present location. In other words it is a peninsula merely "in passing." On April 18, 1906, as reported by Molenaar (1982), Point Reyes suddenly jumped 21 feet (7 m) northwestward; this sudden displacement along a 270-mile-long segment of the San Andreas Fault produced the catastrophic San Francisco earthquake.

The San Andreas Fault, which essentially parallels the northeastern boundary of the Seashore, is actually a broad fault zone composed of badly crushed and sheared Jurassic and Cretaceous marine sedimentary rocks, which are easily eroded. The north half of the fault zone lies beneath Tomales Bay; the remainder forms Olema Valley. Granites underlying the peninsula are exposed only on the inner edge and outer tip of the peninsula; else-

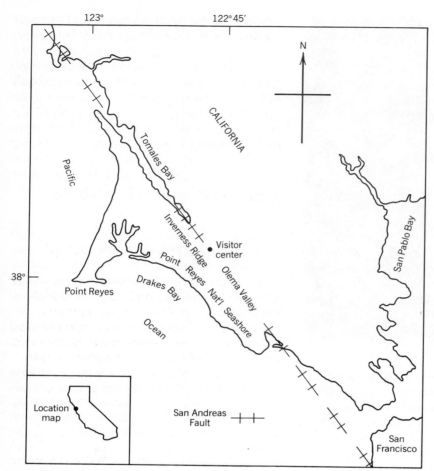

FIGURE 4-24

Map of Point Reyes National Seashore.

where they are buried by gently folded sedimentary rocks of Miocene age.

The Point Reyes peninsula is part of the Pacific Plate that has been moving north-northwestward along the San Andreas Fault for at least 30 million years. The granite on the peninsula was once continuous with the granite in the Tehachapi Mountains, some 370 miles (592 km) to the south. (We will find similar evidence for large-scale movements in Pinnacles

National Monument, the next park area on our trip.) Most visitors will doubtless be intrigued when they learn that they are on an "endless belt"—one that moves by jerks—and that the land could suddenly be jerked out from under their feet, northwestward another 20 feet or so. Minor faulting has recently occurred in several places along the fault; stresses are building up, and the "big one" will soon occur. That it will come and that it will generate a severe earthquake is not guesswork; it is merely the moment when the rocks will snap that is the question.

The rocks of Point Reyes vary greatly in type and in response to weathering and erosion. Rocks that are highly resistant to wave erosion form headlands extending out into the ocean; there, arches and stacks are common features. Where the rocks are alternately hard and soft, as near the Coast Campgrounds, sea caves and tunnels are carved out of the soft rocks. Thus, the variety of features is truly remarkable.

Waves striking the shore at an angle produce longshore currents that move sand laterally along the outer edge of the beaches. The south coast at Point Reyes is better protected from storm waves than the west coast; consequently, the longshore currents build long *spits* along the south coast, like the 2.5-mile-long Limantour Spit near Drakes Bay. Bars built across the mouths of small streams act as dams to impound water in small lakes. Estuaries such as Drakes Estero form large indentations in the coastline. [For a discussion of the complexities of life in an estuarine ecosystem or in the tidepools, see Russell Sackett's book, *Edge of the Sea* (1983).]

Point Reyes has something of historical interest also. When Sir Francis Drake sailed into the bay in 1579, to make repairs on the Golden Hinde, he was greeted by the friendly Miwok Indians. Commemorating his visit are Drakes Bay, Drakes Estero, Drakes Beach and Sir Francis Drake Highway, which crosses the Seashore from Tomales Bay southward to Point Reyes. Oddly, this area, which is only a few minutes' drive from San Francisco, has not been greatly changed since the time of the Miwoks.

Point Reyes has many miles of trails to explore, each with its own fascinating features. Elk, deer, sea lions, harbor seals, and gray whales can be observed from viewpoints on the sand dunes. And at the end of the Bear Valley Trail is Arch Rock, a sea arch tunneled through the rock by the endless pounding of the waves. Crawl through the arch at low tide but be sure to crawl back through before the tide comes in, or you will be marooned until the next low tide! Some of the trails are steep, such as those leading up Mt. Wittenberg (1407 feet; 407 m), high point on Inverness Ridge. Carry a good supply of water because the water in the streams is not fit to drink. And watch out for poison oak and stinging nettles. There are places for swimming and places where the hammering of the surf is extremely hazardous; check your Seashore brochure, and be safe.

PINNACLES NATIONAL MONUMENT

Pinnacles National Monument is small — less than 23 square miles including the additions made since it became a national monument in 1908. Nevertheless, its craggy columns and pinnacles that rise sharply above the rounded mountains have attracted visitors from afar.

The only entrance to Pinnacles is on the east side, about 34 miles (55 km) southeast of Hollister, which is east of California's Monterey

FIGURE 4-25

Sharp pinnacles in volcanics in Pinnacles National Monument. (Photo by V. Matthews III.)

Bay. Only a small part of the monument can be seen from a car; several foot-trails lead you through this fabulous geological area. Condor Gulch Trail in the central section took early visitors through the breeding grounds of the California condor; now the condors are gone. You are likely to see black-tailed deer and raccoons, but not the nocturnal bobcat and gray fox. The cougar is king of the mountain but usually keeps out of sight. Keep on the lookout for rattlesnakes and poison oak, both of which could take the edge off your visit. Carry a canteen on the trails; some of the streams are polluted and they look alike. Parts of the trails are steep, and summer days are likely to be hot.

The long, dry summers are unfavorable for many plants but are evidently ideal for chaparral, the "pygmy forest." Greasewood (chamise) with lesser amounts of manzanita, buckbrush, hollyleaf cherry, and toyon make up the chaparral community within the monument. In the moister areas, sycamore, live oak, willows, and alders abound; yellow pine and digger pine are dominant in the higher areas.

Of the many types of rock, the Miocene volcanics are of special interest because only they form pinnacles. As the lava flows and pyroclastics cooled, shrinkage cracks developed, many in a near-vertical direction. Weathering widened the cracks and formed broad columns which were later narrowed. The streams that developed had sinuous courses and undercut their banks; where the rocks were least resistant the undercutting was at its maximum, and remarkable overhangs or "caves" were formed. You will need to have your flashlight with you when you explore Bear Gulch Cave on the Moses Spring Trail south of the visitor center. Weathering, mass-wasting, and stream erosion worked hand in hand in the fashioning of the fascinating, sometimes grotesque features that entice people into the Pinnacles.

The oldest rock exposed here is the Paleozoic Gavilan Limestone, largely metamorphosed by granitic intrusions (Andrews, 1933). Also extensively exposed are conglomerates

—some of which are older than the volcanics: arkosic sandstones and unconsolidated gravels, including terrace deposits.

The Miocene Pinnacles Volcanic Formation was described by Andrews (1933) and in greater detail by Matthews (1973). A highly complex series of rhyolitic and andesitic lava flows and pyroclastics were piled up, largely beneath the sea, about 23.5 million years ago. Perhaps by now this sounds a bit repetitious; however, what happened to these rocks after they were belched up out of the earth is more than a little exciting. They have migrated 195 miles (315 km) from where they were formed and are still going! Similar rocks were mapped earlier, down in Los Angeles County, and named the Neenach Volcanic Formation (Dibblee, 1967). Matthews determined that 23.5 million years ago the two volcanic masses, now separated by 195 miles (315 km), were one continuous mass. The numerous layers are the same and in the same sequence; moreover, major structures are (or were) continuous. The Pinnacles lie on the edge of the Pacific Plate whereas the Neenach rocks are on the American (Continental) Plate on the opposite side of the San Andreas fault zone.

For years, the most frequently asked geologic question was "How much displacement has taken place along the San Andreas Fault?" Many geologists had labored over the problem, using all known methods; yet, until this discovery there was room for doubt. In all cases, there was more than one possible interpretation of the data. It appears that now the problem is solved — for the period since the early Miocene. When Hill and Dibblee suggested in 1953 that since Jurassic time the displacement has been as much as 350 miles (565 km), a certain amount of skepticism greeted them. With convincing data for the last 23.5 million years, however, many geologists now accept their 350-mile displacement without difficulty.

But as one problem is solved, at least one other appears to be taking its place. Has the

FIGURE 4-26

Sketch map showing displacement of Pinnacles volcanics along the San Andreas Fault. (From National Park Service brochure.)

rate of movement been essentially uniform or has it speeded up or slowed down? Although the average rate of about 1.4 cm per year for the last 23.5 million years must be accepted, the larger problem remains for the future.

Many people will visit Pinnacles National Monument and marvel at the columns, spires, balanced rocks, and caves without particular concern for the fact that the monument is moving; others will want the whole story. We have offered only a few glimpses of the vast amount of work involved in putting this puzzle together.

CHANNEL ISLANDS NATIONAL PARK

Offshore from Santa Barbara in Southern California, a range of mountains rises out of the Pacific Ocean, on the south side of the 25-mile-wide, oil-rich Santa Barbara Channel. Actually, it is only the tops of the mountains that you see — the westward extension of the

Santa Monica Mountains, northwest of Los Angeles. From Santa Barbara they appear as rugged mountains, rising more than 2000 feet (606 m) above the sea.

Channel Islands National Monument was established in 1938 and, with considerable enlargement, became a national park in 1980. The park consists of the islands of San Miguel, Santa Barbara, Anacapa, Santa Cruz, and Santa Rosa (at the time of writing Santa Cruz and Santa Rosa are still privately owned). Park headquarters are in Ventura, California.

The oldest rocks in the park are volcanic and sedimentary, intensely deformed and metamorphosed into schists during Mesozoic time. Locally thick marine and terrestrial sediments accumulated in basins produced by Cenozoic faulting and folding. Organic material trapped in these sediments was converted to oil and gas. Although no large accumulations are known on the Channel Islands, substantial oil fields occur on land and in the Santa Barbara Channel. Oil seeps and associated tar deposits occur on San Miguel, and an oil seep on Santa Cruz Island has burned continuously for hundreds of years.

Southern California, of which the Channel Islands are a part, is broken by faults, major and minor. Studies by various workers have shown that large crustal units were rotated in the zone between the American and the Pacific plates when these two plates came into contact, late in the Oligocene. Luyendyk and his co-workers (1980) subscribe to this general interpretation and present evidence that the Channel Islands and the western Transverse Ranges have been rotated clockwise 70° to 80° since Miocene time. Their paper, supported by paleomagnetic data and more than 40 references, provides an in-depth analysis of this complex area.

The Channel Islands were uplifted above the sea sometime during the Pleistocene Epoch. The rocks are both hard and soft, and in places have been differentially eroded, producing grotesque sculpturing.

The sea has cut marine terraces both above and below sea level, reflecting local tectonic adjustments as well as changes in ocean depth — the result of the waxing and waning of Pleistocene glaciers. Faults cut across the islands, and there the waves have carved out sea caves, arches, and stacks. Some of the caves are quite large — the roof of Painted Cave on Santa Cruz Island is in places 130 feet (40 m) above the sea.

There are many species of animals on the islands, but one former inhabitant is missing —

the pygmy mammoth, about 6 feet (2 m) high. The charred bones found on Santa Rosa Island suggest that Early Man was responsible for its extermination. Some early workers believed that the presence of the mammoth and other land animals indicated that the islands were once joined together and extended as a peninsula westward from the mainland. Howarth (1982), however, assumes that the animals were swimmers and not dependent on a land bridge. It is also probable that they had a shorter distance to swim during the periods when sea level was lower than at present.

Visitors interested in sea life will be drawn to the Channel Islands by the great variety of animals there. The long list is headed by sea lions, which will noisily challenge your invasion of their domain. The sea otters that managed to survive the vicious hunting by Europeans are now safe. The abalones, lobsters, and nudibranches (naked, gilled sea slugs) are not widespread in the park system.

The Chumash Indians inhabited the islands for a long period prior to arrival of the Spanish explorers late in the sixteenth century (Bunnell and Vesely, 1983). With an abundance of food from both the ocean and the land, the Chumash prospered — until they were transplanted onto the mainland by missionaries.

Archeologists find the Indian middens of interest, as well as the occasional pictographs. The "paint" in Painted Cave — splashes of green, orange, pink, and purple — comes from slime molds and algae, however, and not from the Chumash as was once believed (Bunnell and Vesely, 1983).

Exploring the islands — and returning home safely — requires careful planning with guidance from someone well acquainted with their hazards. Sea swells can be violent and unforgiving, as Bunnell and Vesely describe. Remember that the awesome power of a sea that formed these huge caves is there today, and the animals — including man — that survive are the ones who understand and respect it.

FIGURE 4-27

Painted Cave, Channel Islands National Park. (Photo by D. Bunnell.)

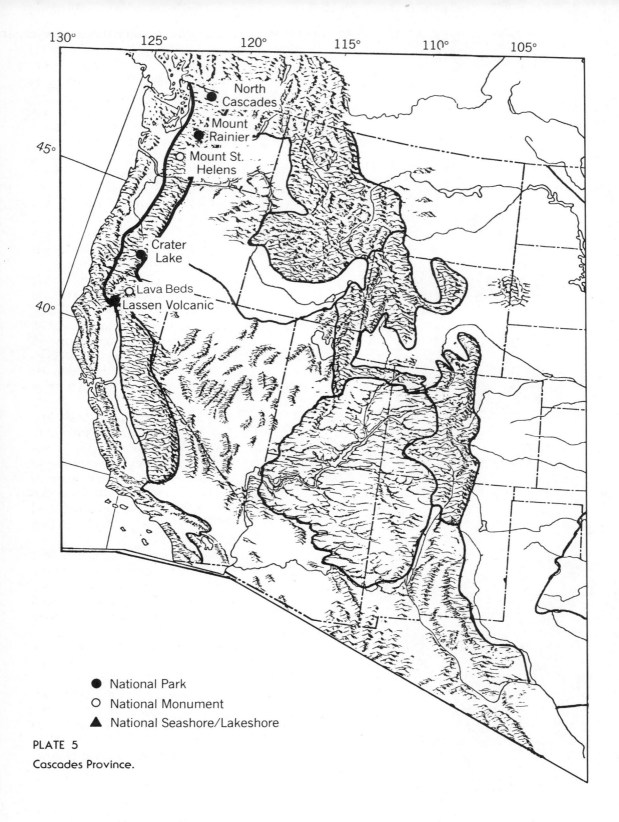

North
Cascades

Mount
Rainier

Mount St.
Helens

Crater
Lake

Lava Beds
Lassen Volcanic

130° 125° 120° 115° 110° 105°

45°

40°

● National Park
○ National Monument
▲ National Seashore/Lakeshore

PLATE 5

Cascades Province.

CHAPTER 5

CASCADES PROVINCE

The quiet of Sunday morning, May 18, 1980, was broken by a thunderous explosion that shook southwestern Washington as it had not been shaken for many years. It was Mt. St. Helens blowing its top, after being essentially quiet for more than a century. Minor eruptions had begun on March 27, but this was the "big one," the one that blew away the mountaintop—the one that killed more than 60 people and devasted a large area.

Mt. St. Helens is one of the youngest and most active of the Cascade volcanoes, any one of which could erupt at any time. Fortunately, it was St. Helens that geologists Crandell and Mullineaux had selected for detailed study, because their findings were alarming. In 1978, they published U.S. Geological Survey Bulletin 1383-C, in which they issued a warning that a major eruption was due to occur in the near future. The 1980 eruptions proved how right they were. Man is unlikely to devise a means of capping an ebullient volcano; however, if sufficient warning is given—as it was at St. Helens—the loss of life and property can be kept to a minimum.

The Cascades Province extends from northern California northward through Oregon and Washington into southern Canada. The southern boundary is a short distance south of Lassen Volcanic National Park, where the Cenozoic volcanics of the Cascades give way to the granites and metamorphic rocks of the Sierra Nevada.

The province is basically a dissected volcanic plateau above which tower many lofty volcanic peaks that were only recently added to the Cascade landscape. They are classic examples of stratovolcanic cones, as illustrated in the Introduction (see Figure I-8). It appears that many of these cones are less than a few hundred thousand years old, and that explosive eruptions producing pyroclastic materials generally dominated over the more quiet, lava-producing eruptions.

The early history of the Cascade region is obscure except in the North Cascades area where pre-Cenozoic gneisses, schists, and granites record a complex history. In places, there are fault-bound masses of exotic rock that are of completely different origin than adjacent rocks, indicating that the exotic rocks are far from "home." Until the subduction zone concept was visualized there seemed to be no solution to this puzzle. Although considerable recent progress has been made, not all investigators are in complete agreement on the sequence of geologic events (see Hammond, 1979; Hamilton and Myers, 1966; Atwater, 1970). The sequence given here is greatly simplified.

The picture now emerging indicates that earlier subduction zones were farther east, permitting widespread deposition of Paleozoic and Mesozoic marine and volcanic rocks in the western geosyncline. As new material was added to the growing North American continent mountain-building occurred and the subduction zone occasionally shifted westward to new positions.

Late Mesozoic mountain-building in the northern Cascades produced metamorphism, granitic intrusions, folding, and large-scale thrust faulting. *Nonmarine* sediments record the final retreat of the seas from the northern Cascades by late Cretaceous time.

In contrast to the northern Cascades where older rocks are abundant, the southern Cascades contain only Cenozoic rocks. Until recently, geologists assumed that the pre-Cenozoic rocks and structures beneath the Cenozoic rocks are similar to those farther north. However, studies by Atwater (1970), McBirney (1978), and others suggest alternative interpretations. There is a general consensus that Cascades volcanism is related to the subduction of the Juan de Fuca plate. As an in-depth study of the tectonic evolution of the Cascades, Hammond's 1979 paper is particularly informative.

Oligocene mountain-building and uplift was accompanied by extensive erosion which was followed during the Miocene by one of the greatest outpourings of basaltic lava that the earth has ever endured. Highly fluid lavas deeply buried the Columbia Intermontane Province to the east and covered much of the eroded Cascade Mountains. In earliest Pliocene time there probably was very little to distinguish the Cascades and Columbia Inter-

FIGURE 5-1

Air view of Mt. St. Helens on April 7, 1980. The crater was formed by volcanic explosions that began on March 27. (Photo by Wes Guderian of the *Portland Oregonian*.)

montane provinces. The Cascade Mountains as we know them began to take shape during the Pliocene. Broad upwarping of the Cascades, in places several thousand feet, separated the mountains from the adjacent Columbia Intermontane Province.

During the Pliocene upwarping, the west-flowing streams were, one by one, blocked off — unable to keep pace with the uplift athwart their courses. The Columbia River was an exception; perhaps already the largest of the streams, it increased its flow and eroding power by gathering the waters of the streams that were blocked off. Moreover, the uplift across its course was less than that to the north; thus the Columbia maintained its course and cut a deep gorge through the Cascade Range. By so doing, it became an *antecedent stream,* one of comparatively few whose antecedence cannot be successfully questioned.

In late Pliocene time, the Cascades Province lacked the main features for which it is famous — the magnificent volcanic cones such as Mt. Rainier and Mt. Shasta. Instead, it was an elongate plateau more than 8000 feet (2454 m) high at the northern end and about 5000 feet (1534 m) at the southern end. Streams were at work cutting deep canyons below the plateau surface, particularly in the northern area where lava flows were apparently thin or absent. Volcanic eruptions began again, and a long line of "sentinel mountains" rose one by one high above the plateau. Their general alignment from northern Washington to California strongly suggests structural control in the basement rocks. Apparently, the magma generated by subduction still feeds the mighty volcanoes that crown the Cascade Range. The magmas have been mainly andesitic in composition — less fluid than basaltic magmas — and most of the eruptions have been explosive; therefore the cones are high and steep-sided.

Glaciers have had a profound effect on the volcanic cones, gouging out deep valleys on their flanks. Glaciers formed on each volcanic cone when it reached sufficient height, in most cases late in the Pleistocene. Most of the sculpturing now in evidence occurred during the Wisconsin glaciation, which reached its maximum only 15 to 20 thousand years ago. However, the youngest cones, such as St. Helens — mere babes in the geologic cradle — were born in time to host only the latest of the Wisconsin glaciers (S. Harris, 1980). In northern Washington where volcanic eruptions only locally interrupted glacial activity, glacial erosion is most advanced. It is mainly this section that justifies reference to the Cascades as the "American Alps."

Certain of the Cascade volcanoes other than St. Helens have erupted or threatened to erupt during historic times. Lassen Peak erupted periodically from 1914 to 1921, and nearby Cinder Cone exploded in 1851. The latest eruption of Mt. Shasta apparently occurred in 1786 (Finch, 1930); now, near the top of the mountain, there are mud pots which are kept active by steam and other gases. Here, John Muir and his companions kept themselves from freezing by lying in the warm mud during a long night's blizzard, as reported by Douglas in his book, *Muir of the Mountains* (1961). In 1975 there was much excitement when steam billowed out of Sherman Crater on Mt. Baker. In addition to the danger of actual eruption, the increase in heat could cause melting of large quantities of glacier ice and trigger devastating mudflows. Therefore, Mt. Baker was carefully monitored for the next few years; however, on the basis of all observations, including the lack of significant seismic activity, the attending scientists concluded that an eruption was not imminent (Kiver, 1978).

Thermal features such as fumaroles and hot mud pots are surface indications of the sleeping "fires" below. The Bumpass Hell area near Lassen Peak is perhaps the most spectacular. Hazard Stevens and P. B. Van Trump made the first documented climb of Mt. Rainier in 1870. It was bitterly cold, but they survived a long night huddled over one of the fumaroles inside

FIGURE 5-2

Mt. Baker, one of Washington's finest volcanic mountains.

FIGURE 5-3

Telephoto of Mt. Shasta and parasitic cone, Shastina, in Northern California. (Photo by E. Kiver.)

a geothermal ice cave in the summit crater. The release of even small quantities of heat and gas serve as reminders that beneath each of these active volcanoes lies a restless magma that will probably erupt again.

An extension of James Hutton's pronouncement that "the present is the key to the past" — one of the basic principles in geology — is that the past is the key to the future. This was the basis for a study by Kiver (1982) in which he determined that the number of significant eruptions of Cascade volcanoes thus far in the twentieth century is considerably less than during the nineteenth century. This should not be interpreted to mean that volcanic activity is slowing down, however; the twentieth century is not over! If your traveling companions think that you are taking too many pictures of Mt. Rainier or some of the other volcanoes, remind them that these pictures may prove to be valuable "before" pictures, and that the "after" may come sooner than most people are aware.

At present, there are four national parks and one national monument in the Cascades Province; beginning at the north end, they are: North Cascades, Mt. Rainier, Crater Lake, Lava Beds National Monument, and Lassen Volcanic National Park. Mt. St. Helens, now a National Volcanic Monument administered by the U.S. Forest Service, is being considered for transfer into the park system. For this reason and because Mt. St. Helens has provided a wealth of geologic information on Cascades volcanism, it is not only included but is discussed first.

MT. ST. HELENS
NATIONAL VOLCANIC MONUMENT
(United States Forest Service)

Mt. St. Helens was until 1980 one of the truly beautiful mountains of the world — symmetrical, with glistening glaciers streaming down its flanks. Now it is only a vestige of its former self — an empty shell, the shattered remains of the colossal explosions of May 18, 1980.

The explosions blew away about 99 billion cubic feet of rock material and roughly 3.5 billion feet of ice (Brugman and Post, 1981). At the beginning of 1980 the mountain was 9671 feet (2931 m) high; after the eruption, it was only 8358 feet (2533 m) high. Within a few minutes, Washington's fifth highest mountain became its thirty-seventh highest.

Is St. Helens destined to remain a jagged stump or will it rise again? Geologists point out that St. Helens is one of the youngest of the Cascade volcanoes, and that the records of the events involved in the building of Mt. Rainier and the other high cones indicate that each had one or more major setbacks before reaching its present height. Therefore, it is almost a certainty that St. Helens will sometime in the future take her place among the "high ones." Sadly, we will not be here to admire her majesty!

Meantime, the paroxysms of the past few years have provided a wealth of information that will assist us in predicting what will likely take place when the next Cascade volcano erupts. We must pay tribute to the geologists and geophysicists who risked or gave their lives assembling every conceivable type of information about active volcanoes.

The Indians of the several local tribes were well aware that St. Helens was an active and dangerous volcano long before the "new Americans" arrived. Their names for the mountain — for example, Tah-one-lat-clah — mean "Fire Mountain." Their legends are included, along with a beautiful pictorial record, in Williams' *Mount St. Helens: A Changing Landscape* (1980). Spirit Lake, at the base of the cone, was for many years a favorite retreat for thousands of people; to the Indians, it was "evil spirit lake." A more recent legend, dating back only to 1924, got its start when some miners claimed that they had killed a "large, hairy ape" on the rim of a canyon — Ape Canyon — on the east flank of the mountain.

Although they were unable to produce the evidence, it was enough to give birth to "Bigfoot," the elusive giant who will probably never be apprehended!

The hairy figment of imagination also gave its name to Ape Glacier and Ape Cave, both on the southern flank of the mountain. The cave is actually a lava tube similar to Thurston Lava Tube in Hawaii Volcanoes National Park. Here also, the tube is in a basalt lava flow, and it was similarly formed when lava drained out from under the solidified crust, about 1900 years ago. Ape Cave has the distinction of being the longest known lava tube in North America — about 12,500 feet (3788 m) long. Take a reliable flashlight and a warm jacket for the 42°F year-round temperature.

The 1980 eruptions brought Mt. St. Helens into the limelight and onto television screens throughout the country. Many articles and books have been written about its behavior, both recent and ancient; for an excellent summary, read Stephen Harris' *Fire and Ice* (1980). Lipman and Mullineaux edited an impressive volume (U.S. Geological Survey Professional Paper 1250, 1981) that contains significant papers by scientists in several different fields. These and other publications that inform the public of the hazards of volcanic outbursts contribute much to human welfare. But without question, the greatest contribution was the little 26-page booklet that appeared in 1978, in which Crandell and Mullineaux issued a warning that St. Helens was soon to erupt. Without that warning, and the resulting preparedness plans, the loss of life would without doubt have been much greater.

Geologic Events
(36,000 to 300 Years Ago)

Mt. St. Helens is geologically very young, much younger than most of the Cascade volcanoes. According to Hyde (1973); Mt. St. Helens was born at least 36,000 years ago; but much of the upper part of the cone was added only recently, within the past 400 years (Crandell and Mullineaux, 1978).

The early events are but little known; the evidence is largely buried deep inside the cone. The record of the past 4000 years is more nearly complete. Studies by Crandell and Mullineaux (1978) indicate that 10 major eruptions occurred between 4000 and 3600 years ago, several took place during the period 3200 to 2600 years ago, and since then, the spacing has generally been between 100 and 200 years. Most were eruptions of ash and other pyroclastics, accompanied in most cases by volcanic mudflows; a few were dominantly lava flows, notably the Cave Basalt flow (occurring about 1750 years ago) which contains Ape Cave, Ole's Cave, and other lava tubes. About

FIGURE 5-5
Mt. St. Helens prior to the destructive 1980 eruptions. (Photo by C. Mahoney.)

FIGURE 5-6
Jagged "stump", the remains of Mt. St. Helens after the 1980 eruptions. Note the growing plug-dome in the crater. (Photo by Dr. William Halliday.)

300 years ago thick, pasty lava was gradually pushed up out of the vent to form a plug dome that capped the mountain.

Precisely when the cone reached sufficient height for glaciers to form has not been determined. It is probable that they developed sometime during the cold period known as the Little Ice Age, which began about 4000 years ago. An ice cap glacier covered the peak and 13 glaciers streamed down the valleys on all sides. Almost 2 square miles of the mountain were covered with about 5 billion cubic feet of ice (Brugman and Post, 1980). The largest glaciers — Wishbone, Loowit, Lesch, and Forsyth — were on the north and northeast flanks where the direct effect of the sun was least. The two longest glaciers, Forsyth and Shoestring, extended down to about 4500 feet (1500 m). This was the mountain prior to the 1980 eruptions.

Geologic Events
(300 B.P.[1] to May 17, 1980)

The explosive eruption that, according to Indian legend, occurred about 1802 was the first documented major event. The tree-ring method was used to demonstrate distinct growth-retardation of trees that survived the heavy ashfall in 1802 (Lawrence, 1941; see also Williams, 1980). (These trees, on the south shore of Spirit Lake, did not survive the 1980 eruptions.)

Intermittent volcanic activity continued until early in the 1840s when the Goat Rocks plug dome was emplaced on the north flank of the cone. Violent eruptions preceded dome emplacement, as shown by the large volume of ash. The last well-documented eruption of the 1800s occurred in 1898 when billowing clouds of steam and ash were observed.

To the local people, and to many from out-side, the Cascade volcanoes were there to admire and enjoy. The powder kegs beneath the volcanoes were generally ignored until 1969 when Dwight (Rocky) Crandell addressed an emergency preparedness conference in San Francisco, urging immediate action in assessing potential dangers. That same year a cooperative study involving the U.S. Geological Survey and the University of Washington was initiated (Time/Life, 1982). Then in 1978, Crandell and his Geological Survey co-worker Don Mullineaux issued a specific warning that Mt. St. Helens was about due to erupt and that it was time to prepare for it. Thus the groundwork was laid for what soon became necessary — perhaps a bit sooner than even they anticipated.

The monitoring system was in operation when, on March 20, 1980, the seismographs suddenly went wild, registering an earthquake of 4.2 magnitude on the Richter scale. When the quakes continued, the seismologists alerted the Geological Survey team and the local U.S. Forest Service in Vancouver, Washington.

On March 27, a thunderous explosion was heard and when the clouds lifted there was a large hole in the mountaintop and volcanic ash blackened the cone; the volcano was operational once again. The preparedness plan was now in effect, and only scientists and certain officials were allowed in the Red Zone — the zone of highest danger. Local residents were evacuated, some with reluctance; one long-time resident, Harry Truman, a feisty 84-year-old operator of a lodge on Spirit Lake, decided to stay with "his" mountain.

Geologic Events
(May 18, 1980, 8:32 A.M. — present)

The morning was clear, ideal for picture-taking. Geologists Keith and Dorothy Stoffel were up there in a plane, with cameras shuttering. Of particular interest was the huge bulge that was hanging precariously out over the north

---

[1] B.P. — before the present.

flank. Suddenly, at 8:32 A.M. (they learned later) they were transfixed in horror; directly below them, the bulge shuddered violently and disintegrated into a gigantic debris-flow that quickly gained momentum and rushed down the mountainside. Almost immediately, a huge mass of superheated steam and ash exploded from the wound (see Frontispiece). The hot, searing *nuee ardente* roared down the north flank, overtaking and then racing ahead of the still-moving debris avalanche.

Expansion of the hot, highly compressed gases gave the ground-hugging death cloud ever-increasing momentum — far in excess of any recorded hurricane force — sufficient to carry it up and over high ridges and down into valleys, devastating a 230-square-mile area north of the mountain. Millions of trees were uprooted or snapped off, involving about 3.2 billion board feet of lumber.

Mudflows began to develop soon after the blast, destroying houses, bridges, and logging operations far downstream, along the Toutle and Cowlitz rivers, for as much as 45 miles (75 km). Also, on the southeast flank, smaller mudflows swept down Pine, Swift, and Muddy creeks, fortunately stopping short of Swift Reservoir (see Figure 5-9). (A mudflow — which has the consistency of concrete as it pours out of the cement mixer — is much more forceful and destructive than a normal flood.) The mudflows were generated when large sections

FIGURE 5-7
Trees flattened by the volcanic blast of Mt. St. Helens on May 18, 1980.

of glaciers and snow-fields were melted by the heat from the blast wave and from subsequent eruptions.

Spirit Lake, a favorite recreation spot, was formed perhaps a thousand years ago when a mudflow blocked off the valley. This time, the debris-flow buried the old mudflow dam and raised the level of the lake by about 200 feet (60 m). Harmony Falls, on the east side of Spirit Lake, is now submerged beneath lake water.

The full extent of the loss of life will never be known. Countless trees, shrubs, flowers, song-birds, deer, elk, and bear were destroyed. At least 60 people died in the blast, mercifully almost instantaneously — from asphyxiation. Some had climbed the beautiful mountain many times through the years and had re-turned to pay their last respects. Harry Truman had said, "That mountain is part of me"; now he is a part of the mountain, buried under the volcanic debris, which is perhaps the way he wanted it.

Geologist Dave Johnston was at his post about five miles north of the ominous bulge on the mountain. At 8:32 he saw it all and radioed his last words: "Vancouver, Vancouver, this is *it!*" Unlike many who lost their lives, he was entirely aware of the danger, but he was deter-mined to do his share in obtaining information that would improve our understanding of ex-plosive volcanoes — perhaps it would help save lives in the future.

Future Activity

Since May 18, 1980, there have been many eruptions, several quite spectacular. Domes of thick, pasty dacite have welled out of the vent, but each time, unable to "cork the bottle," they have been blown to bits. Likely the vent will be plugged securely one day, conceivably preparing for a truly gigantic explosion. About 4000 years ago, St. Helens blew up with a vio-lence several times that of the 1980 eruption. And, about 6900 years ago, approximately 15 cubic miles of Mt. Mazama were blown off,

FIGURE 5-8

Trojan Nuclear Power Plant about 35 miles west of Mt. St. Helens.

forming the 6-mile-diameter caldera called Crater Lake (Findley, 1981).

If St. Helens's next major eruption blows out the south flank, the three reservoirs on the Lewis River could be destroyed, with colossal flooding downstream. Or, a blast on the west flank would head toward the Trojan Nuclear Power Plant, which is located much too near the volcano.

When the new dam at Spirit Lake is over-topped it will probably be eroded away, thus draining the lake. Perhaps Harmony Falls, now submerged, will be resurrected and restored to its former beauty.

The future of the remaining glaciers is un-certain, depending on how soon or whether the volcanic cone cools down. Those on the

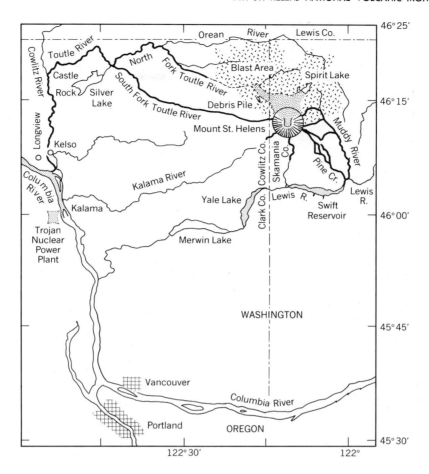

FIGURE 5-9

Map of Mt. St. Helens area, showing areas affected by the May 18, 1980 volcanic blast (shutter pattern) and areas covered by volcanic debris (stipple pattern). Heavy lines indicate areas of mudflows and flooding in stream valleys. (Adopted from Brugman and Post, U.S. Geological Survey Circ. 850-D.)

top and on the north flank were blown away in 1980, and several of the others were beheaded. Melting of the remaining ice may result in devastating mudflows and highly dangerous outburst floods, if large quantities of water suddenly break loose (Brugman and Post, 1981). A few glaciers may survive; some of the others may remain as stationary icefields.

The future of Mt. St. Helens is impossible to predict and definitely beyond our control, but the future of the Mt. St. Helens *area* is ours to determine. At present, it is a National Volcanic Area administered by the U.S. Forest Service,

FIGURE 5-10
Map of North Cascades National Park.

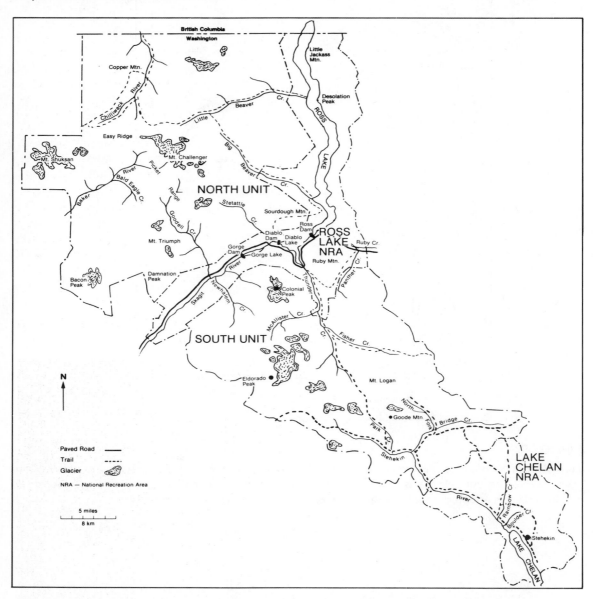

an agency dedicated to multiple use of lands. Already, many of the downed trees have been salvaged, defacing the mountainside — altering the natural scene. Unquestionably, the 1980 eruption was the most significant natural event to take place in the United States in recent times. Clearly, the volcano and the entire area of devastation should be preserved in its natural state — to be meaningfully interpreted for the millions who will come to view it. In order to prevent further damage to its scientific values, the Mt. St. Helens area should be designated a national park or a national monument.

NORTH CASCADES NATIONAL PARK

''Sharp as a cougar's fangs'' is N. T. Kenney's description of the Picket Range in the heart of North Cascades National Park. His *National Geographic* article (1968), supplemented by James P. Blair's excellent color photography, shows that the North Cascades clearly deserve a place in the National Park System. This broad section of the Cascades is one of the largest areas of uninterrupted spectacular scenery in the 48 states, and much of it is in the park. Glaciers — cirque glaciers, valley glaciers, and shelf glaciers — are too numerous to count, as are the tarns and other glacial lakes. Only a flight over the area affords awareness of the great expanse of this geologic spectacle.

Ice-clad Mt. Baker, to some the most beautiful of the Cascades volcanoes, stands high just west of the park near the Canadian border. In 1975, Mt. Baker was ''smoking'' and threatening to erupt; now, it appears to have gone back to sleep — temporarily! The peak is buried beneath a small ice cap that divides into several valley glaciers on the flanks of the cone. Coleman Glacier is noteworthy because it was the first glacier in the Pacific Northwest known to advance after a century-long period of retreat. This discovery alerted scientists to make observations elsewhere, and by the mid-1950s a significant number of glaciers were known to be advancing. As a result of these world-wide studies a new glacier regime has been recognized. During the century prior to about 1945,

FIGURE 5-11

The jagged Picket Range in northern section of the North Cascades National Park, as pictured from Skagit River below Diablo Dam.

glaciologists rarely had the opportunity to observe an advancing glacier; since 1945, at least a few in each glacier area have been advancing, reflecting a new climatic regime more favorable for glaciers. This climatic change is not fully understood; however, Flint (1973) presents a summary of the available information in his book, *The Earth and Its History*.

North Cascades National Park, consisting of about 790 square miles, extends southward from the Canadian border for about 50 miles (81 km). It is made up of two parts separated by a strip along the Skagit River which has been developed by the Seattle Light and Power Commission. Gorge, Diablo, and Ross dams harness the Skagit; Ross Dam, 540 feet (165 m) high, backs water for more than 20 miles (32 km), even into the edge of Canada. The Ross Lake area has been established as a national recreation area. North Cascades Highway, which follows the Skagit upstream to Ross Dam and continues eastward out of the mountains, affords unparalleled views of the Cascades.

North Cascades is a "walk-in" park, with no roads or trails for motorized vehicles. Later, an aerial tramway on Ruby Mountain in the Recreation Area will enable visitors to see some of

the remote backcountry. The Park Service should be supported in its policy to exclude the automobile. If park officials can withstand public pressure to build highways, and if they can persuade park visitors to behave like animals and not desecrate the forest, the park will continue to be unspoiled.

In a number of parks and wilderness areas the question arises as to whether mineral deposits should be developed. The law prohibits mining in such area, except for mining claims valid at the time of official designation. The law, however, does not prohibit arguments about the justification for the law! Copper and silver deposits in and near North Cascades National Park and in Glacier Peak Wilderness Area to the south have given rise to such arguments.

David Brower, modern-day John Muir, is a preservationist of wide reputation; even the thought of defacing a beautiful mountain is repulsive to him, regardless of purpose. If you are interested in hearing a lively discussion between Brower and a mining geologist, a man with equally strong but opposing views, read McPhee's account (1971) of their encounter in Glacier Peak Wilderness Area. As you will see, a strong case can be made for each stand; hearing both sides may help us decide what we want our parks and wilderness areas to be.

Geomorphic History

The geologic events involved in the building of the Cascade Mountains were outlined earlier in this chapter; here, only material directly applicable to the North Cascades is discussed.

The bedrock geology is now known to be much more complex than it was earlier thought to be. The Chilliwack batholith was believed to be a single large granitic intrusion; now, as McKee (1972) points out, it is known to consist of several intrusions, ranging from granite to granodiorite in composition, which were emplaced at different times, from Late Cretaceous to Miocene. Still older rocks,

FIGURE 5-12

Gneiss laced by quartz veins, at Diablo Lake Overlook on Highway 20. Vertical lines are manmade drill holes.

largely metamorphic gneisses, may be as old as Precambrian; definitely they are older than mid-Paleozoic. Crystalline rocks — schists, gneisses, and granite — form the backbone of the range.

The spectacular scenery of the North Cascades is clearly the work of glaciers. Guided by stream-cut valleys, they dug deeply and carved out U-shaped canyons, many of which are more than a half-mile deep. Locally, where rock resistance was less, the glaciers dug deeper and formed rock-basin lakes. Where glaciers surrounded a peak, they developed cirques with steep headwalls; as the cirques were enlarged, the mountain was eaten away, leaving only a sharp spine or horn, like the Matterhorn in the Alps. Many of the high, jagged peaks are composed of granite gneiss. Near-vertical structure in the layered gneiss is largely responsible for the jagged features; locally, homogeneous massive granites are cut by vertical fractures which, if the spacing is favorable, also lead to the development of sharp features.

Today's Glaciers

Glaciers today are much smaller than those of the great Wisconsin advance which reached its maximum 15 to 20 thousand years ago. Whether any of today's glaciers in the North Cascades are remnants of Wisconsin glaciers is doubtful. Conceivably those in sheltered locations were able to survive the higher temperatures during the Thermal Maximum about 6000 to 4000 years ago. Many, however, were born during the Little Ice Age when temperatures were lower. More than 300 glaciers are alive and well today in the park, and a view from the air almost compels one to believe that the Ice Age is not over!

Within the northern section of the park, Mt. Shuksan (9127 feet; 2782 m) with its snowfields and glaciers is a beautiful sight, both in reality and in reflection in Picture Lake. Farther east, in the Picket Range, Challenger Glacier

FIGURE 5-13

Le Conte Glacier in Glacier Peak Wilderness, south of North Cascades National Park. (Photo by E. Kiver.)

on the north side of Mt. Challenger is one of the larger glaciers and does in fact present a challenge for the climber. Near the Canadian border, cobalt-blue Glacier Lake lies in a large cirque on one of the Glacier Peaks in the Cascades, recording the size of the Wisconsin glacier which was responsible. (The other Glacier Peak [10,568 feet; 3222 m] is in Glacier Peak Wilderness Area which lies to the south of the park.) In the southern section of the park, Colonial Glacier on Colonial Peak, Neve Glacier on Snowfield Peak, and those on Eldorado Peak and on Buckner Mountain (9200 feet; 2805 m) are a few of the many existing glaciers. The largest, Boston Glacier, is located a few miles northeast of Cascade Pass, near the southern boundary of the park.

The North Cascades are not exceptionally high mountains, yet they support many glaciers, some of which are large. In contrast, the mountains of comparable height in Glacier National Park, some 350 miles to the east, support only small glaciers. The latitude is the same, but the amount of snowfall is much less. When the moisture-laden westerly winds lift to get over the Cascades, cooling causes precipitation — in the form of heavy snows during the winter. When these winds lift to

FIGURE 5-14

Map of Mt. Rainier National Park.

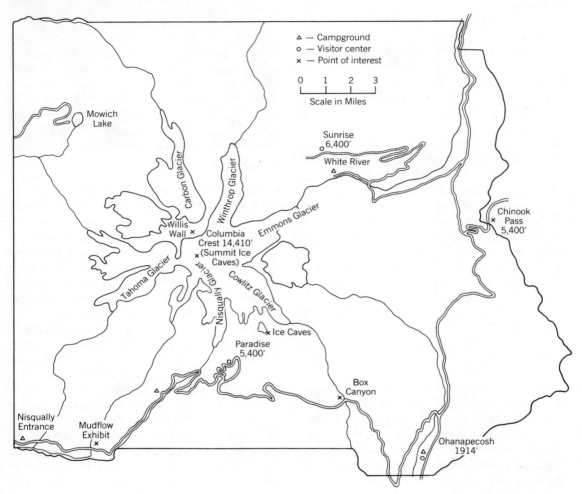

pass over the Rockies, they drop a significant amount of their remaining moisture, but the remainder is not great. It is clear that precipitation is the "life-blood of glaciers."

Go to the North Cascades National Park; it is truly unique. It might be your most unforgettable trip.

MT. RAINIER NATIONAL PARK

Mt. Rainier is high (14,410 feet; 4367 m) — the highest mountain in the Pacific Northwest. Buried beneath snow and ice it is a sight to behold, especially when seen in the setting sun from Puget Sound. It was from here that Captain Vancouver saw it in 1792. Lest anyone who has visited Mt. Rainier during a rainy period get the impression that it is rainier there, Captain Vancouver named the mountain after a fellow officer, Peter Rainier.

Mt. Rainier National Park, located near the northern boundary of the Middle Cascades in south-central Washington, was established in 1899. It consists of 378 square miles, all of which lies on the west flank of the Cascades and drains into the Pacific. Extending out from the base of the cone are benches, remnants of lava flows, called *parks*. Yakima Park, where Sunrise Campground is located, and Grand Park, also on the northeastern side of the mountain, are examples.

With its variety of attractions, Mt. Rainier is a popular spot. Skiers, botanists, volcanologists, and glaciologists find it of particular interest. For hikers, the 90-mile-long (145 km) Wonderland Trail encircles the mountain, leading through one wildflower garden after another if the season is right.

Geology and Geologic Sequence

A glance at a geologic map of Mt. Rainier indicates that the geology is highly complex, even though pre-Tertiary rocks, widely exposed in the North Cascades, are absent. Early reconnaissance by S. F. Emmons (1879) and others established that Mt. Rainier is a composite cone (stratovolcano) perched on a plateau of eroded volcanics and granite. Detailed geologic analysis of the area was slow in coming, however, because of the physical environment. Below 5000 feet most areas are covered by dense forest and underbrush, and the area above 5000 feet is mainly glaciers. There are extensive outcrops in the higher areas, but many are on steep slopes in unstable rocks. The Willis Wall, the headwall of the Carbon Glacier cirque, is an example. At the top of the headwall, 3600 feet high, are shelf glaciers that push out over the edge of the shelf. The resulting avalanches and rockfalls are not conducive to prolonged and detailed geologic examination. Thus, credit is due Fiske, Hopson, and Waters (1963) for their work under difficult conditions. Much of the following summary of bedrock geology is based on their U.S. Geological Survey Professional Paper 444.

The oldest rock in the park is the Ohanapecosh Formation, more than 10,000 feet thick, which is composed mainly of late Eocene andesitic volcanic debris, with lesser amounts of lava, mudflows, and waterlaid clastics. It rests on early Eocene sedimentary rocks, which in turn rest on pre-Tertiary rocks.

The Ohanapecosh is overlain by Oligocene volcanic ash and pumice, which underlie Miocene andesite and basalt flows.

During the Oligocene orogeny, Ohanapecosh rocks were folded, faulted, and eroded, and the younger rocks were then laid down on the erosion surface. Additional folding and faulting affected all of these rocks, probably during the latter part of the Miocene. At this same time magmas intruded and formed the Tatoosh batholith and also stocks, sills, and dikes.

Pliocene uplift accelerated stream erosion which uncovered the Tatoosh pluton and cut canyons as deep as 4000 feet into the granodiorite. It is on this irregular surface that the Mt. Rainier volcano was built, beginning in Pleistocene time.

Mt. Rainier is a stratovolcano composed of lava flows interlayered with pyroclastics and volcanic mudflows. However, its overall configuration has been modified significantly; its flanks have been steepened and its top has been removed. Glaciers carved deep gorges into the flanks of the cone and paved the way

FIGURE 5-16

Bridge built after the 1947 mudflow buried Kautz Creek Valley; Mt. Rainier in the background.

for large mass movements of the unstable materials, as rockfalls, avalanches, landslides, and mudflows. Thus, the moderate slopes of the cone were markedly steepened.

Mt. Rainier was until rather recently much higher than now, as its distinctly rounded top indicates. Some workers suggested that the upper thousand or fifteen hundred feet of the cone was removed by explosion and collapse, but Crandell (1971) presents convincing evidence that avalanches, rockfalls, and landslides of highly unstable, hydrothermally altered materials lowered the summit. He found much of the summit material in the huge 5700-year-old Osceola mudflow deposits on the northeast and north flanks of the mountain. Probably steam explosions triggered the mass movements of the unstable material at the summit.

Two summit craters now occupy the rounded mountaintop; the younger of the two, Columbia Crest is the higher. Thus we see today a magnificent volcano, albeit without a peak and with its flanks deeply scored by glaciers, avalanches and mudflows.

Mt. Rainier Today

Is Mt. Rainier a dead volcano? The 1820–1854 eruptions, although minor, and the steam that issues forth from both of the summit craters suggest that the volcano is sleeping, but restlessly. Moreover, by remote sensing, Moxham and others (1965) have detected "hot spots" on the flanks of the craters. Whether the hot spots have been hot for a long period or whether they are now heating up is moot. Regardless, it is clear that Mt. Rainier has not cooled off completely and that it is safer to regard it as a dormant or quiescent volcano, one that could erupt again, conceivably in our time.

A more immediate threat than an actual volcanic eruption, however, is the hazard of mudflows. In 1947 a huge mudflow rushed down Kautz Creek Valley and destroyed everything in its path. The old bridge over Kautz Creek is now buried in mud about 50 feet (15 m) beneath the new one. In 1963 a large rockfall avalanche broke loose on Little Tahoma Peak and roared down the east flank of the mountain. It was headed for and seemed intent on wiping out the Park Service's White River Campground; fortunately it "ran out of steam" a short distance up-valley. On the other side of the mountain the Tahoma Creek Campground was located on a 400- or 500-year-old mudflow, according to the alert sounded by Crandell and Mullineaux (1967). Their warning was heeded and the campground was relocated outside the path of future mudflows.

Conditions are ideal for mudflows in volcanic areas where snow and ice cover steep, highly unstable pyroclastic slopes. Meltwater penetrates readily into the permeable materials, causing them to become highly mobile, and on steep gradients these mudflows travel at high velocities. If the hot spots on Mt. Rainier are in fact becoming hotter, more snow and ice will be melted and the frequency and magnitude of mudflow activity will increase. There appears to be nothing that will prevent mudflows from developing, and nothing will likely stop them once they are on their way. Campgrounds and certain other installations, however, can be relocated out of their reach, on benches high above the valleys. Crandell and Mullineaux's environmental studies provide a basis for sound planning, and in a National Park where large numbers of people congregate, such guidance has particular significance.

Snow and ice fill the two summit craters, but fumarolic activity has melted more than 6500 feet (2000 m) of ice tunnels along the crater floor (Kiver and Steele, 1975). In the east crater a small crawlway leads from the large main cave to a huge grotto more than 340 feet (104 m) below the surface of the ice fill. A similar passage in the west crater leads to Lake Grotto where, beneath a thick cover of ice (160 feet; 50 m), there is a crater lake, believed to be the highest on the continent, at 14,113 feet (4329 m). Mt. Rainier is one of only a few volcanic peaks in the world to possess extensive geothermal ice caves, or "steam caves." These caves and their warm fumaroles have saved the lives of many, including Stevens and Trump, who made the first documented climb in 1870. In December of 1979, a small plane crashed near the summit in subzero temperatures. The pilot and his passenger survived the cold by huddling around an ice cave fumarole during a long night.

Mt. Rainier's Glaciers

There are 26 named glaciers on the slopes of Mt. Rainier; of these, Carbon, Emmons, Paradise, Nisqually, and Tahoma are perhaps the best known. A small ice cap covers the top of the cone and divides into the individual valley glaciers high on the slopes. These valley glaciers deepened and widened their valleys, leaving narrow, jagged ridges called *cleavers*. The Cowlitz Cleaver between Cowlitz and Nisqually glaciers is an example. East of the summit, Gibraltar Rock sticks up above the surrounding ice as a nunatak. Other glacial features such as horns, aretes, and moraines

FIGURE 5-17

Entrance to Mt. Rainier's Paradise Caves. (Photo by E. Kiver.)

are also found within the park. Of special interest are the Paradise Ice Caves. Although tunnels and caves in most glaciers are hazardous because of the potential danger of collapse of the ice, those in Paradise Glacier have been found to be *relatively* safe. Snowbound most of the year, they are generally accessible toward the end of the summer.

FIGURE 5-18

Nisqually Glacier (arrow) has cut deeply into volcanic tuff and flows exposed in far cliff.

Many glaciers in the mid-latitudes did not survive the higher temperatures of the Thermal Maximum, but most of Mt. Rainier's glaciers merely retreated upslope into the "arctic" of the mountaintop (Crandell, personal communication, 1974). Advances and retreats occurred in response to climatic fluctuations during the Little Ice Age. The latest retreat is the one of greatest concern to humans. In many places glacial meltwater is a valuable resource — as a source of water for human consumption, irrigation, and hydroelectric power. Meltwater from Nisqually Glacier on Mt. Rainier is vital to the city of Tacoma. According to Matthes (1942), records of Nisqually's retreat extend back to 1857. Even with brief readvances, the average rate of retreat increased until by 1920 it was truly alarming. During the 1940s, however, a change took place throughout the Pacific Northwest, and by 1950 Nisqually and certain other glaciers had begun to advance. An excellent pictorial record of Nisqually's retreat and advance is presented by Veatch (1969). Nisqually's re-advance has been less than spectacular, with halts and minor retreats interspersed, but by 1965 the ground lost after 1890 had been regained. Although a climatic change is clearly established by glacier expansion, the problem of the cause or causes of the climatic change remains unsolved. During the 1930s, while the warming trend was still in progress, it was assumed by certain workers that the increase in air pollution was the principal if not the sole cause; if this is true — and it should not be so assumed — it is difficult to use this same explanation for the cooling trend, particularly since air pollution has been increasing at an even greater rate. It becomes obvious that we are dealing with a complex system in which the role of humans is difficult to determine.

In summary, Mt. Rainier is typical of most of the Cascades stratovolcanoes. They are essentially sizable blisters that developed recently atop a major range of folded and faulted mountains which had been blanketed by Miocene basalt flows. Mt. Rainier was born in latest Pliocene or earliest Pleistocene time. Throughout the Pleistocene, eruptions of andesite lava alternated with violent eruptions of pyroclastic materials. During construction, erosion by streams and glaciers modified the cone by stripping away sufficient material from its flanks to significantly steepen the sideslopes. Periodic eruptions interfered with the glaciers and obliterated certain glacial features. About 5700 years ago, the top of the cone was lowered perhaps as much as 1500 feet (458 m) to its present height of 14,410 feet (4367 m). Volcanic mudflows have occurred repeatedly and present a real hazard today. Glaciers have fluctuated throughout; from about 1850 to 1945 they were receding, but now they are making a minor but significant advance.

When will Mt. Rainier erupt again? Although geologists now have a reasonably complete record of the recent past, they are not yet in a position to make such predictions. On the other hand, with the constant watch over the volcano, it is unlikely that a major eruption will occur unannounced.

CRATER LAKE NATIONAL PARK

Crater Lake is without question one of the beauty spots of the world. The deep blue of the lake with varicolored cliffs rising 2000 feet (610 m) above it presents an unforgettable picture. The lake is 1932 feet (589 m) deep and about 6 miles across. Until 1869, it was called Blue Lake and Deep Blue Lake, names more appropriate than Crater Lake; Crater Lake's basin is not a crater but a caldera!

The national park, established in 1902, occupies an area of about 250 square miles in southern Oregon. It includes the caldera, the lake with the volcanic cone Wizard Island rising above it, low mountains such as Mt. Scott, the Pumice Desert, and cinder cones at several locations around the caldera. The Rim Drive affords good views of the Palisades, Phantom Ship, Eagle Crags and other fascinating features. Boat trips around the lake and to Wizard Island are available. Visitors hike down the mile-long Cleetwood Trail, take the boat trip, then get some good exercise struggling back up the steep trail to the rim. Tiny Wizard Island, which from the rim appears a mere impudence, rises above the boat almost 800 feet (242 m); given that two-thirds of the cone is submerged, Wizard Island begins to command respect. From the boat, a large dike called Devils Backbone, the Phantom Ship, and many other features can be viewed at close range.

Although much of the park area is dry, precipitation in the higher areas is sufficient to support a good stand of evergreens, including pines, firs, and hemlock. Wild animals in the park include ground squirrels, foxes, porcupines, marmots, deer, and bears.

Geology and Geologic History

The rocks of Crater Lake National Park are all late Tertiary and Pleistocene volcanics except for a few small intrusives. Andesites are the dominant rocks; however, basalts are fairly common and there are several areas of more highly siliceous rock — rhyolites, *dacites,* and obsidian, of late Pleistocene age. Much of the area north of Crater Lake is blanketed by pumice and ash that was deposited only about 6900 years ago (Bacon, 1983).

In general the early geologic history of Crater Lake parallels that of Mt. Rainier and other stratovolcanoes of the Cascades. Pleistocene eruptions of mainly andesitic materials, both lavas and pyroclastics, built a large stratovolcano. At times, during highly explosive eruptions, fiery clouds of ash, pumice, and volcanic gases roared down the flanks of the cone and blanketed large areas. Heat welded the materials together, particularly around the vents through which the hot gases escaped. Later, as streams eroded the area, the rocks that had been thoroughly welded remained as

FIGURE 5-20

Map of Crater Lake National Park.

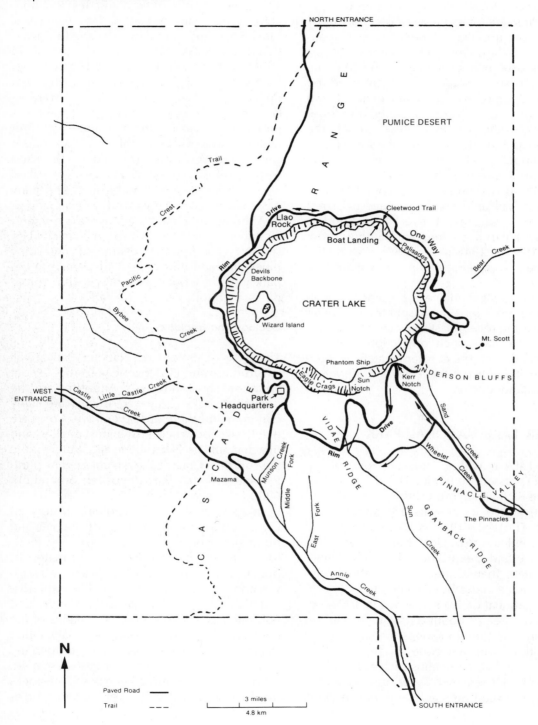

columns and pinnacles, as in the Pinnacles area near the southeast corner of the park.

During cone-building, the vent was at times so firmly sealed with solidified lava that the gases broke through elsewhere, even well out beyond the base of the cone. Eruptions at various points built up small satellite cones such as Red Cone, Crater Peak, and Mt. Scott.

Valley glaciers developed on the cone and scoured out deep valleys on its flanks. Periodically, lavas poured out over the lip of the summit cone, dispatched any glaciers that were there, and filled certain of the valleys with lava. The lower end of one of the U-shaped valleys extends down from Kerr Notch on the southeast side of the caldera rim.

Then came the truly cataclysmic event, the destruction of the cone, named Mt. Mazama, and the formation of the caldera. The destruction of Mt. Mazama has been the subject of many discussions. At one time it was believed that a gigantic explosion was solely responsible. To have a huge mountain literally blown to bits certainly lacks nothing of the sensational, particularly when we become aware that about 15 cubic miles of mountain must be accounted for (Williams, 1942). The surrounding area is buried with pumice, ash and other materials, and an unknown quantity of dust was carried long distances by the wind. Moreover, as pumice floats in water, significant amounts were probably carried to the ocean and thus cannot be measured. Even so, it appears unrealistic that all of the 15 cubic miles of material were removed solely by explosion. For this reason, Williams and others favor the subsidence or collapse theory. Although these terms suggest that the mountain simply fell in and that the material was somehow engulfed into the depths of the earth, actually the theory involves explosions, followed immediately by collapse of the walls to enlarge the depression. The explosions blow the mountaintop to bits and at the same time eject large volumes of material from beneath, thus providing space for infalling material. Ensuing flank eruptions

FIGURE 5-21

Devils Backbone, a large dike rising above the north shore of Crater Lake.

lower the level of the magma and cause additional collapse of the walls, until a depression of large diameter, a caldera, is formed. This general interpretation appears to satisfy the evidence at Crater Lake.

FIGURE 5-22

Llao Rock on north rim of Crater Lake Caldera; glacial valley on the flank of old Mt. Mazama was filled to overflowing with lava. (Photo by E. Kiver.)

Since that time, three cinder cones have been built up from the floor of the caldera; Wizard Island stands 764 feet (233 m) above present lake level. Wizard Island is recent enough to be only a little eroded; however, a still more recent eruption broke through the west flank of Wizard Island and built a lava platform almost to the edge of the lake.

The restoration of the original cone, Mt. Mazama, was done by projecting upward the dipping strata of the caldera, in keeping with other stratovolcanoes of the area. As Matthes (1956) reasons, "As it was as large around at the base as that peak (Mt. Rainier) and rose from about the same level, it may be conceived to have had similar proportions, except that it probably tapered upward to a smaller summit crater." He further points out that Mt. Mazama had to be a lofty mountain in order to support large glaciers; large U-shaped valleys at Kerr Notch and Sun Notch are the lower sections of long glaciated valleys. These valleys are larger than those on Mt. Rainier, and since Mt. Mazama was almost 300 miles (484 km) farther south, in an area where precipitation was less than at Mt. Rainier, Mt. Mazama must have been about as high. Recently, however, Mazama has been under attack! It wasn't that high, possibly less than 12,000 feet high, and it was highly asymmetrical, so say those who would humble majestic Mazama. Perhaps they are right, but until more convincing evidence has been published, Mt. Mazama will remain in the eyes of these writers a lofty mountain.

Crater Lake obtains its water from the precipitation that falls within the caldera rim. The lake level is fairly constant, at about 6160 feet (1878 m) above sea level. The minor fluctuations are related to climatic cycles, with precipitation and evaporation as the controlling factors. Measurements clearly show that the input of water far exceeds that which is lost by evaporation; it is clear, therefore, that a significant amount of water seeps out of the basin through the porous rocks. It appears that the system is essentially in equilibrium; should increased precipitation begin to raise the lake level, the increased water pressure would probably cause sufficient seepage to offset it. Consequently we can anticipate future fluctuations to be relatively minor, as in the past.

Glaciers are not present today, but the evidence of former glaciers is unmistakable. In addition to the glaciated valleys, glacial grooves and striations are abundant, and moraines are found at several places around the caldera.

Stream erosion, supported by weathering

and mass-wasting, is changing the landscape today. In places away from the caldera, streams have cut youthful canyons below the general surface. Of the many erosional features, perhaps the Pinnacles in the southeastern part of the park are the most spectacular.

In summary, Mt. Mazama was a stratovolcano of great height, perhaps as much as 14,000 feet (4269 m). It was deeply scored by glaciers during the latter part of the Pleistocene. The destruction of Mt. Mazama, by explosion and collapse, occurred about 6900 years ago, leaving the 6-mile-wide caldera which contains Crater Lake. Later eruptions built three cinder cones, of which Wizard Island was high enough to reach above lake level.

If you care even a whit for things beautiful or things geological, you must visit Crater Lake. The deep blue of the water you see in pictures is no exaggeration. Take the boat trip. As you see the different features, one by one, in leisurely fashion, Crater Lake and its exciting past begin to take shape.

FIGURE 5-23

Schematic diagram showing the downfall of old Mt. Mazama following explosive activity.

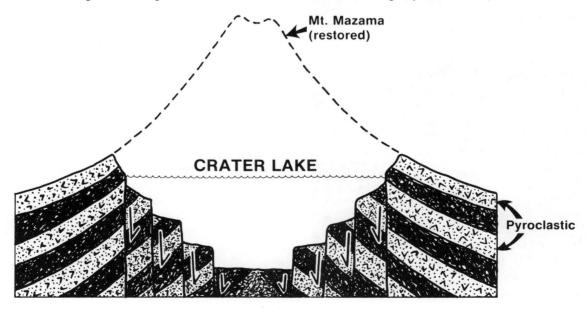

LASSEN VOLCANIC NATIONAL PARK

On May 30, 1914, the quiescent volcano, Lassen Peak, came to life. This should have been no surprise, because in 1851 Cinder Cone, about 10 miles northeast, erupted violently and steam issued from nearby Chaos Crags in 1857. Lassen erupted repeatedly in 1914, but it did not really blow its stack until in May of 1915, when a five-mile-high column of volcanic ash and dust shot up into the sky. There was no problem generating interest in establishing Lassen Volcanic National Park, and in 1916 an area, later increased to 165 square miles, was so designated.

Visitors entering the park from the south will be impressed first with intriguing glimpses of Lassen Peak; shortly, however, their attention will be diverted by an unpleasant odor, namely, hydrogen sulfide, from the Sulphur Works — a *solfatara*. The Sulphur Works and a larger solfatara, Bumpass Hell, indicate that the "fires" beneath the Lassen Peak area are not dead.[2]

Geology and Geologic Sequence

The Lassen Peak area is a partially dissected volcanic plateau, above which rise many volcanic hills and mountains, the highest of which is Lassen Peak (10,457 feet; 3188 m). Unlike Mt. Rainier, Mt. Shasta, and other stratovolcanoes, Lassen Peak is a huge volcanic dome, or *plug dome,* which rises some 2500 feet above its base.

How, then, is Lassen Peak related to Crater Lake and Mt. Rainier? As a result of detailed geologic studies, the following general sequence has been established. First, as at Mt. Rainier, a large stratovolcano developed, mainly during the Pleistocene. Mt. Tehama, at least 11,000 feet (3354 m) high, was then demolished by explosion and collapse to form a caldera generally similar to Crater Lake. Brokeoff Mountain, about five miles southwest of Lassen Peak, is a part of the rim of the caldera. Later, highly viscous magma was forced up through various vents in and adjacent to the caldera, thus forming several plug-domes, the largest of which is Lassen Peak. Therefore in Lassen Volcanic National Park is recorded a similar but more complex sequence than at either Crater Lake or Mt. Rainier.

FIGURE 5-25

Lassen Peak in eruption, May 22, 1915. (Photo by National Park Service.)

---

[2] Additional information is available in Schulz, 1950, 1959; Macdonald, 1966; and Harris, 1980.

FIGURE 5-26

Map of Lassen Volcanic National Park.

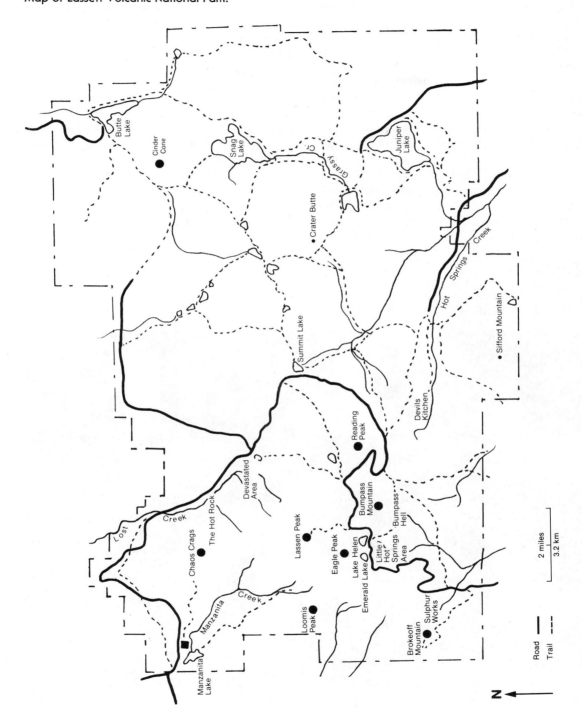

How and why do plug-domes form? The first essential is highly viscous magma. Gases, mainly steam, are always present, but here the high viscosity prevents sufficient gas concentrations to cause violent explosions; consequently, the viscous magma is squeezed up through the vent somewhat like toothpaste squeezed from the tube, as Schultz (1959) aptly describes the process. As extrusion takes place, rapid cooling produces rocks that are mainly glassy. As the process continues, the structure is built higher and higher; a "mushrooming" effect enlarges the dome and causes large amounts of rock to break off from the sides to form huge taluses around the solid plug. The solid plug, essentially buried in its own talus, is a steep-sided dome, in the case of Lassen Peak almost half a mile high. Nearby Eagle Peak, Chaos Crags, Bumpass Mountain, and Reading Peak are prominent although small plug domes.

The rocks in Lassen's plug domes are largely dacites, rocks that are similar to rhyolites. The rocks in old Mt. Tehama, of which Brokeoff Mountain is a small remnant, are mainly andesites as in the other Cascades stratovolcanoes. Basalts of recent origin also occur in parts of the park. Cinder Cone in the northeastern part is an example; here, lavas and pyroclastics were extruded during the last few hundred years, the latest occurrence being in 1851.

FIGURE 5-27

Looking down into Lassen's newest crater, formed by the 1915 eruptions. (Photo by National Park Service.)

FIGURE 5-28

Cinder Cone, about ten miles northeast of Lassen Peak. Basal eruption of lava occurred as late as 1851. (Photo by National Park Service.)

Lassen's Latest Eruptions

The eruptions, which began in 1914, are well documented. They were recorded in pictures by B. F. Loomis, a local photographer, and in words by J. S. Diller, geologist with the U.S. Geological Survey, and by others. Diller's first climb to the crater rim was in 1883 when he began his geological survey of the region. His *Lassen Peak Folio,* published in 1895, includes a geologic map that serves as the basis for "before and after" pictures of what happened during the period 1914–1921. By then Diller was in his 70s, but on hearing that Lassen was on a rampage, he lost little time in getting to the scene. Again he climbed to the crater rim, but it was not the same one; Lassen had blown out through a new vent this time. His studies during the next several years were published in part, but his long report was still unfinished at the time of his death in 1928. It was for this reason that Howel Williams undertook his investigations, which were published in 1932. Later, Macdonald (1966) prepared a more detailed geologic map as a part of his regional studies of the Cascades.

The May 30, 1914, eruption was the first of about 150 relatively small but significant outbursts during the first year. In May of 1915, lava filled the crater and poured down the southwest and northeast flanks of the mountain. Melting snow triggered a large mudflow which rushed down the northeast side and buried an area more than five miles long. Large blocks of lava were carried downslope; one, called Hot Rock, weighed about 30 tons. The main event, however, occurred on May 22, when the most violent of Lassen's outbursts sent a "mushroom cloud" about five miles above the vent and showered volcanic ash over a large area of northern California and Nevada. This eruption generated a *nuée ardente* (fiery cloud) which roared down the northeast flank and devastated everything in its path. This "Great Hot Blast," as it is frequently called, destroyed an estimated 5 million board feet of lumber. Today, although in the Devastated Area significant vegetative recovery has been made, the destructiveness of that "Great Hot Blast" is still clearly marked. During this same eruption, the countryside was pelted by volcanic bombs, some of which were large. Vulcan, the god of fire, had apparently vented his wrath and most of his energy in this one demonstration; later eruptions, which occurred periodically until 1921, were relatively mild.

To our knowledge, Lassen Peak is the only place where a man who was felled by a volcanic bomb survived to tell his story. Curiosity compelled three men to have a look down into the crater on June 14, 1915. Ominous noises caused two of the three to withdraw rather hurriedly, but the third, Lance Graham, was apparently transfixed by the unearthly sounds. The blast occurred as he started off the rim, and it was then that volcanic bombs rained down on him. Unconscious and believed dead by his companions, who ventured back up to rescue him, Lance eventually came around and later recounted his experience. Those interested in climbing around on live volcanoes may wish to read Hill's article in the November, 1970 issue of Mineral Information Service, California Division of Mines and Geology. It is mainly a reprint of newspaper clippings on the Lassen eruption, but it also contains historical data on the area, including material on the life of Peter Lassen, early pioneer, for whom Lassen Peak was named.

Lassen's Thermal Areas

Momentarily, the thermal areas provide the only direct evidence that the Lassen volcanic area is not dead. The Sulphur Works, Bumpass Hell, Devil's Kitchen, and Boiling Springs contain such features as solfataras (vents or fumaroles that emit sulfur fumes), *mud pots, mud volcanoes,* and hot springs. Bumpass Hell, at the end of a "long mile" trail, shows most of the above features in an impressive manner, and the appropriateness of the terms *"Big Boiler"* and *"Steam Engine"* is immediately apparent. Here and in other thermal areas, you must use extreme care to stay on marked trails if your visit is to be an entirely pleasant one. Both boiling water and steam are dangerously hot! Moreover, the "water" in certain of the hot springs is in reality a dilute solution of sulfuric acid.

Hydrothermal alteration is widespread in and around the thermal areas. In addition to sulfur, such minerals as opal, tridymite, kaolinite and pyrite are being formed today (Allen, 1925, and Anderson, 1935). Geochemical studies have shown that where the acidity of the water is relatively high, nearly pure opal is formed; lower acidity and lower temperatures produce mainly kaolinite.

Of possible significance is the fact that many of the thermal areas are not far from the centers that were active when old Tehama was abuilding. This does not pinpoint this spot for the next eruption; then again, magma at shallow depth is the most probable source of the heat that operates these systems.

Glaciers, Lakes, and Swamps

Glacial features are abundant in the park. Along what is now the Bumpass Hell trail, many of the rocks were polished, grooved, and striated as debris-laden ice moved over them. Also, a huge glacial boulder was left perched high on a ridge overlooking Little Hot Springs Valley. Lake Helen and Emerald Lake south of Lassen Peak are two of the many glacial lakes. High on various peaks, Loomis and others, there are many cirques and cirque lakes (tarns). According to Schulz (1959), in the area near the saddle between Lassen and Eagle peaks, the rock hills were reshaped by glacial erosion, thus forming elongate, asymmetrical roches moutonnées. It is clear that glaciers were once widespread; although there are many small icefields, no true glaciers are here at the present time.

FIGURE 5-29

Erratic boulder resting on polished and striated rock.

Most of the lakes in the park are of glacial origin, but Manzanita Lake in the northwestern corner of the park was formed when a landslide dammed Manzanita Creek about 1000 to 1200 years ago. The landslide is contemporaneous and genetically related to the formation of Chaos Crags, the park's youngest group of plug domes. About 300 years ago, steam explosions in the Chaos Crags domes, or perhaps the intrusion of another small dome, sent some 150 million cubic yards of rock hurtling down the north face of the mountain with such force that it crossed a broad valley and rushed 400 feet (123 m) up the side of Table Mountain.

Chaos Crags, high on the slope above Manzanita Visitor Center and associated tourist accommodations, is highly unstable and capable of producing another destructive avalanche. In January, 1974, geologists of the U.S. Geological Survey pointed out that a rockfall-avalanche could take place without warning and that it would travel at high velocity — too high to provide time to warn the people in the area downslope. Accordingly, on April 26 the Park Service issued a news release stating that, because of the geological hazard, the facilities at Manzanita Lake would be moved to a safe place. Environmental geology is multifaceted; one important facet is the evaluation of volcano-related hazards, particularly in the Cascades and specifically in the parks where large numbers of people congregate.

Certain lake basins are the direct results of volcanic activity. In many places depressions are left in the surface of lava flows; in others, as at Snag Lake, lava flows dam valleys. There are also the crater lakes; here the craters are reasonably watertight, as at Crater Butte about three miles east of Summit Lake.

Destruction is the fate of all lakes. Some are destroyed by streams which cut channels through the outlets; more obvious, however, is the filling of the lake basins which gradually reduces their capacities. Delta building can be observed in Manzanita Lake and many others in the park. In 1959 Schulz pointed out that Hat Lake, 44 years old at that time, had already been about half filled. But as lakes are destroyed by filling, wet meadows and swamps are born, thus creating another ecological system in Lassen Volcanic National Park.

LAVA BEDS NATIONAL MONUMENT

Lava Beds National Monument may be the only place in the National Park System where geology played a key role in an Indian war. The Modoc Indians may not have known how the lava tubes and the collapse features were formed, but they knew how to make the most of the resulting terrain. In the Modoc War (1872–1873) Captain Jack's 53 braves fought — and eluded only to fight again — about 1000 U.S. soldiers. Finally, after five months of embarrassment (not to mention casualties) the soldiers had the Indians surrounded and were ready to round them up. However, as related by Waters (1981), they were a bit downhearted the next morning when they could find not one Indian! By knowing every inch of the terrain, 160 women, children and warriors had slipped through the lines and escaped, temporarily. The story has a sad ending; they were captured a short time later, and the U.S. Army passed up the opportunity to use Captain Jack to train its officers in the applications of military geology!

Located on the north flank of the Medicine Lake shield volcano in northernmost California, Lava Beds National Monument consists of many lava flows, 17 cinder cones, 200 known lava tubes, and other volcanic features (Heiken, 1981), several of which were formed during the past few hundred years. Apparently in 1910, there was a small ash eruption and flame from vents in the top of the shield (Finch, 1928); therefore, the probability of renewed volcanic activity is high. This unique area, consisting of about 72 square miles, became a national monument in 1925.

Geologic History and Features

Only a mere instant of geologic time is recorded in the rocks and landforms in Lava Beds. The oldest rocks are shield-forming basalts that are 1.0 to 1.4 million years old; these rocks are exposed in Gillem's Bluff, a prominent fault scarp along one of the park roads. A series of these north-south trending faults extends into the area from the nearby Basin and Range Province. Many of the faults are marked by lines of cinder and spatter cones. As the 3000-foot (914-m) shield volcano was building, basaltic eruptions continued in Lava Beds and elsewhere on the flanks, and increasing amounts of silicic (andesite, dacite, rhyolite) volcanism occurred in and near the summit caldera. The extreme variation of lava types led Eichelberger (1975) to postulate the existence of both rhyolitic and basaltic magma chambers beneath the area, with different degrees of mixing to account for the extreme compositional range of volcanic products.

Lava Beds experienced basaltic eruptions that produced a wealth of geologic features for the visitor to discover. Fleener Chimney, an impressive 130-foot (40 m) hole, helped feed the 500- to 1000-year-old Devil's Homestead Flow; Mammoth Crater is a large collapse crater on a small lava shield. The Calahan flow is the youngest lava flow in the monument, younger than the nearby 500-year-old lavas. Pockets of pumice on the flow surface record even more recent activity that originated at the summit area of the Medicine Lake volcano.

About 200 lava-tube roofs have collapsed sufficiently to permit entry. Fourteen caves are readily accessible from Cave Loop Road near park headquarters, and trails from other roads lead to Merril Ice Cave and other interesting features. There are also numerous speleothems such as basalt stalactites ("lavacicles"), frozen lava cascades, horizontal lines ("bathtub rings"), and lava shelves.

A walk through this maze of tunnels and passageways gives the visitor an understanding of how the Army let Captain Jack and his people escape. Remember that the noon-day sun beating down on the black lava can make it seem almost as hot as when the lava was flowing!

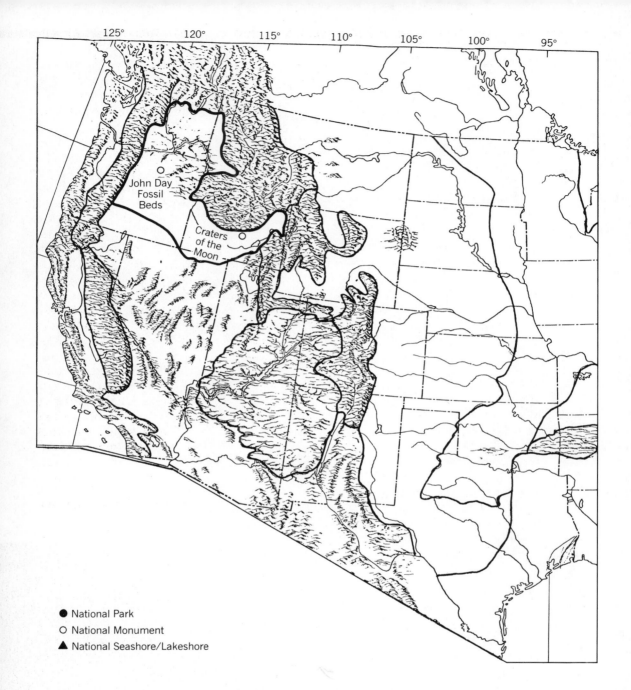

PLATE 6

Columbia Intermontane Province.

CHAPTER 6

COLUMBIA INTERMONTANE PROVINCE

The Columbia Intermontane Province has generally been called the Columbia Plateau. When examined, however, the province is found to include large areas of plains, hills, and mountains, in addition to plateaus. Therefore, as the province is essentially surrounded by mountains, it is logical to use the term *intermontane,* as Freeman et al. (1945) and Thornbury (1965) have done. The province extends eastward from the Cascades to the Northern and Middle Rocky Mountains. The southern boundary with the Basin and Range Province is less definite but is ordinarily drawn across southern Oregon and Idaho, including a narrow strip along northeastern Nevada.

The rocks of the province are mainly basaltic lava flows ranging in age from Eocene to Pleistocene, with 15(±)-million-year-old Miocene flows most abundant (Hooper, 1982; Swanson et al., 1979). The latest activity was the building of a line of cinder cones and spatter cones.

Older rocks occur around the borders of the province and, in places, as *steptoes* — hills that, like kipukas, were surrounded by lava. In John Day Fossil Beds National Monument in northeastern Oregon, streams cutting through the Miocene basalts have exposed older Cenozoic continental deposits containing remarkable assemblages of plant and animal fossils.

Although the results of volcanism are most obvious, other great changes that took place during the Pleistocene resulted from glacia-tion, both directly and indirectly. The Okanogan lobe of a Canadian glacier pushed southward into eastern Washington, damming and displacing the Columbia River; while temporarily displaced, the river — repeatedly in flood — cut the Grand Coulee, a 50-mile-long gorge that is in places nearly 1000 feet deep. A glacier also blocked the outlet of a large basin north of Spokane, Washington, forming huge Lake Missoula. In a short time the water broke through the ice dam, causing the largest flood ever documented (Bretz, 1923, 1969). Actually, there were several, perhaps many floods — 13 according to Kiver and Stradling (1982), 40 according to Waitt (1980), and 70 according to Atwater (1983). After each flood, probably 30 to 50 years were required for ice to again block the outlet and for the impounded water to reach the depth necessary for another breakthrough.

These catastrophic floods cut huge channels and scars, thus forming the massive erosional features in the famous Scablands of eastern Washington. Erosion went on at a rate difficult to comprehend; imagine what took place when more than 9 cubic *miles* of water

FIGURE 6-1

Twin Craters, Sheep Trail Butte. (Photo by J. Wiebush.)

FIGURE 6-2

Dry Falls. Diversion of the Columbia River combined with catastrophic floods from glacial Lake Missoula created this now-abandoned giant cataract, Sun Lakes State Park near Grand Coulee National Recreation Area, Washington. (Photo be E. Kiver.)

per *hour* discharged through the broken dam (Pardee, 1942).

Giant ripple marks—ridges of gravel—confused geologists regularly until a theory of colossal floods was suggested and then substantiated. When 500 cubic miles of water are suddenly released, some gigantic features—erosional and depositional—can be formed in a short period of time.

The Columbia Intermontane Province has been divided into subprovinces, some of which have been divided again into sections. Here, the discussion is confined to the Snake River Section, where Craters of the Moon National Monument is located, and to the Blue Mountains Section, which contains John Day Fossil Beds National Monument in northeastern Oregon.

The eastern Snake River Plain is a classic area for studying basaltic volcanism and is the subject of a significant volume edited by Greeley and King (1977). Hot, highly mobile lavas issued forth from a series of fissures and spread over the adjacent area as gently sloping, coalescing shield volcanoes (Greeley, 1977), ex-

cept in Craters of the Moon where pyroclastic eruptions built a number of cinder cones. The volcanic rocks become younger eastward along the Snake River Plain, and the northeastern trend of the trough leads directly to the Yellowstone caldera, where molten material probably lies beneath the surface. There was either a stationary hot spot with the North American plate moving approximately westward over it (Eaton et al., 1975; Smith and Christiansen, 1980), or a rift with upwelling magma as a plate carried the Northern Rockies northward (Hamilton and Myers, 1966).

In either case, hot, highly fluid, and highly mobile lavas poured out of long fissures during the Miocene, covering an area of about 200,000 square miles. In at least one place the total thickness of all the flows exceeds 10,000 feet (3067 m).

The older rifts or fissures are buried beneath younger flows, but they can be reconstructed by observing the latest rift system (Prinz, 1970). The Idaho Rift System extends northward and then northwestward across the eastern Snake River Plain. In the southern section several of the rifts are open; King's Bowl Rift has been descended to a depth of 800 feet where an ice floor ended the exploration. The northwestern section, called the Great Rift, is largely within Craters of the Moon National Monument.

The groundwater system in the Snake River Plain is unusual. In this dry area underlain by fractured lava flows, essentially all of the water in the streams flowing out of the mountains disappears into the fractures in the basalt and are therefore "lost rivers." Eventually this water reappears in the form of large springs along the north wall of the gorge cut by Snake River. Of these, Thousand Springs, about 25 miles downstream from Twin Falls, are most widely known. Here, more than 150 feet above the river, large volumes of water gush forth; the quantity is sufficient for the irrigation of large tracts of valley land. The rate of flow of these underground waters is abnormally rapid, a puzzling problem until detailed geological investigations were made. It was discovered that there are continuous layers of highly permeable volcanic materials between certain of the flows; the water moves rapidly through the porous materials and reappears as springs, especially the Thousand Springs.

CRATERS OF THE MOON NATIONAL MONUMENT

This area of about 83 square miles in southern Idaho was set aside as a national monument in 1924. The Snake River Plain in which the monument is located is floored with Quaternary basalts, some of which are too recent to have a good soil cover. The lavas consist of both the rough, jagged aa type and the smooth, sometimes ropy pahoehoe flows. In addition there are cinder cones, spatter cones, and *squeeze-ups*. Most of the cones form a line along the 34-mile-long (55 km) Great Rift which extends southeastward across the monument. The Great Rift is the northwestern part of the Idaho Rift System. Carbon-14 analyses of wood indicate that the Great Rift opened up last about 2000 years ago. Lava flows poured out in places, but the explosive development of cinder cones along the Rift dominated the scene.

Lava tubes, formed by the draining out of lava from beneath a solid roof, are plentiful in and around the monument. In places the roofs have collapsed, thus forming entrances into the tubes, as at Boy Scout Cave and Great Owl Caverns. Visitors are generally surprised after a hike across the sun-baked lava to find themselves on ice in the bottom of Great Owl Caverns. The ice probably dates back to the time when a cooler climate prevailed; once formed, the air circulation is apparently too poor to cause the ice to melt. Interesting also are the stalactites that hang from the roof. They are not composed of calcite as in limestone caves; neither are they icicles, as might be expected. They are basalt stalactites formed immediately

FIGURE 6-3

Rumpled pahoehoe flow. The almost solid "skin" was folded into rolls as the flow moved forward.

after the lava drained out of the lava tubes. Also of special interest are the tree molds — fossils — which were formed only a few thousand years ago. The trees were encased in the lava and their charred trunks left an unmistakable imprint in the rock.

The Craters of the Moon National Monument fascinates many people; geologists and

FIGURE 6-4

Tree mold, the mold of a charred log.

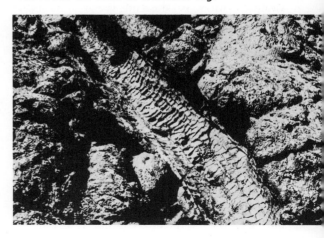

volcanologists are enthralled there. Others may have reactions of their own. Perhaps Washington Irving was restraining his enthusiasm for lava when he wrote in 1868 that it was an area "where nothing meets the eye but a desolate and an awful waste; where no grass grows nor water runs, and where nothing is to be seen but lava." This is a beautiful area, like no other on earth; it is somewhat bleak and lonely, but so are most areas on the moon!

If on a hot day, when the heat waves are dancing on the lava, you see a boy carrying a sled, don't assume that the sun was too much for you; he is going sledding on the ice in one of the caves. Exploring in the sections where the roof is only two or three feet above the ice is much easier on a sled while lying on your back. By so doing, you can avoid being speared by a stalactite. Wear a hard hat and take a flashlight.

JOHN DAY FOSSIL BEDS NATIONAL MONUMENT

John Day was a member of the party sent into the Pacific Northwest to develop the fur-trading industry for John Jacob Astor. The "John Day Country" lies in the northeastern part of Oregon, in the upper reaches of John Day River, a tributary of the Columbia. The national monument, established in 1974, consists of three widely separated units—Sheep Rock, Painted Hills, and Clarno—with monument headquarters in the town of John Day. The area is famous for the extraordinarily complete record of the plant and animal life of the Cenozoic Era.

Geologic History

Within the John Day Country, rocks as old as Devonian record mid-Paleozoic deposition, and late Paleozoic-Mesozoic volcanism and intrusive igneous activity. Probably Early Cretaceous magmas gave rise to the gold veins in the Canyon Mountain area—the gold that caused the rush of 1862 and the settlement of the area. During the latter part of the Mesozoic there was marine and nonmarine deposition; marine deposition ended with the retreat of the sea late in the Cretaceous, with the uplifting of the Rocky Mountains to the east (Thayer, 1974; Baldwin, 1979).

Within the monument's boundaries, the record begins with the Eocene Epoch and the rise of the mammals. The "dawn horse" (Eohippus in some publications, Hyracotherium in others) lived here, but he was anything but impressive; when full grown he was no bigger than a fox terrier! Diminutive ancestors of the modern camels and rhinoceroses also got their start here in North America, along with other mammals that, like the oreodons, are now extinct.

During late Eocene and early Oligocene time, the warm, humid climate intensified chemical weathering and produced an iron-rich red soil. This red material was eroded and redeposited as the basal beds of the John Day Formation (Oligocene). On top of these red layers are volcanic materials of many colors—delicate shades of green, buff, and cream. These varicolored beds are best displayed in the Painted Hills Unit and are also exposed in the Sheep Rock Unit of the monument.

During the Oligocene, a warm, temperate climate prevailed and forests of birch, oak, and chestnut replaced the earlier tropical species. Saber-toothed cats, giant pigs, and other mammals roamed the John Day Country. Mesohippus, the three-toed horse, was a forest dweller about the size of a sheep.

The Miocene was a time of violence in this section of the country. Long cracks or fissures developed in the earth's crust, and highly fluid basaltic lavas poured out and flowed for long distances, as much as 100 miles (Thayer, 1974). The many flows, one on top of another, make up the thick Picture Gorge Basalt which, with associated dikes, features prominently in the John Day Country.

During the Pliocene, volcanoes erupted violently and thick ash flows filled John Day Valley, burying gravels that contain bones of horses, camels, bears, and rabbits. By this time the horse had reached the size of a pony and had feet and teeth like those of the modern horse. Plant fossils indicate that the climate had changed to one similar to that of today. Associated with the volcanic eruptions were tectonic adjustments that produced folds and faults that are readily observed, particularly in the Painted Hills Unit and in the spectacular Picture Gorge near John Day.

Most of today's landscapes were carved out during the Pleistocene, by glaciers above 5000 feet (1515 m) and by streams in the lower areas. Several areas of badland topography were developed, thus placing the colorful John Day Formation on display, particularly in the Painted Hills Unit. Landslides are unusually abundant, due mainly to the instability of the John Day beds that underlie the thick Picture Gorge Basalt, well exposed along the river north of Picture Gorge.

Here, we have merely glimpsed a few of the many fascinating features that you will see when you visit the John Day country. You will probably want to obtain Thayer's book, *The Geologic Setting of the John Day Country* (1974), before you start exploring the area. In addition to his discussion of the changes in the life forms and of the geologic changes that have taken place during the past 50 million years, it contains a road log that explains what you are seeing at the many points of interest on the John Day "loop." Perhaps the only thing that he neglected to mention is that you should keep an ear open for the "buzz" of a less-than-friendly rattlesnake!

PLATE 7
Sierra Nevada Province.

CHAPTER 7

SIERRA NEVADA PROVINCE

Massive and majestic are the words that describe the Sierra Nevada, which rise boldly above the adjacent Basin and Range Province. To get the full impact of the Sierra Nevada, make your approach from the east. Stop in Owens Valley and look up, up, and up to the top of Mt. Whitney, the highest peak in the conterminous United States. Then drive north to Lee Vining and up and over Tioga Pass. In the days before the road was paved, the trip up the east face was an adventure; now it is disgustingly simple! Early motorists made many stops and enjoyed the fabulous scenery, and certain ones pondered the fact that they were on the edge of a huge fault-block, the Sierra Nevada, which had been tilted toward the west—all while their car radiators were cooling down! Now, all too many zip up the mountain, eager to get over it and down to the lodge in the valley in time for dinner. Take a little time. There will be a tomorrow. Stop at a few of the overlooks and let a bit of it sink in! On Tioga Pass, tarry longer than just to show your Golden Eagle Pass at the entrance into Yosemite National Park. Look to the west. You can see the Pacific, or at least you could back in the pre-pollution days. But you can still see the High Sierra with its mountain meadows and open parks and the snow-capped peaks rising high above them. Finally, cruise down the gradual west slope of the fault-block, past the many rounded granite domes and then into the main valley.

Two other park areas are within the province, Sequoia-Kings Canyon National Parks, about 100 miles (162 km) to the south, and Devils Postpile National Monument, a short distance southeast of Yosemite. The Sierra Nevada presented an irresistible challenge to geologists, beginning with early explorations and the discovery of gold. The roster of names of the "greats" who worked in the Sierra is essentially a "Who's Who in Geology"—J. D. Whitney, G. K. Gilbert, Clarence King, Adolph Knopf, F. L. Ransome, Waldemar Lindgren, Francois Matthes, and John Muir. (For a summary of the work done by these and other geologists, refer to Bateman and Wahrhaftig, 1966.)

Bedrock Geology and Structures

"The Sierra Nevada is a huge block of the earth's crust that has broken free on the east along the Sierra Nevada fault system and been tilted westward. It is overlapped on the west by sedimentary rocks of the Great Valley and on the north by volcanic sheets extending south from the Cascade Range"—so say Bateman and Wahrhaftig (1966). In addition, the Sierra Nevada block is chopped off at the south end by the Garlock fault. The trend of the 400-mile-long (648-km-long) range is north-northwest, reflecting the dominant direction of the Sierra Nevada fault system.

Most of the rocks are granites, quartz diorites and granodiorites of the huge Sierra Nevada batholith. As King (1959) and others have pointed out, the "batholith" consists of several large igneous masses that intruded at different times rather than as a single gigantic intrusion. Many inclusions of metamorphic rocks, some of large size, are found within the batholith, and extensive areas of metamorphics flank the western side of the batholith, particularly in the northern section. There are also extensive areas of young volcanic rocks along the eastern border, to the east and north of Yosemite National Park.

Geologic and Geomorphic Sequence

The area that is now the Sierra Nevada, like the Cascades area, was a part of the western geosyncline during much of the Paleozoic and Mesozoic time. During the latter part of the Mesozoic, deformation occurred and batholithic magmas began to intrude in the Sierra Nevada area. Then there was uplift and much erosion, which stripped all the roof rocks off the deep-seated batholith.

It was not until late in Cenozoic time that the Sierra Nevada began to take shape. Faulting along the east side and tilting of the fault-block to the west are basically responsible for the Sierra (see Figure I-18). Fault scarps, many now much eroded, are abundant along the steep east face; there are, however, certain essentially unmodified fault scarps that cut across recent volcanics and glacial deposits, indicating that parts of the fault system are still active. More convincing to some is the faulting that caused the Lone Pine Earthquake in 1872; a 13-foot-high fault scarp remains today as a reminder. As a combined result of the faultings during a 10-million-year period, there exists a difference in relief of about 11,000 feet (3354 m) between the top of Mt. Whitney (14,495 feet; 4419 m) and the floor of Owens Valley to the east. Therefore, the Sierra Nevada rises higher above the adjacent area than any other mountains in the conterminous United States.

Faulting also took place within the fault-block, notably in the northwestern section. In the southern part, one major fault, the Kern Canyon Fault, has special geomorphic significance, which is discussed later.

Volcanic eruptions accompanied the faulting in certain areas, particularly in the boundary zone between the Sierra Nevada Province and the Basin and Range Province. East of Devils Postpile National Monument near the resort town of Mammoth Lakes, the dominant structure is a large (10 by 19 miles; 17 by 32 km) caldera, the Long Valley Caldera. Geologists of the U.S. Geological Survey determined that the caldera was formed about 700,000 years ago, by explosions that would dwarf Mt. St. Helens' 1980 paroxysms (Kerr, 1983).

The Long Valley area has particular significance because the intense earthquake activity that began in 1978 is continuing today. According to Ryall and Ryall (1982), many geophysicists are convinced that magma about 5 miles (8 km) beneath the surface is moving slowly upward. The caldera floor rose 8 inches (20 cm) from 1975 to 1982, and steam vents developed above the magma chamber.

The Survey's 1981 report indicates that, based on the frequency of past activity, an eruption will probably take place somewhere in this Long Valley area within the next 50 years. Perhaps Mt. St. Helens may have to share its publicity with a full-blown Long Valley Volcano — conceivably in our time.

The recent faulting along the eastern boundary of the Sierra Nevada and recent faulting in the Basin and Range strongly suggest a direct tectonic relationship between the two provinces. Apparently this large section of the earth's crust has been stretching, with many of the blocks being rotated (tilted), a common occurrence in the Basin and Range Province. Possibly the massive rigid Sierra Nevada block maintained its integrity while the lesser rocks farther east were fragmented, forming smaller mountain ranges.

Glaciers: Today and Yesterday

The glaciers of today are small, yet they were large enough to intrigue John Muir into studying them and eventually other, much larger ones in Alaska. Muir was the first to contend that Yosemite Valley was shaped by a glacier, at a time when State Geologist Whitney incorrectly interpreted it as a *graben,* a downfaulted section of the earth's crust.

FIGURE 7-2

Yosemite Valley. Rock floor of valley is about 1600 feet (488 m) below stream level (see Figure 7-9).

FIGURE 7-3

Glacial polish and erratics in Center Basin, looking east to Mt. Bradley (13,289 feet) in southeastern part of Kings Canyon National Park. (Photo by Anthony Morse.)

According to the Geological Survey's latest inventory, there are about 80 glaciers in the Sierra. They are all cirque glaciers that were reborn during the Little Ice Age, according to Matthes (1930), and are located on the "cool" (east or northeast) side of the divides at elevations of 10,500 feet (3201 m) and greater.

When glaciers were at their maximum, an ice cap 20 to 30 miles (32 to 48 km) wide covered the higher mountains from just south of Mt. Whitney north for about 270 miles (435 km) (Wahrhaftig and Birman, 1965). Valley glaciers extended out from the ice cap and relandscaped the valleys, deepening and widening them. Yosemite Valley is one of many large glacial gorges in the Sierra. There were several advances but it was the Wisconsin glaciers that finished the reshaping of the mountains.

The problem of the length of the Pleistocene Epoch has plagued geologists for a long time. The original estimate of a million years was made more than a century ago, before any of the modern methods of dating were developed. Although we have long recognized the fact that the Pleistocene began considerably earlier than one million years ago, specifically how much earlier remained a mystery until recently. In 1966, Curry made a significant discovery, one that required a distinct revision of the Geologic Time Scale. On Deadman Pass, above Mammoth Lakes, California, Curry found glacial till sandwiched between two layers of volcanic ash. By radiometric dating of the two ash beds, he determined that the glacial materials are approximately 3 million years old, the oldest Pleistocene glacial deposits which have been dated.[1]

---

[1] The glacial origin and the Pleistocene age of these deposits have been questioned by Dr. N. K. Huber of the U. S. Geological Survey (personal communication, September 10, 1979).

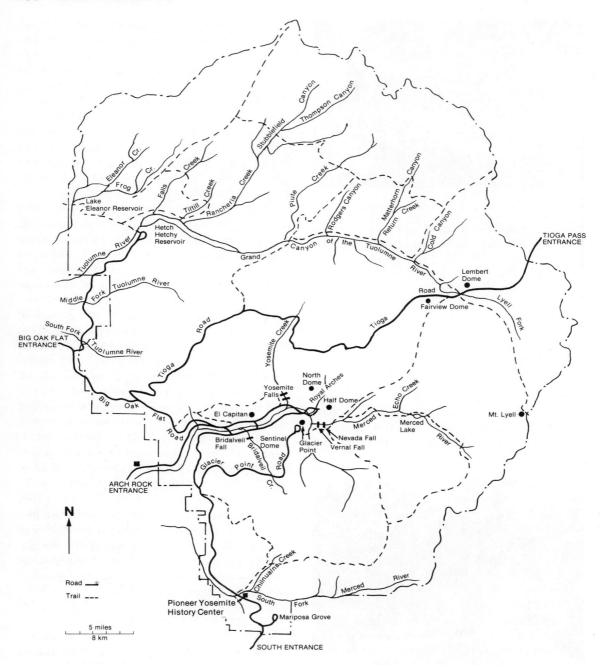

FIGURE 7-4

Map of Yosemite National Park.

FIGURE 7-5
President Theodore Roosevelt (left) and
John Muir at Glacier Point, May 1903.
(Photo by National Park Service.)

YOSEMITE NATIONAL PARK

Yosemite was established as a national park in 1890. It now consists of approximately 1200 square miles of the mid-section of the Sierra, extending westward from the crest of the range. Yosemite was John Muir's. He tramped over much of it; he climbed its mountains; he climbed out under Yosemite Falls and felt its power; he knew the animals, the birds, the trees and the flowers; he knew the glaciers and how Yosemite was born. He knew what environment was about. He did not merely glance at it and talk knowingly about it; he lived it — for months at a time — alone. He was much disturbed to see the wanton destruction by timbermen and sheepmen, especially in Yosemite Valley. Around the turn of the century there was no effective control within park areas (the National Park Service was not established until 1916). Muir was alone at first in his defense of Yosemite; he made speeches and wrote many articles appealing for support, and one who responded was President Theodore Roosevelt. The President went to Yosemite and camped out with Muir, and the two climbed up and stood together on Glacier Point, high above the Big Valley, where they discussed their mutual interest — conservation. As a result of their long talks, Yosemite was expanded to its present boundaries, and soon thereafter the Park Service was established to protect all of the park areas.

Yosemite Valley has prompted many to the use of superlatives. Francois Matthes says of Yosemite that "no other valley holds within so small a compass so astounding a wealth of striking and distinctive scenic features," and that "El Capitan is the most majestic cliff in the Yosemite and perhaps in the world." Yosemite Valley was the "Incomparable Valley" to him and it was his intention to describe it in a book for the use of the general public. Dr. Matthes, geologist with the United States Geological Survey, retired after 51 years of service; he died the following year and his book was not complete. His friend and co-worker, Fritiof Fryxell, undertook to complete it as editor, using Matthes' Professional Paper 160 and his other writings; with Mrs. Matthes' assistance, he published a delightful book in 1956. The title is *The Incomparable Valley: A Geologic Interpretation of the Yosemite;* its remarkable photographs and drawings alone would make it an outstanding book. Of Dr. Matthes' many contributions in glacial geology and glaciology, the greatest is his concept of "the Little Ice Age," discussed in the Introduction. The latest reactivation of glaciers, which began about the time of Matthes' death, fits readily into his concept of glaciers and their behavior.

Many other worthwhile general-interest books on the Sierra have appeared; the Sunset book, *National Parks of the West* (1970), is one. It includes striking photographs, both in color and black-and-white, some by Ansel Adams, David Muench, and Forrest Jackson. For those

FIGURE 7-6

Wire rope ''trail'' up to the top of Half Dome in Yosemite National Park. (Photo by E. Kiver.)

readers who know John Muir, *The Yosemite* need not be mentioned; for those who do not know his writings, this book, published in 1912, is the best introduction.

The High Sierra section of the park is a ''must''; some visitors, carried away by the unbelievable in the Incomparable Valley, forget that there is more. See the High Sierra first, and then fit the Valley into it, as one piece of a gigantic jigsaw puzzle. Go around and up to Glacier Point; from here you see the High Sierra in the distance, Half Dome nearby, Vernal and Nevada falls below you, and across the valley the big one — Yosemite Falls. Then go over to the railing and look almost straight down to the floor of the valley more than 3000 feet (910 m) below you. After a long look, go down (by road!) into the valley and look up to Matthes' ''most majestic'' cliff, El Capitan, to Yosemite Falls, Glacier Point, and Half Dome.

Then, when you must go, leave by the south entrance, visit the Pioneer History Center, see the Grizzly Giant sequoia in the Mariposa Grove. But do not plan to drive through the Wawona Tunnel Tree, perhaps the most photographed tree in the park. This old tree, its heart ruthlessly cut out in 1881, was blown down during a severe winter storm in 1969.

Yosemite has problems. So many people visit the park that it is almost impossible for anyone to see and enjoy what is there. Traffic became so congested that the Park Service was forced to establish a new policy of phasing out the use of private cars in the main valley. Visitors are encouraged to use the shuttle buses that are available in the more crowded sections of the park. Such restrictions are, of course, infringements on the constitutional right of Americans to drive their cars where and when they like! On the other hand, Yosemite must be reclaimed and its wounds healed.

Yosemite's Landscapes

The huge granite domes, the deep gorges, and the other features that make Yosemite famous are young geologically; they were shaped by glacial action, weathering, mass-wasting, and stream erosion. Even so, the conformation of these features was predetermined long ago — when the rocks were formed and later when the mountain block was uplifted. The great rounded domes of Yosemite could not have been developed in the rocks at Mount Rainier or in Glacier National Park. Massive rocks with widely spaced joints are essential.

The Sierra Nevada batholith was formed several miles below the surface; therefore, with slow cooling of the magma, the crystals formed are large and the rocks are coarse-grained. A few are true granites; others are quartz diorites, granodiorites, and diorites. (Here, the general term *granite* will be used except where it is necessary to be more specific.)

Most of the granites are equigranular, but porphyritic rocks are fairly common. In many of the porphyries — rocks in which large crystals (phenocrysts) are embedded in a finer-grained groundmass — the phenocrysts are in some cases as much as 3 to 4 inches long. On weathered surfaces, the feldspar phenocrysts stand out in relief and are handy for climbers, assuming that the phenocrysts are still firmly embedded in the rock.

The first interpretation of the granite domes was that a thick ice sheet had moved across the entire range and rounded off the sharp peaks. Matthes, however, found that Half Dome, El Capitan, Sentinel Dome, and many others had been surrounded but not entirely covered by glaciers; hence, they were nunataks and their rounded tops had been shaped in another way. They are *exfoliation* domes formed by the loosening and spalling-off of concentric layers

FIGURE 7-7

Porphyritic granite — large feldspar crystals embedded in coarse-grained groundmass.

of rock, similar to the shelling-off of the layers of an onion. The granites, as they were being formed at depth under enormous pressure, stored up potential energy; when pressures were lessened by unloading (removal by erosion), the rock expanded and formed concentric fractures spaced a few feet to several feet apart. Mainly by the expansion of ice in the fractures, the concentric layers were pried loose and the domes, as we see them today, were formed.

There are a number of domes, however, which are distinctly elongate and asymmetrical. In the park, the steep ends are on the west, and their gradual east slopes are polished and striated. These domes — such as Lembert Dome and Fairview Dome — were overridden and reshaped by the ice. The up-glacier end was ground down by abrasion and the down-glacier end was quarried away, leaving a steep, stair-stepped cliff. Thus, the generally symmetrical domes are the result of exfoliation, and the distinctly asymmetrical domes — *roches moutonnées* — were formed by glacial abrasion and quarrying.

Arches are among Yosemite's remarkable features. They are formed on the steep sides of exfoliation domes by the spalling-off of large sections of an exfoliation sheet. Almost invariably, the roof section forms a graceful inset arch. In some cases smaller arches are recessed back of the main arch. The Royal Arches in upper Yosemite Valley, as classic examples, stand a thousand feet above the base. Here, Matthews (1968) surmises that the glacier removed the lower sections of the exfoliation shells, leaving the upper sections without support. (Just in case, these arches are not the "see-through" type of arch such as those in Arches National Park.)

The main valley was carved by the Merced Glacier. A V-shaped stream-cut valley at the beginning of the Pleistocene, it is now a deep U-shaped valley with near-vertical walls in places, as at El Capitan, which rises abruptly 3000 feet. Here a large, essentially unjointed monolith of granite is responsible for the high cliff. This sheer wall presents a very real challenge to professional climbers; persevering subprofessional climbers are likely to become professional here. Regardless, they all soon develop a profound respect for El Capitan.

Actually, the valley is not U-shaped but ⊔-shaped, with its flat floor extending almost to the base of the cliffs. The glaciated rock floor of the valley is as much as 2000 feet (610 m) below the present floor, as Gutenberg and others (1956) determined by geophysical methods. Therefore, the valley cut by the glaciers is almost a mile deep, but a significant part of it is now below the valley floor. This deep glacial basin, scooped out of bedrock, later impounded water—Lake Yosemite; still

later the lake was completely filled with outwash materials.

Striations high on valley walls have been used by some as reference points in measuring the thickness of the ice. But this method is not valid because the position of the bottom of the glacier (and of the valley) at the time the striations were made remains unknown. What is now known is that glaciated valleys were not cut during a single advance but during perhaps a dozen or more advances, each cutting deeper than the one before.

Half Dome is a mountain of moods, depending on the lighting; it is thus intriguing to photographers and artists. It is also intriguing geologically. Was it a complete dome sheared off by a passing glacier, as was once thought?

FIGURE 7-8
Lembert Dome, a roche moutonnée; glacier rode over it from right to left.

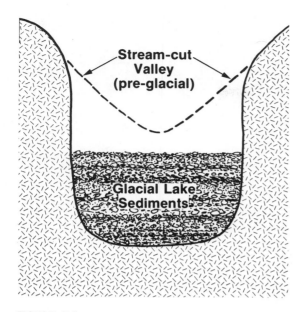

FIGURE 7-9

Generalized cross section of Yosemite Valley, showing stream-cut valley (dashed lines) and thick lakebed deposits.

Surely it was once a fairly complete dome, but inasmuch as the top of the glacier was 700 feet below the top of the dome, as Matthes determined, that early theory is untenable. However, the glacier did have a role in the deepening of the valley and in the steepening of its side at this point, thus rendering the rock face unstable. Collapse along well-defined vertical fracture planes exposed the existing sheer wall.

Waterfalls, large and small, are abundant in Yosemite. Those along the sides of main valleys fall out of hanging valleys — tributary valleys that lacked glaciers large enough to downcut as rapidly as the main glacier. Upper Yosemite Falls deserve first place in the discussion, because they are unusual in more than one respect. First, they are "free-leaping"— that is, the water is in free fall from top to bottom. In addition, they are 1430 feet (436 m)

high, nine times the height of Niagara Falls, and are among the highest of the free-leaping falls in the world. After cascading down another 675 feet (206 m), the water plunges over the Lower Falls another 320 feet (98 m). On the opposite side of the canyon across from El Capitan, is Bridalveil Falls. Also a free-leaping falls from a hanging valley, Bridalveil is but 620 feet (189 m) high; nevertheless it is one of the truly beautiful features in the park. During most of the summer the falls are relatively small, and up-valley breezes cause the filmy

FIGURE 7-10

El Capitan — supreme challenge for rock climbers.

FIGURE 7-12

Bridaveil Falls; here Bridalveil Creek pours out of a hanging valley. (Photo by R. S. Creely.)

FIGURE 7-13

Erratic boulder, moved by a glacier, high up on Pothole Dome. (Photo by C. K. Harris.)

green garden; hence the name. Upstream about one-half mile, Nevada Falls, 594 feet (181 m) high, is a twisting turmoil of white water; therefore the name Nevada is appropriate.

Glacial deposits — end, lateral, and recessional moraines — and outwash deposits are found in many parts of the park. (For details of the glacial deposits and the different glacial stages they represent, Matthes' Professional Paper 160 [1930] is recommended.) Several frontal moraines may be inspected in roadcuts in the valley, one at El Capitan Bridge. Glacial erratics, large boulders transported and deposited singly by the glaciers, can be found in many places. Particularly striking are the erratics that were transported and deposited high on the roches moutonnées where they rest on polished abraded surfaces, as in a number of places along the Tioga Road.

Today's Glaciers

Of the 80 glaciers in the Sierra, only three are in Yosemite National Park. None of the 80 is more than a half-mile in length, puny indeed compared to the huge ice streams that fashioned the Sierra Nevada. In fact, they were thought to be merely stagnant icefields by

spray to float veil-like back and forth along the wall. Frequently, rainbow colors are vivid — an impressive sight. Many of the falls are at their best during the first half of the summer while stream flow is substantial. Vernal Falls and Nevada Falls, on the Merced River, are reliable the year round. These and similar falls mark points where the stream crosses major vertical joints, below which the Merced Glacier quarried out large blocks of granite. The result is a stair-step valley profile. Vernal Falls, 317 feet (97 m) high, spray a large area at the base — a some of the early day geologists. It was John Muir who contended that they were true glaciers, and to prove his point he made measurements of the rate of movement of one of them, the Maclure Glacier. Lyell Glacier near the top of 13,114-foot Mount Lyell is about a half-mile long and a mile across, the largest one in the park. These glaciers are young; they were born during the Little Ice Age.

John Muir said, "No temple made with hands can compare with Yosemite."

SEQUOIA-KINGS CANYON NATIONAL PARKS

Sequoia and Kings Canyon national parks, an area of about 1314 square miles, are located in the southern part of the Sierra Nevada. The two parks have a common boundary and are administered together; therefore, they are discussed in large part as one.[2]

In most of Sequoia-Kings Canyon the rocks are very similar to those in Yosemite; consequently many of the landforms are similar. In contrast, however, metamorphic rocks that are insignificant in Yosemite are present in a number of places in Sequoia-Kings Canyon, particularly in the western area of Sequoia. Here, among other metamorphics, are several large exposures of marble, one of the less common rocks in the park system.

The "Big Trees," *Sequoia gigantea,* are prominent attractions in both sections, in Sequoia's Giant Forest and in Kings Canyon's General Grant and Redwood Mountain groves. The largest tree in the world, General Sherman Tree in the Giant Forest, is believed to be more than 3000 years old.

Landscapes of Sequoia-Kings Canyon

As is readily observed from photographs and maps, the High Sierra are generally more

---

[2] Recommended references are Matthes, 1950, 1965; Sunset, 1970; Bateman and Wahrhaftig, 1966.

rugged here than in Yosemite, particularly along the Great Western and Kings-Kern divides. Serrated ridges, horns, and cirques are prominent features; rounded exfoliation domes are, however, both numerous and well-developed. Examples cited by Matthews (1968) are Beetle Rock, Moro Rock, Alta Peak, and Tehipite Dome. There are well-developed roches moutonnées but apparently they are not as numerous here as in Yosemite.

One feature, somewhat unusual in most areas, is very common, particularly in Sequoia. These are avalanche chutes — smooth, clean-swept troughs that seem to stand essentially on end along steep mountain sides. In places, as in the vicinity of Mt. Whitney, Matthes (1956) reports that the chutes are so closely spaced as to give a fluted appearance.

One of the more striking geomorphic features in the Sequoia-Kings Canyon area is the anomalous drainage pattern in the southern part. Unlike all other large streams on the west side of the divide, Kern River flows almost due south, parallel to the crest, instead of westward down the regional slope. Moreover, Kern River Canyon is much straighter than most canyons. Here the reflection of structure in the topography is obvious; Kern River has developed its canyon along one of the major faults, the Kern Canyon Fault, within the Sierra Nevada fault-block. Kern River is therefore an excellent example of a subsequent stream.

Mt. Whitney, unlike lower peaks nearby, has an essentially flat top. It is not a remnant of an exfoliation dome; it is not a surface smoothed off by a glacier. It is, according to Matthes (1956), a tiny remnant of the Whitney erosion surface. Formed at a low elevation, it is now the highest point in conterminous United States. Need there be more convincing evidence that we are living on a restless earth?

Although Sequoia is famous for features that stand out in bold relief, there are also fascinating features underground, namely caverns. Most caverns are formed in limestone but these are in the marble referred to earlier. These large masses of marble take us back to

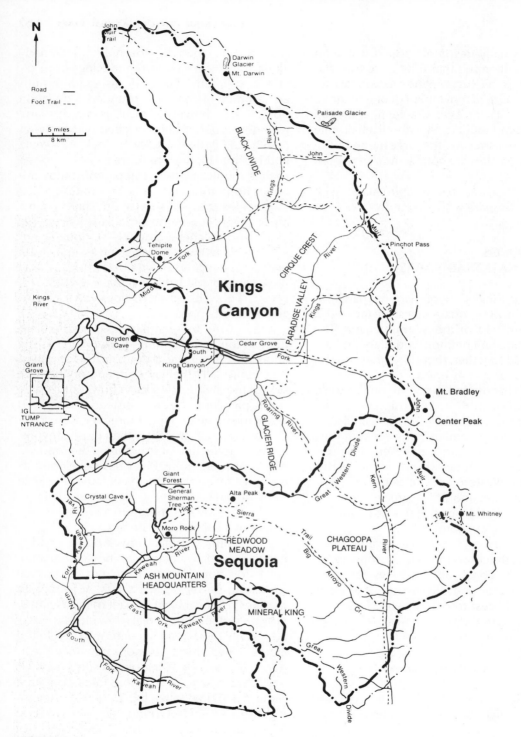

N

Road —
Foot Trail ---

5 miles
8 km

John Muir Trail

Darwin Glacier
Mt. Darwin

Palisade Glacier

BLACK DIVIDE

Kings River

John

Tehipite Dome

Middle Fork

Kings Canyon

CIRQUE CREST

PARADISE VALLEY

Kings River

Muir Trail

Pinchot Pass

Kings River

Boyden Cave

South Fork

Cedar Grove

Kings Canyon

Grant Grove

IG TUMP NTRANCE

GLACIER RIDGE

Roaring River

Mt. Bradley

John

Center Peak

Kaweah River

Crystal Cave

Giant Forest
General Sherman Tree

Alta Peak

Great Western Divide

Kern River

Muir Trail

Moro Rock

Sierra High

REDWOOD MEADOW

Sequoia

Trail

CHAGOOPA PLATEAU

Big Arroyo

Cr.

Kern River

Mt. Whitney

North Fork

ASH MOUNTAIN HEADQUARTERS

East Fork Kaweah River

MINERAL KING

South Fork

Kaweah River

Great

Western

Divide

FIGURE 7-14

Map of Sequoia-Kings Canyon National Parks.

FIGURE 7-15

Center Peak (12,760 feet), with the peaks along King-Kern Divide in the distance. (Photo by Anthony Morse.)

FIGURE 7-16

Ice pinnacles (Nieve Penitentè) in Mt. Whitney area. (Photo by Peter Barth.)

FIGURE 7-17

Dollar Lake with granite domes and horns, from Muir Trail south of Pinchot Pass, Kings Canyon. (Photo by Peter Barth.)

events early in the geologic sequence outlined for the Sierra. Until recently, the age of the limestone which later became marble was not known because of the lack of diagnostic fossils. However, in 1973 Jones and Moore reported the discovery of an Early Jurassic ammonite from Boyden Cave near the park. Although some restraint is suggested here, it is probable that this is the same stratigraphic unit as the one in which the caves in Sequoia were formed. The limestone was recrystallized mainly by the heat from the batholithic magmas that intruded in Jura-Cretaceous time. Much later, after the overlying rocks were eroded off, the caverns began to form, by solution of the calcite in the marble. Enlargement of the solution openings eventually resulted in large caverns such as Clough Cave, Palmer Cave, Paradise Cave, and Crystal Cave, all in the Kaweah Basin in Sequoia National Park.

Glaciers of Today

The work of yesterday's glaciers is well documented in Sequoia-Kings Canyon. Today's glaciers are difficult to find. Darwin Glacier, nestled in a protective cirque on the north side of Mt. Darwin, is located along the divide at the north end of Kings Canyon. Along that same divide but just outside the park there are cirque glaciers; one, the Palisade Glacier, appears to be more than a mile wide. These glaciers are on the northeast side of the divide, generally out of reach of the afternoon sun.

DEVILS POSTPILE NATIONAL MONUMENT

Devils Postpile, established in 1911, is one of the smaller of the park areas. It is located almost in the shadow of Yosemite National Park, far below on the east side. It is reached at the end of a long, dusty road across pumice slopes, above Mammoth Lakes, California. This road tends to discourage some people; consequently, Devils Postpile is not overcrowded and is therefore a delightful place to visit. (At least, this was true in 1971!)

Devils Postpile was established as a national monument primarily because it contains classic examples of geologic features, such as columnar jointing in a lava flow (Figure 7-18). As is readily seen in the picture, the columns are near-perfect.

But how is Devils Postpile related to the Sierra Nevada, a range of fault-block mountains? Early geologic events in the Sierra Nevada are not directly related to the formation of Devils Postpile; however, when faulting began late in the Cenozoic, magmas were formed at depth and intruded into the fault zone along the east side of the Sierra block. Extrusions occurred in the Mono Lake area, at Mammoth Mountain and Devils Postpile, among others. About 630,000 years ago, according to Huber and Rinehart (1967), an andesite (now determined to be basalt) lava flow buried the area that is now Devils Postpile. The flow came to rest and the lava cooled and solidified without further movement within the lava. As the cooling progressed, vertical cracks were developed because of the horizontal shrinkage in volume. In most cases the cracks were so arranged as to form 6-sided columns, in others 4-, 5-, or 7-sided, which extended from bottom to top of the flow. The columns in Devils Postpile are mainly from 2 to 3 feet in diameter and 50 to 60 feet long.

In the late Pleistocene, during the Wisconsin advance, a glacier moved across the area, removed by abrasion the tops of the columns, and polished, grooved, and striated their upper ends. The glacier also deepened the valley through the lava flow — the valley now occupied by the Middle Fork of the San Joaquin River. Since the glacier melted away, many of the columns have fallen over, forming the large talus at the base.

Also in the monument, near the south end, are Rainbow Falls, 140 feet (43 m) high, where the Middle Fork of San Joaquin River plunges over a precipice. The white water against the black cliffs presents a striking contrast, enhanced by a rainbow during the middle of the day.

For long-distance hikers, Devils Postpile is one of the stops along the Muir Trail, which extends from Yosemite down to Sequoia National Park. Although small and somewhat isolated, Devils Postpile reveals significant geologic events not recorded in any other park.

FIGURE 7-18
Devils Postpile.

FIGURE 7-19
Polished and striated tops of posts, Devils Postpile.

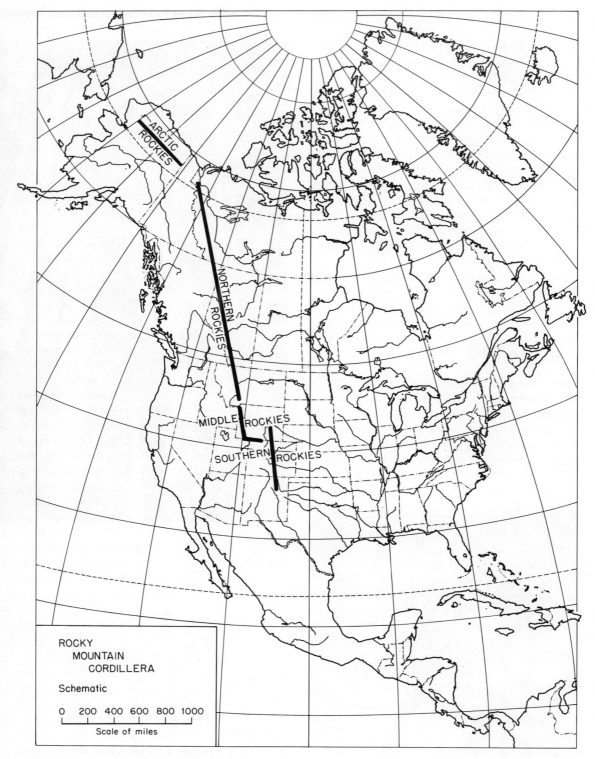

ARCTIC
ROCKIES

NORTHERN
ROCKIES

MIDDLE ROCKIES

SOUTHERN ROCKIES

ROCKY
MOUNTAIN
CORDILLERA

Schematic

0   200 400 600 800 1000
Scale of miles

PLATE 8   Rocky Mountain Cordillera.

CHAPTER 8

ROCKY MOUNTAIN CORDILLERA; NORTHERN ROCKY MOUNTAIN PROVINCE

THE ROCKY MOUNTAINS

The Rocky Mountains, as outlined by Thornbury (1965), consist of the mountain ranges that extend northward from New Mexico through Canada and westward across northern Alaska. The general sequence of geologic events is much the same throughout, but the more recent events are sufficiently different to justify the recognition of three geomorphic provinces.

The Southern Rockies extend from north-central New Mexico northward through Colorado into southern Wyoming. The Middle Rockies, an odd-shaped province, includes the Yellowstone Plateau, the nearby Beartooth and Absaroka mountains, and the Bighorn, Teton, Wind River, Wasatch and Uinta ranges. As here defined, the Northern Rockies extend northward from Yellowstone Park through western Canada and then westward across northern Alaska. When more work has been completed in northwestern Canada and in Alaska, this elongate province may deserve to be divided into two or more provinces. On a tentative basis, the "Arctic Rockies" of Alaska are designated here as a separate section of the Northern Rockies Province.

The Building of the Rockies

The material that follows pertains only to the Rocky Mountains in conterminous United States. The Arctic Rockies in Alaska involve a more complex sequence; therefore, they are discussed in a separate section.

All geologic events beginning with the early Precambrian have contributed in one way or another to the conformation of the Rocky Mountains; however, Precambrian history is fragmentary, and only certain of the fragments can be related with confidence to the building of the Rockies. Directly related and therefore a logical starting point is the development of a geosyncline in mid-Mesozoic time. The Rocky Mountain Geosyncline, which extended from the Gulf of Mexico area to Alaska, began to take shape during the Jurassic and became a continuous seaway early in the Cretaceous. It included not only the area that is now the Rocky Mountains but also the areas adjacent on both the east and the west. Sediments — clays, silts, sand, and gravel — were carried in by streams from land masses on both sides of the geosyncline. Limestones were laid down at certain times, by chemical and biochemical precipitation. With continued deposition the geosyncline began to sink, thus providing space for additional sediment. Deposition was more rapid in some parts of the geosyncline; there, thicker sequences of sedimentary rocks were laid down. Deposition continued through much of the Cretaceous period, and in parts of the geosyncline as much as 30,000 feet (9146 m) of sedimentary rocks were laid down. Thus the stage was set for mountain-building.

Mountain-building began with the folding and faulting of the rocks in the geosyncline. This deformation, the Laramide Orogeny, began near the end of the Cretaceous. As compressional forces continued to be applied, deformation became more widespread and intense. Tighter folds developed and displacement along thrust faults increased. The deformed sections were forced upward, thus forming the mountains.

As land appeared above the sea, streams began to erode the newly formed mountains; erosion continued after uplift ceased, near the end of the Eocene epoch. Thus the Laramide Orogeny continued over a long period of time, many millions of years. Near the end of the Laramide, magmas formed near the base of the earth's crust and, mainly by the melting of the roof rocks, intruded toward the surface in various places. Large intrusives, mainly granitic in composition, formed batholiths and stocks, and small intrusives formed sills, dikes, and *laccoliths*. Erosion continued both during and after the intrusions were emplaced, and by the end of the Eocene the mountains were essentially destroyed and a surface of low relief was developed, a surface referred to by some as a peneplain.

The removal of the vast amount of rock material lightened the load and set the stage for *isostatic adjustment* to begin. Uplift of mountains took place at intervals during the Oligocene and Miocene. Magmas formed once again, and in places volcanoes and lava flows developed on the surface. The uplift steepened stream gradients and initiated a new erosion cycle, and near the end of the Pliocene epoch a second erosion surface, a pediplain,[1] was formed, leaving remnants of the older surface standing, bench-like, high above it. Again, for the third time, a major uplift occurred, lifting the mountains to their present position; again a new erosion cycle was initiated — the cycle that is in progress today.

Stream erosion was interrupted in the higher mountains when, several times during the Pleistocene, glaciers became widespread. Early Pleistocene glaciers were generally broad lobate tongues or lobes, according to Richmond (1965), but later, after distinct canyons were cut, valley glaciers became the dominant

type; in the northern section certain glaciers extended out onto the edge of the plains.

The general sequence given above is greatly simplified and condensed; not all of the igneous activity was confined to the times specified, and the uplifts in various areas were probably not perfectly synchronized. The sequence, suggested originally by Lee (1923) and Bevan (1925), has been modified in places because of radiometric dates that are now available. Although there is not complete agreement as to the number and age of the erosion surfaces, the sequence as outlined will serve as a general framework for our discussion of each of the three geomorphic provinces. For more detailed treatment, see King (1959) and Thornbury (1965).

NORTHERN ROCKY MOUNTAIN PROVINCE

The mountains of western Montana and Idaho and the northeastern corner of Washington make up the Northern Rockies, as the term is generally used. (The Arctic Rockies are discussed later in this chapter.) It includes a number of individual mountain ranges, each with its own rock types and structures; consequently, each has its own assemblage of landforms. The Rocky Mountain Trench, a *graben* about 60 miles west of the eastern front of the mountains, separates the front ranges from those to the west. Much of the area west of the trench is occupied by the Idaho batholith, some 16,000 square miles in extent, mainly in west-central Idaho. In this area of homogeneous rocks, mainly quartz monzonites and closely related granitic rocks, there is little of the north-south lineation that is pronounced in the remainder of the province; instead, this is a dissected mountain area in which dendritic drainage patterns are dominant (Thornbury, 1965). The front ranges, such as the Lewis and Livingston ranges, were formed mainly by thrust faulting. The Lewis Thrust is a major structure, recognized to be at least 135 miles

---

[1] A pediplain is generally similar to a peneplain, but pediplains are developed under dry rather than humid conditions.

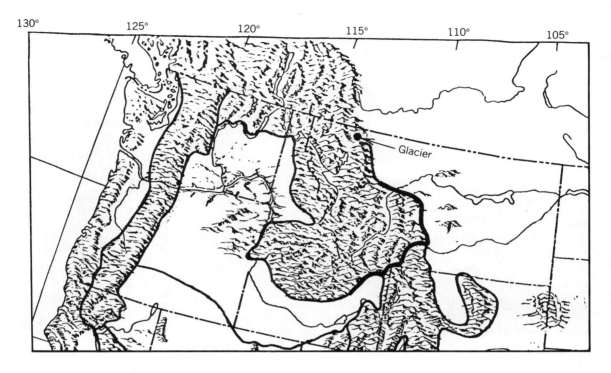

Glacier

● National Park

○ National Monument

▲ National Seashore/Lakeshore

PLATE 8A

Northern Rocky Mountain Province.

long, in Montana and Alberta, Canada. Proterozoic rocks were thrust perhaps 40 miles eastward and up on top of Cretaceous rocks (Ross, 1959).

Overthrusting was until recently the accepted interpretation. Now, however, as Alt and Hyndman (1973) point out, some geologists reject overthrusting in favor of gravitational sliding as the mechanism. They picture a huge slab of the earth's crust sliding gradually down a fault plane toward the east; later, the area was uplifted and tilted toward the west. They argue that overthrusting would have crumpled and broken this thin layer of rocks. But when we consider that, as shown in Figure 8-2, the slab that was faulted was much thicker than now, their argument appears less convincing. Those interested in the controversy are referred to Mudge (1970), Price (1971), and Drake and Maxwell (1981).

Later work by Mudge and Earhart (1980) indicates that the Lewis Thrust may be as much

FIGURE 8-1

Lewis Thrust Fault at the base of a steep cliff of Precambrian rocks; Cretaceous shales lie beneath the fault. (Photo by E. Kiver.)

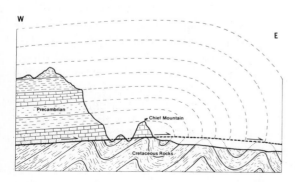

FIGURE 8-2

Generalized cross section of Northern Rockies in the Glacier National Park area.

as 280 miles (452 km) long. Also, it has been suggested that there was a collision of plates that initiated the Laramide Orogeny. There is no consensus, however, and at the moment we are not sufficiently arrogant to predict when or how the controversy will be settled. Attempting to keep such things in perspective, we think it entirely possible that the outcome may not be of major consequence to many who are admiring the scenery in Glacier National Park.

GLACIER NATIONAL PARK

Glacier National Park and Canada's Waterton Lakes National Park are adjacent, and together they form the International Peace Park, a natural monument to friendship between neighbors. About 200 miles farther north are Banff and Jasper national parks where jagged mountains, glaciers and glacial features are on display. Those planning a trip to Glacier should include Canada's parks and sufficient time to see them the right way; they are truly outstanding.

Glacier National Park, established in 1910, contains 1600 square miles of Montana's best mountain and glacial scenery. Gigantic cirques and deep glacial troughs, as well as the name Glacier National Park, suggest that large glaciers are there. Unfortunately, this is not the case, for although there are about 50 glaciers, they are all relatively small. Fortunately, the magnificent glacial features are mainly in full view instead of being buried beneath glacial ice.

Although the glacial features are immediately apparent, geologic structures much older than the Pleistocene must not be ignored. Outstanding in this category is the Lewis Thrust Fault without which there would have been no mountains to be glaciated.

Astride the Continental Divide, the park sheds its waters in three directions: those on the west side flow into the Pacific; those on the east flow into either the Gulf of Mexico or Hudson Bay. Triple Divide Peak in the southern section of the park has the distinction of providing water to all three.

The mountains, although rugged and spectacular, are not exceptionally high. Mt. Cleveland (10,468 feet; 3192 m), near the Canadian boundary, is the highest; many of the peaks along the divide are 9000 to 9500 feet (2744 to 2899 m) high.

Two main highways cross the mountains, offering panoramic views for those who must see the park on wheels. Going-to-the-Sun Road extends from West Glacier past Lake McDonald, over Logan Pass (6646 feet) and down to St. Mary Lake on the eastern side; U.S. Highway 2, the other main highway, skirts around the southern end of the park over Marias Pass (5216 feet). For hikers, the park provides excellent trails into remote areas. For information regarding trails, roads, accommodations and geologic features, George Ruhle's *Guide to Glacier National Park* is a good reference. Of specific interest also is Alt and Hyndman's recent book *Rocks, Ice and Water: The Geology of Waterton-Glacier Park.*

Glacier's Rocks

The rocks in the mountains are Proterozoic (Late Precambrian) in age; the rocks under the mountains and along the eastern base are Cretaceous. (See Figure 8-2.) Several thick *argillites* and limestone formations compose the Proterozoic series. The Grinnell Formation shines out above all others because of its brilliant red color. It is an argillite, an exceptionally well-indurated shale, and as such it weathers into slabs that accumulate as conspicuous taluses easily seen from miles away. The red color is due to hematite, present in small but sufficient amounts to give the Grinnell its striking appearance. A second red formation is the Kintla, but as it is much younger than the Grinnell, it forms the tops of mountain peaks and is not likely to be mistaken for the Grinnell.

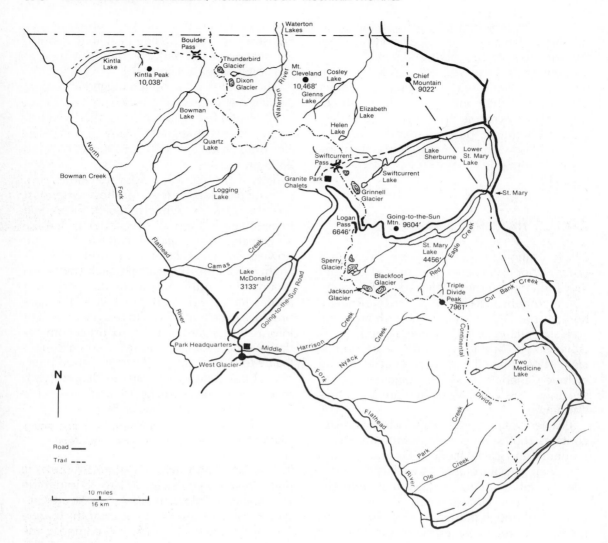

FIGURE 8-3

Map of Glacier National Park.

Overlying the Grinnell is the 4000-foot-thick Siyeh Limestone which forms many high cliffs, including the famous Garden Wall (Figure 8-4) that rises above Grinnell Glacier.

An eye-catching feature on this drab wall is a conspicuous dark band extending from one side to the other; below and above the dark band are narrower bands of white. The dark-colored band is a sill formed by the cooling of magma that intruded along bedding planes in the Siyeh Limestone. Clearly it is a sill and not a lava flow that was later buried, because the limestone both above and below it was metamorphosed to marble, the white bands. The sill averages about 100 feet (30 m) in thickness and is mainly coarse-grained. Dyson (1963) refers to the rock as diorite, but as Alt and Hyndman suggested in 1973, diabase is a better general designation.

The Siyeh Limestone contains fossil algae, colonies of algae that lived in the Proterozoic sea. Excellent exposures are readily seen along the Going-to-the-Sun Road about 25 miles east of West Glacier. Another limestone formation, the Altyn, is the oldest rock in the park. Some of the waterfalls plunge over ledges of the Altyn; Trick Falls near Two Medicine Lake hold particular interest. You may visit Trick Falls a second time and observe that the falls are more than twice as high as they were when you first saw them. They are at full height when streamflow is strong; when it is not, the water is seen pouring out of a cavern in the limestone, an opening that is hidden during periods of high flow.

Igneous rocks are not widespread in the park. The diabase sill which intruded along bedding planes in the Siyeh Limestone has already been mentioned. Dikes that cut across bedding planes are also present but are generally less conspicuous than the sill. The reason is that in places the dike rock weathers more rapidly than the adjacent rocks, thus forming recesses and little valleys that are snow-filled much of the time. Dyson (1962) points out that some of the dikes have small copper deposits

FIGURE 8-4

Garden Wall, showing thick diabase sill (arrows).

FIGURE 8-5

Proterozoic colonial algae in Siyeh Limestone along Going-to-the-Sun Road, Glacier National Park.

associated with them and that there was a momentary mining boom around the turn of the century. However, none of the deposits was commercial; otherwise, the mountains would have been marred by the scars inherent in mining operations.

In several places, basalt lava flows are exposed, some of which contain pillow structures, indicating that some of the lavas poured out into the sea. Lavas are found in the Boulder Pass area, near Swiftcurrent Pass and in Granite Park. (If a geologist had done the naming, it would be Basalt Park rather than Granite Park!)

Glacier's Mountains

Chief Mountain, a spire that stands alone east of the mountain front, is a "mountain without roots." Unlike most mountains, which have as their "roots" the oldest rocks at the base, Chief Mountain's oldest rocks lie above much younger rocks. Proterozoic rocks at least a billion years old form the mountaintop, and Cre-

FIGURE 8-7

Chief Mountain, about 5 miles (8 km) east of mountain front. (Photo by M. E. McCallum.)

taceous rocks, a mere 100 million years old, are underneath, separated by the Lewis Thrust. Chief Mountain's roots are somewhere beneath Flathead Valley, some 30 miles (48 km) to the west. Chief Mountain serves as a reminder that the mountain front was at one time many miles to the east. It was eroded back to its present position, and Chief Mountain was left as a remnant, or *klippe*, an outlier of rocks surrounded by the younger rocks that lie below the fault.

About 60 million years ago, after the Laramide Orogeny was well under way, thrust faulting began to take place here on a grand scale. The low-angle Lewis Thrust, dipping toward the west, brought deep-seated rocks high up in the mountains. The Siyeh Limestone in which the colonial algae were fossilized was, prior to faulting, deep beneath what is now the Flathead Valley; after the faulting and much erosion, the Siyeh Limestone is exposed high in the mountains.

Glacier's Glaciers

Glaciers developed along the Continental Divide first as ice caps with the lobes extending down the slopes; later, they were confined to the canyons that were cut during the interglacials. It was at this time that the deep troughs characteristic of the park began to form. Lakes now occupy many of the troughs and some, like Lake McDonald and St. Mary Lake, are many miles long. Dyson (1962) notes that at one time Two Medicine Glacier extended out beyond the canyon mouth and spread out in the form of a piedmont glacier.

Hanging valleys with waterfalls tumbling from them can be seen along several of the main valleys. The trunk glaciers deepened their valleys more rapidly than the smaller tributary glaciers; when the glaciers melted away, the tributary valleys were left "hanging" high above the main valley floor. Grinnell and Virginia falls are two that pour out of hanging valleys.

Large cirques developed at the head of the valleys, and with prolonged cirque enlargement, horns were formed. From one vantage point, Kinnerly Peak in the northwestern corner of the park closely resembles the famous Matterhorn in the Swiss Alps. Also formed by glacial erosion—abrasion and quarrying—elongate roches moutonnées, large and small, are abundant. On these and other abraded surfaces, grooves and striae can be observed in many places.

Glacial deposits, abundant in and at the ends of the valleys, clearly represent a number of advances. Several techniques that have been used to determine the relative ages of the deposits are topographic location, degree and depth of weathering, and carbon-14 dating.

Frost action also left its unmistakable imprint in the park, in the form of *stone stripes* and *stone rings*. Flat slabs weathered from the bedrock are lifted and tilted by frost heaving. Long stripes or bands of bare rock slabs extend down the steeper slopes; circular patterns are also formed, mainly on flatter slopes. These stone rings are also called *rock polygons* when the "ring" has straight sides.

The large glaciers that left their deep imprints on the park area could not survive the heat of the Thermal Maximum. Today's glaciers formed later when the climate cooled, sometime during the Little Ice Age or Neoglacial, therefore during the last 4000 years. Glacial advances, minor compared to those of the Wisconsin, were separated by retreats as the climate fluctuated. When Glacier National Park was established, the glaciers were more numerous and slightly larger than at present. They were retreating even at that time, and a number disappeared completely. However, Dyson (1962) made measurements of several of the glaciers and observed a slowing of the retreat during the late 1940s, and by 1950 the fronts were essentially stationary; in 1951 Grinnell Glacier and certain others began a feeble advance. In 1972, Alt and Hyndman reported that an essentially stable situation obtained. The effects of the cooling trend are not as pronounced here as in the Pacific Northwest and Alaska, where precipitation is much greater. Nevertheless, if the cooling trend continues for an extended period, even these glaciers will begin to advance.

Although today's glaciers are a bit less impressive than some elsewhere, the opportunities for climbing around on a live glacier are good—even better if under the guidance of an experienced person such as the park naturalist. Crevasses are common on most active glaciers, certainly on Grinnell and Sperry, two of the larger and more frequently visited glaciers in the park. At times, crevasses are bridged over by snow too thin to support a person's weight. The depth of crevasses varies but is never more than 200 feet; at about that depth the ice is rendered plastic by the weight of the overlying ice.

Glacier National Park is time-consuming, so numerous are the colorful and fascinating features. Allow plenty of time, perhaps guided by John Muir's suggestion to "Give a month at least to this precious preserve."

Perhaps John Muir is fortunate; were he

FIGURE 8-8

St. Mary Lake, impounded behind thick outwash sediments.

alive today, he would be saddened by developments to the west and north of the park — yet close enough to affect profoundly the environment of the park itself. As reported by Gene Albert in the November, 1975 issue of *National Parks and Conservation Magazine,* the recent developments of an oil field west of the park and of Canada's coal field north of the border have already damaged the North Fork of Flathead River, an excellent trout stream. And many of the grizzly bears, elk, and wolves have been driven away.

Visitors to Glacier National Park now have an aid available to them; in 1983, Raup and his co-workers published a book in which they described and explained the geologic features along the Going-to-the-Sun Road across the park. Watch for changes in the names of certain formations, such as Helena instead of Siyeh Limestone.

THE ARCTIC ROCKIES (BROOKS RANGE)

The informal name "Arctic Rockies" was suggested to the senior author by Gil Mull (personal communication, 1983) as an appropriate designation for the series of mountain ranges collectively known as the Brooks Range, named for early geologist Alfred H. Brooks.

FIGURE 8-9

Arrigetch Peaks, Gates of the Arctic National Park. (Photo by R. Belous, National Park Service.)

These ranges, beginning on the west, are the DeLong, Baird, Waring, Schwatka, Endicott, and Philip Smith mountains; and, to the east, the Romanzof, British, and Davidson mountains near the Canadian border—all north of the Arctic Circle. Although they vary considerably in height and in configuration—from subdued to highly exciting—Mull (1977) sees a common thread that ties these mountains together, namely, geologic history and evolution.

When compared to the Alaska Range or the St. Elias, the Arctic Rockies (Brooks Range) come up short, with the highest peaks only slightly more than 9000 feet (2727 m) high; even so, they have the distinction of being the highest mountains north of the Arctic Circle. They are high enough to have been intensely glaciated, with horns, cirques, moraines, and glacial lakes of all kinds well displayed. But don't expect to find any large glaciers there now; with relatively low snowfall the existing environment is not favorable for glaciers. There are many small glaciers hiding in sheltered sites (Post and Meier, 1979) but at least most of them appear to be slowly wasting away.

The Arctic Rockies are composed in large part of Paleozoic sedimentary rocks—many of which have been hardened by metamorphism—and largely unmetamorphosed Mesozoic sedimentary rocks (Mull, 1977). There are also a few small granitic batholiths which intruded into the core of the range during the Devonian; when the mountains were uplifted and eroded much later, these hard granites formed extravagantly spectacular scenery, such as the Arrigetch Peaks and Mt. Igikpak in the Schwatka Mountains.

Apparently, the Laramide Orogeny began at some time during the Cretaceous. The highly complex sequence—in which the plates and microplates, some from distant points including even from the ocean floor, were telescoped and uplifted into mountains—is presented by Mull (1979). It seems appropriate to

mention that the progress that these Alaskan geologists have made in "putting the pieces together"—in this vast and remote area—has earned the respect of many.

The Rockies of Canada's Yukon Territory are not as well documented as those farther south, in British Columbia and Alberta, where thrust faulting was the primary mountain-building process. In this latter area Canada has a number of outstanding national parks, including Jasper, Banff, and Waterton Lakes. In the Arctic Rockies there are two national parks, Gates of the Arctic and Kobuk Valley.

GATES OF THE ARCTIC NATIONAL PARK AND PRESERVE

The Gates of the Arctic, set aside in 1978, became a national park and preserve in 1980. It consists of about 8.0 million acres of the Endicott and Schwatka mountains, in the central section of the Arctic Rockies.

This was the land that Bob Marshall explored and mapped during the 1930s. He named many of the mountains and other features, often using Eskimo names. Mt. Doonerak (Devil Mountain to the Eskimos) he first named the Matterhorn of the Koyukuk, suggesting major problems for the climber. On the west side of the North Fork of the Koyukuk River, Frigid Crags rear up abruptly above the valley. Directly across the valley, Boreal Mountain forms the other of his "Gates of the Arctic." Marshall's book *Alaska Wilderness* is a remarkable account of his experiences—also explaining the arts of mushing, geepoling and siwashing—using the town of Wiseman as his taking-off point. His earlier haunts in the Flathead and Lewis and Clark National Forests are now known as the Bob Marshall Wilderness Area.[2]

[2] See Marshall, 1970.

The Arctic Continental Divide enters Gates of the Arctic National Park near Atigun Pass and rides the ridge westward across the Endicott Mountains, then southward out of the park. Rivers such as the Killik, Chandler, and Anaktuvuk are tributaries of the Colville, which flows down the north slope to the Arctic Ocean. The Koyukuk and other rivers on the south side of the divide are tributaries of the mighty Yukon.

There is no lack of wild scenery anywhere in the park but it is at its best in the Schwatkas, in the southwestern part, where Mt. Igikpak rises to 8570 feet (2600 m). Nearby, the Arrigetch Peaks (Figure 8-9) leave everyone breathless, especially the rock climbers who can't resist the challenge.

Weather variations are extreme, and dramatic changes may take place at a dangerously rapid pace. As Mull (1977) points out, long, warm summer days may change overnight, bringing very low temperatures and even blizzards. It is not a place for the unprepared or the careless.

The climate, particularly the short summers, limits vegetative growth, which in turn has an impact on certain of the animals. Grizzly bears are there, but they are distinctly smaller than those farther south. Although the precipitation is low, many of the tundra areas are wet because of widespread permafrost. Caribou herds migrate through Anaktuvuk Pass in the eastern part of the park and north of one of the important winter ranges. Moose, Dall sheep, and a few wolves and wolverines are there but the most numerous and vicious animals are the mosquitoes! Unlike in many other places, here the first killing frost is most welcome.

The National Wild and Scenic Rivers are appropriately represented here in this wilderness park. In addition to the John River, the Alatna, and the North Fork of the Koyukuk, there is the Noatak River, which heads in the park and flows westward across the Noatak National Preserve, adjacent on the west.

The geology of the area is extremely complex, and although geologic work continues apace, particularly by the U.S. Geological Survey, certain detailed problems remain unresolved. Those who feel compelled to get deeply involved are referred to Chapman et al. (1964), Tailleur and Brosge (1970), Mull (1976, 1979), and Roeder and Mull (1978). These and our references indicate that considerable progress has been made since A. H. Brooks wrote his *Geography and Geology of Alaska* in 1906.

The consensus is that most of the rocks were not formed here, but were deposited in sites well to the south; and as the plates were thrust northward, they were telescoped together, metamorphosed, and uplifted to form the Endicott Mountains, early in the Cretaceous. During the same orogenic period the Arctic Alaska Plate was moving southward and thrusting its way underneath the Brooks Range. In time, magmas generated deep in the subduction zone, forming batholiths and stocks that uplifted the Schwatka Mountains (Mull, personal communication, 1983). As in other mountain ranges in Alaska, without plate tectonics this collage of rocks would have forever remained a mystery.

Glaciers began to relandscape the Gates of the Arctic 2 to 3 million years ago. There were several major advances separated by recessions during interglacial periods. Naturally, the latest major advance was responsible for essentially all of the present scenery, having obliterated most of the earlier glacial features.

Cirque glaciers formed first, high in the mountains, and pushed down the valleys and out on the flats, where they built up end moraines along their fronts and lateral moraines along the sides. When the glaciers melted away, perhaps 8500 years ago, lakes formed in the basins hemmed in by the moraial deposits; Walker Lake is an excellent example (Figure 8-10).

In the park, the handiwork of the glaciers is best displayed in the Arrigetch Peaks. In the hard granites of the Schwatkas, when sharp

FIGURE 8-10

Walker Lake, one of many glacial lakes in the Arctic Rockies. (Photo by G. Arehart.)

FIGURE 8-11

Mt. Doonerak, a jagged horn along the North Fork of the Koyukuk River. (Photo by J. Kaufman, National Park Service.)

horns and knife ridges are formed they last "forever." In the eastern part of the park, Mt. Doonerak is a prominent landmark—an isolated horn carved out of Paleozoic volcanic rocks.

KOBUK VALLEY NATIONAL PARK

Kobuk Valley National Park, near the western end of the Arctic Rockies, is about 50 miles (83 km) west of the Gates of the Arctic. The park occupies a 50-mile-long segment of the broad Kobuk Valley, the southern flank of the Baird Mountains to the north, and the Waring Mountains on the south. This 1.7-million-acre wilderness was established as a national monument in 1978 and upgraded to park status in 1980. The Kobuk's main tributaries within the park flow out of the Baird Mountains; from east to west, they are the Akillik, Akiak, Tutuksuk, and the Salmon, the latter being one of Alaska's many Wild and Scenic Rivers—fine for canoeing and kayaking.

Kotzebue—on Kotzebue Sound about 75 miles (125 km) west of the park—is the usual taking-off point, by light plane, for Kobuk Valley. While in Kotzebue you may have the opportunity to visit the jade shop where the Eskimos cut and polish the beautiful jade mined in the Jade Mountains.

The park's wildlife includes the grizzly and black bear, lynx, wolves, moose, and caribou; there are grayling, arctic char, and several species of salmon in the streams.

The Baird and Waring mountains are westward extensions of the Arctic Rockies, and the geologic sequence is generally similar to that outlined for the province. They are considerably lower than the mountains farther east: the Baird Mountains along the north side of the park rise only to about 5000 feet (1515 m), and the Waring Mountains along the south side are even lower, about 2000 feet (606 m) high. Evidently they were not uplifted as high originally; also, the Paleozoic schists of the Baird Mountains and the Cretaceous shales and sandstones of the Waring Mountains are poorly resistant and were eroded down at a fairly rapid rate.

Glaciers were active in much of the area, and Kobuk Valley is floored with glacial deposits and glacial outwash materials. Deposits of Wisconsin age are most widespread, but deeply weathered pre-Wisconsin and essentially unweathered Little Ice Age deposits are also present (Fernald, 1964).

The silts and sands that underlie Kobuk Valley provide ideal conditions for the development of *permafrost*. This term is likely to be misleading; actually, it is frozen ground that is tens or even hundreds of meters thick. Permafrost probably began with the cooling of the climate early in the Pleistocene. As evidence, Flint (1971) cites the entombment of now-extinct mammals in the permafrost in Alaska and the Soviet Union. Since then, it has fluctuated in thickness with the fluctuations in climate.

With the warming of the climate since the latest Little Ice Age advance, the upper layer of permafrost has thawed, in some places more rapidly than in others; consequently, the surface is now pitted with depressions, some small and others more than a mile across. The meltwater cannot drain away because it is contained in shallow "ice-bowls"; the lakes thus formed are called *thaw lakes*. Although thaw lakes are common in many permafrost areas, within the park system they are especially well represented in Kobuk Valley.

Other striking features that have been developed recently in Kobuk Valley are sand dunes. Many people associate sand dunes, at least large dune areas, with hot desert regions such as the Sahara. Nevertheless, the Great Kobuk Sand Dunes are here, and some are as much as 100 feet (30 m) high. The dunes, both active and stabilized, cover about 25 square miles. This area—including the nearby Little Kobuk Dunes—lies only a short distance south of the river. Actually, the occurrence of dunes here is not anomalous; the two requi-

sites for dune accumulation are present: an abundance of unstabilized sand and strong prevailing winds. Loose sand is readily available, especially on the large sand bars in the river channel, and the polar easterlies are frequently strong during much of the year. According to Fernald (1964), the summer westerlies cause very little movement of the dunes. Dunes are not confined to Kobuk Valley; Pewe (1975) points out that they are fairly common in the northern part of Alaska.

Unlike many dunes that are composed mainly or entirely of quartz sand derived from nearby sandstone cliffs, the Kobuk dunes consist of sand derived by glacial abrasion from rocks of many kinds that outcrop in the surrounding mountains. Fernald has identified sand composed of quartz, feldspars, garnet, magnetite, pyroxenes, and several other dark-colored minerals. The glaciers mixed the mineral fragments from one area with those from other areas traversed; the streams flowing out from the glaciers rounded the fragments and deposited the sand in the channel and on the floodplain where it was picked up by the wind.

The finer materials, mainly silt-size particles, that were picked up by the wind were carried over a larger area and more gradually deposited as a blanket of varying thickness. Such wind-laid fine deposits, called *loess* (less), cover large sections in Kobuk and other valleys in northern Alaska. Elsewhere, in the Mississippi and Missouri river valleys for instance, soils formed on loess are some of the most productive agricultural soils in the United States.

Archeologists have a particular interest in northwestern Alaska because they are convinced that the first Americans lived here, after hiking across the Bering Land Bridge from Siberia. Just who they were, when they arrived, and where their first settlement was, remain a mystery—so the search goes on. Perhaps the original occupation site is in one of the newly established parklands; at least, that was the main reason for setting them aside—so that priceless archeological materials would not be destroyed.

In Kobuk Valley, near the eastern boundary of the park, Onion Portage may be the most significant of the sites found thus far. There, layer upon layer, are the records of continuous occupation over a very long period. [For an 80-page report on Onion Portage, see Anderson (1970).]

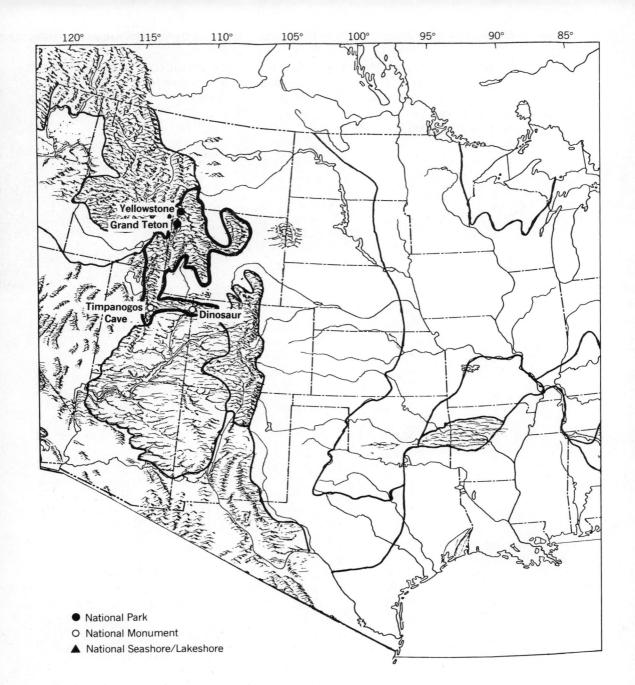

● National Park
O National Monument
▲ National Seashore/Lakeshore

PLATE 9

Middle Rocky Mountain Province.

CHAPTER 9

MIDDLE ROCKY MOUNTAIN PROVINCE

FIGURE 9-1

The Middle Rockies are unstable, as indicated by the fault scarp that appeared on the night of August 17, 1959, west of Yellowstone National Park.

The Middle Rocky Mountain Province is made up of the mountains and plateaus of western Wyoming and northern Utah. In addition to the various mountain ranges, there are two large plateaus — one is a volcanic area in Yellowstone National Park, the other is the Beartooth Plateau immediately northeast of the park.

In the Bighorn Mountains of northern Wyoming, in the Beartooth Plateau, in the Tetons south of Yellowstone, and in the Wind River Mountains to the southeast, the Precambrian basement complex is extensively exposed. The ranges of westernmost Wyoming and easternmost Utah and Idaho are the results of overthrusting of sedimentary beds. To the south, in northeastern Utah, the north-south Wasatch Range has been interpreted to be a large fault-block, generally similar to the Sierra Nevada. The Uinta Range, which extends eastward across northeastern Utah, is a large asymmetrical anticline.

The general pre-Tertiary sequence outlined for the Rocky Mountains is recorded in the Middle Rockies; however, certain of the later events — such as the volcanism in the Yellowstone region — were localized.

The erosion surfaces — peneplains and/or pediplains — are recorded by extensive remnants, particularly in the Bighorn Mountains, Beartooth Plateau, and the Uintas. As Thornbury (1965) points out, their precise origin and age have been variously interpreted; regardless, the general consensus is that these flattish upland areas are the results of long periods of erosion.

Pleistocene glaciers — valley glaciers and piedmont glaciers — relandscaped extensive areas within the province, as will be shown in the discussions of the four parks and monuments.

YELLOWSTONE NATIONAL PARK

Enthralled by the performances of Old Faithful and other geysers and by the beauty of Yellowstone Canyon and Yellowstone Falls, park visitors may leave without becoming aware of other attractions, less spectacular but actually more important in the geologic story of Yellowstone. The variety of geologic features is perhaps greater here than in any other park.

As outlined in the Introduction, Yellowstone was where it all began — where, around a campfire, the National Park concept was en-

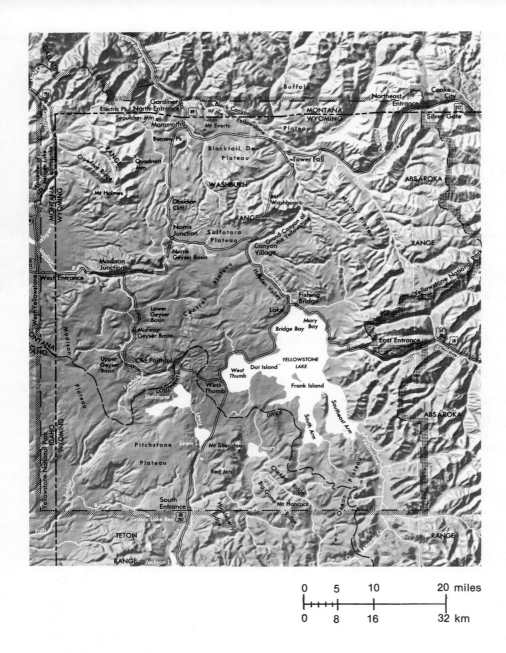

```
0     5     10        20 miles
├──┼──┼──┼──┤─────────┤
0     8     16        32 km
```

FIGURE 9-2

Map of Yellowstone Park, showing rivers, lakes, landforms, roads, town, settlements, and major geyser basins (stippled). The park includes 3472 square miles (2,221,770 acres); its boundaries traverse a distance of nearly 300 miles. (Reprinted with permission of the USGS.) Yellowstone Lake, with an irregular shoreline of 110 miles and surface area of 137 square miles, is one of the largest natural mountain lakes in the United States. (From U.S. Geological Survey Bulletin 1347 by W. R. Keefer. Reprinted with permission of the USGS.)

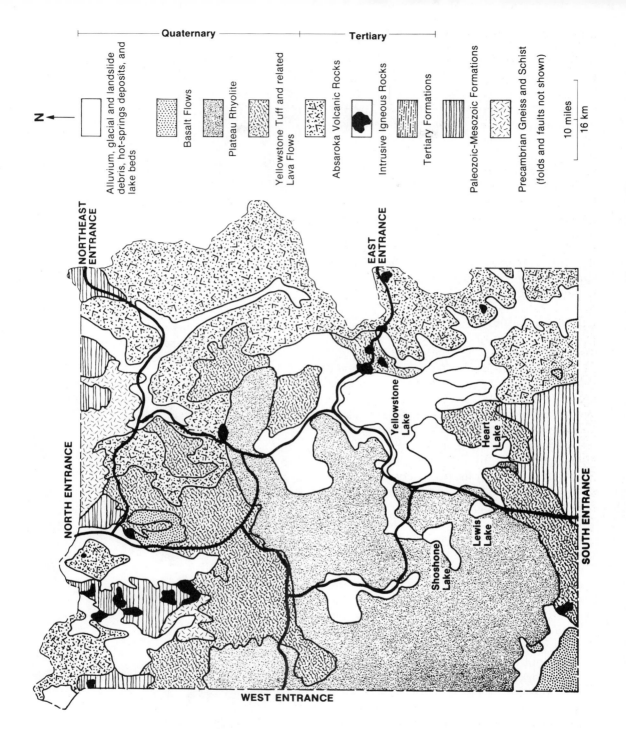

FIGURE 9-3

Geologic map. (Adapted from Keefer, Plate 1, U.S. Geological Survey Bulletin 1347.)

visioned. In 1872, about 3472 square miles of Wyoming, Idaho and Montana were officially set aside as our first national park.

The early years were hectic ones. The Yellowstone country had belonged to the Indians and, as should have been expected, there were periodic conflicts. More serious by far were the depredations by park visitors; they damaged the geysers and hot pools and the beautiful Hot Springs Terrace formations. Finally, in 1883, the Army was given responsibility for the administration of the park, and it was saved from complete destruction. Now it is seriously threatened again, this time by the large numbers of tourists — too many for any park to endure.

Geomorphic Setting

Most of the park is a broad, undulating volcanic plateau about 8000 feet above sea level. Near the eastern boundary there is an abrupt rise from the plateau to the Absarokas, a high range of volcanic mountains. The Gallatin Range, which extends northwestward out of the park, is actually part of the Northern Rockies Province; thus Yellowstone occupies parts of two geomorphic provinces. The Teton Range extends southward from the southwestern corner of the park, into Grand Teton National Park. There are several large lakes on the plateau; by far the largest is Yellowstone Lake in the southeastern part of the park. Yellowstone River flows northward into the lake and northward out of the lake down through the Grand Canyon of the Yellowstone and into the Missouri River. South of the Continental Divide, which crosses the southwestern part of the plateau, the Snake River flows southward into and out of Jackson Lake in Grand Teton National Park south of Yellowstone, eventually to flow into the Pacific.

The Yellowstone Earthquake

The Yellowstone area is tectonically active. In 1871, Hayden's survey party experienced

FIGURE 9-4

Fault scarp, which developed suddenly on the night of August 17, 1959. This building at Blarneystone Ranch was built across the fault.

FIGURE 9-5

The squeezing together of the alluvium near Hebgen Lake resulted in this serpentine section of fence, during the 1959 earthquake.

FIGURE 9-6

Landslide that dammed Madison River and formed Quake Lake in 1959; excavation (lower left) is the spillway, outlet of Quake Lake.

earth tremors, but no severe earthquakes occurred until 1959 when on August 17 a 7.5-magnitude earthquake rocked the region. The earthquake center was near the park boundary, about 12 miles (19 km) north of the town of West Yellowstone. Farther west, in Madison River Canyon, the quake triggered a huge landslide that buried a forest service campground and dammed the valley, forming Quake Lake.

Yellowstone's geysers became more active for a few years, and a new one, Seismic Geyser, was born. At first, its vent was small, but it was soon enlarged when it blasted large chunks of rock out of the vent (Keefer, 1971).

Earthquakes may be anticipated periodically, and without warning. One occurred in

1975 but, with a Richter scale reading of only 5.7, it was less damaging than the 1959 quake.

Geologic Sequence

The oldest rocks in the park are in the Gallatin Range, where small areas of early Precambrian metamorphic rocks form the backbone; also present here along the northern boundary are Paleozoic and Mesozoic sedimentaries. The oldest volcanic rocks are the andesitic volcanic breccias and basalt lava flows of Eocene age (Keefer, 1971). These Absaroka Volcanics make up the Absaroka Range and probably lie beneath the younger flows of the plateau. During the Eocene this area was extensively forested. At least 27 times volcanic material rained down on and buried the trees (Dorf, 1960). Each time, new trees grew on the new volcanic surface; the evidence is there to see in the Lamar Valley in the northeastern part of the park. By Eocene time the trees were generally similar to those of today, and fossilized pines, redwoods, walnuts, elms, and about 100 other trees and shrubs have been found.

But not all of the Absaroka magmas broke the surface; instead, they intruded into pre-Tertiary sedimentary rocks and formed stocks, sills, dikes, and laccoliths. These intrusive igneous bodies are well exposed in the Gallatin Range in the northwestern section of the park.

Precisely what occurred within the park during the Oligocene and Miocene is not clear. If volcanic eruptions occurred in the park, as they did south of the park, the evidence has either been destroyed by erosion or buried by younger volcanics.

During the Pliocene, faulting took place and certain of the fault-blocks were uplifted. In this way, according to Keefer, the Gallatin Range and the Teton Range south of the park were formed.

But even at the beginning of the Pleistocene, the latest epoch, the area bore little resemblance to the Yellowstone of today. Lacking were the rhyolites and volcanic tuffs that cover the vast Yellowstone Plateau. Lacking also were the geysers, hot springs, and mudpots. And of course none of the glacial features were there. Clearly, colossal changes took place during the past few winks of geologic time.

First, magma forced its way up toward the surface, arching up and fracturing the overlying rocks. This magma was high in silica — a rhyolitic magma — in which large quantities of gases were trapped. Finally the eruption occurred, with violence far in excess of anything experienced by humans, as reconstructed by Keefer and others. Huge quantities of ash, pumice, and blocks of rock were hurled into the air. Propelled by hot, expanding gases, ash flows spread rapidly across the Yellowstone area, filling the valleys and smoothing the topography. The heat from the gases fused the material together, forming a "welded-tuff," the Yellowstone Tuff.

With the instantaneous ejection of many cubic miles of rock material, collapse of the roof formed a huge caldera. When the infall of the wall rocks ceased, the caldera had an area of about a thousand square miles! Subsequently, however, large volumes of rhyolite were periodically extruded into the caldera, filling it and spilling out over the rim in places.

There was also extrusion from fissures outside the caldera; according to Keefer, the Obsidian Cliff flow is one example. Some of the rock is rhyolite, but most of the flow cooled very rapidly and formed obsidian — black volcanic glass.

Although rhyolitic lavas were dominant, a few basalt flows are known. Near Tower Falls a buried basalt flow exhibits excellent columnar jointing, as exposed in the side of the canyon.

The above is merely a skeletal sequence; many more details are presented by Keefer (1971). Those for whom Yellowstone is their favorite park will want more — to learn, for example, that there is a "baby" caldera, perhaps only about 125,000 years old, within the main caldera. The West Thumb of Yellowstone Lake now fills the baby caldera.

With the cooling of the climate at the beginning of the Pleistocene, glaciers developed in the mountains, particularly in the high Absarokas. Valley glaciers moved downvalley and at times extended out onto the plateau where they coalesced to form large piedmont glaciers. The early Pleistocene history is buried beneath lava flows in the plateau area, but one pre-Wisconsin advance has been documented by Richmond (1969). He further reports that during the Wisconsin advance, glaciers originating in the Absarokas formed a piedmont glacier which moved across the low continental divide and down the Snake River Valley to Jackson Lake. We will pick up its trail at this point when Teton National Park is discussed. Along the way, the glacier deposited its materials, including erratics such as the one

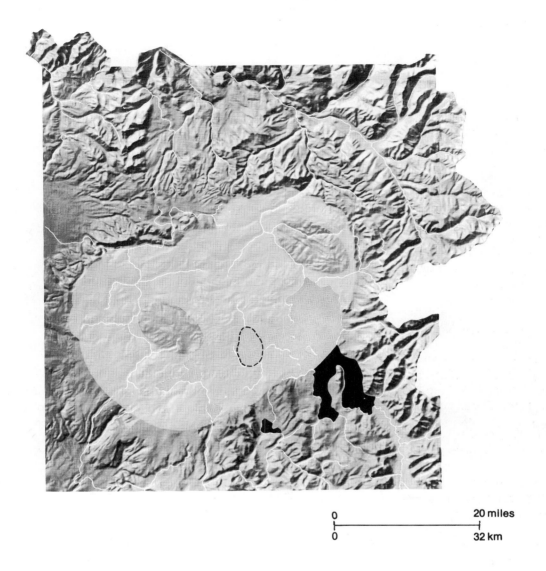

```
0                                    20 miles
|----------------------------------------|
0                                    32 km
```

FIGURE 9-7

Outline of Yellowstone Caldera produced by an enormous volcanic eruption 600,000 years ago. The two oval-shaped areas are resurgent domes that arched the caldera floor over twin magma chambers after the eruption. The margins of the resurgent domes are surrounded by ring-fracture zones which extend outward toward the edge of the caldera. Numerous fractures in these zones provided escape routes through which lava of the Plateau rhyolite oozed to the surface and poured out across the caldera floor. Today these zones also provide underground channels for the circulation of hot water in the Yellowstone thermal system. The area outlined by the dotted line shows the smaller and younger inner caldera now occupied by the West Thumb of Yellowstone Lake. (Based on information supplied by R. L. Christiansen and H. R. Blank, Jr. The existence of a caldera in Yellowstone National Park was first recognized by F. R. Boyd in the late 1950s.) (From U.S. Geological Survey Bulletin 1347, by W. R. Keefer. Reprinted by permission of the USGS.)

FIGURE 9-9

Erratic boulder, about 24 feet (8 m) long and weighing more than 500 tons, near Inspiration Point on the north rim of Yellowstone's Grand Canyon.

FIGURE 9-8

Obsidian Cliff. Part of the flow exhibits poorly developed columnar jointing.

near the rim of Yellowstone Canyon (Figure 9-9). The waning stages of this advance are too complex to detail here, but glacial deposits modify the outline of Yellowstone Lake basin, basically a lava-bound basin.

During the Thermal Maximum the glaciers disappeared; Neoglacial glaciers were confined to the higher mountains, mainly the Absarokas (Richmond, 1969). Thus we see that although Yellowstone is noted mainly for its geysers, glaciers left their imprint over large areas of the park.

The Grand Canyon of the Yellowstone

Yellowstone Canyon is a youthful canyon that heads about 15 miles downstream from Yellowstone Lake. Whether the lake drained northward by this route immediately after the Lamar Glacier melted out of the northern area is questionable. In any event, Yellowstone River cut rapidly down through thick rhyolite flows and tuffs that had been highly altered by hydrothermal solutions. The "Yellow Stone" is yellow, hydrothermally altered rhyolite. Other hues — red, pink, orange and brown — add variety and beauty.

The head of the canyon is marked by the 109-foot-high Upper Falls; the Lower Falls, 308 feet high, have been photographed by millions. Without question, a picture of the multi-colored canyon with the foaming white of the Lower Falls in the background is likely to be a prized possession.

The canyon is more than a quarter of a mile deep and about 20 miles long. The falls originated where the rhyolite was more resistant — less altered — than downstream; however, they are slowly eroding back upstream. Given time, they will reach the lake and it will be drained, leaving this part of the huge caldera exposed for the first time.

Geysers, Hot Springs, and Mud Volcanoes

Old Faithful is but one of a large number of geysers, but it is, as the name suggests, one of the most reliable. Whereas the other geysers erupt infrequently and at their own convenience, Old Faithful can be counted on to go off again within an hour and a half, and generally within about an hour. To date, the interval has varied from about 35 to about 95 minutes. The question "What makes it do that?" is often heard as the steam is clearing away. Actually, the system is not extremely complex, and is detailed by Keefer (1971) in his Yellowstone booklet. Obviously, heat beneath the surface is one essential; water in an underground "plumbing system" is the other necessary component. Consequently, geysers are confined to areas of recent igneous activity.

Geysers are many and varied in Yellowstone, each with its own habits that are related to peculiarities in its plumbing system. The geyser basins are located in the western part of the park, from south of Old Faithful north to Norris Junction. Grand, Grotto, Riverside, Castle, and Lone Star are among those which are likely to perform for the visitor who spends several hours in the geyser area.

As indicated above, most geysers are unpredictable. After a long period of inactivity a geyser may perform in a spectacular fashion. On March 27, 1978, Steamboat Geyser blew its stack, sending water and steam more than 300 feet (91 m) into the air, reportedly higher than Old Faithful's best effort. Those lucky enough to be in Norris Geyser Basin witnessed a play that they will long remember. Prior to this eruption, Steamboat had been quiet for several years; two weeks later, it was resting peacefully once again.

Siliceous sinter or *geyserite* is deposited around geysers and hot springs. Silica is dissolved from the rhyolite down below and gradually deposited in the form of rings and mounds around the vent. By this process the "pipes" may be so enlarged as to reduce the violence of the eruptions; eventually, the water merely overflows as a hot spring. Morning Glory Pool, shaped like a giant morning-glory and filled with blue water, is a thing of great beauty.

The Hot Springs Terrace area at Mammoth Hot Springs near the north entrance is a "must" at Yellowstone. These huge and beautiful terraces rise step-like high above the valley. They are composed of *travertine* or calcareous tufa rather than siliceous sinter. Here, the hot waters are charged with calcium bicarbonate which is deposited as calcium carbonate in the form of travertine. Many of the terraces are beautifully colored, apparently by different colored algae; when the hot waters shift to another place the abandoned section soon turns white. Liberty Cap is a striking feature in this travertine area. It stands high, like a 37-foot drinking fountain. The hot spring that formed it is now dead, but evidently it maintained a small flow that deposited all of its dissolved material around the pipe rather than out beyond the base.

Mudpots and mud volcanoes are common features in the hydrothermally active areas of the park. Here, the vents are filled with mud that is kept active by steam and other gases. Blisters are constantly built up by the rising gases, and as they break endlessly, the "bloop—bloop—bloop" is entrancing. Certain of the mudpots are periodically choked with mud until the gases concentrate sufficiently to

FIGURE 9-11

Old Faithful Geyser, hallmark of the National Park System, inspired those who gave birth to the National Park concept more than a century ago. Photo by B. R. McClelland

FIGURE 9-12

Tavertine terraces building in Mammoth Hot Springs area of Yellowstone National Park. (Photo by E. Kiver.)

cause an explosion; in places small spatter cones are built up by these mud volcanoes. Over all these hangs a pall of sulfurous fumes — John Colter's "smell of brimstone."

As we go from geyser to geyser and then to the hot springs and the mud volcanoes, it becomes increasingly clear that it is *hot* in the basement. What is less than clear is what geologists have been intrigued with for a long time. Is the Yellowstone caldera resting uneasily on top of live magma? Might we see another caldera come into being in our time, or perhaps at least a volcano erupting?

Yes, Yellowstone is full of the weird, the wonderful, and the unpredictable. It had to be so to inspire the concept of National Parks.

GRAND TETON NATIONAL PARK

Established in 1929, Grand Teton National Park was confined to the major peaks of the Teton Range; in 1950 it was enlarged to include Jackson Hole to the east. It is still a small park, 485 square miles, and on the map appears dwarfed by mighty Yellowstone almost adjacent on the north. Love and Reed (1968) come to its defense by stating flatly that whereas other mountain ranges are "longer, wider and higher, but few can rival the breath-taking Alpine grandeur of the eastern front of the Tetons." Anyone standing on the shore of little Jenny Lake looking up to the high peaks is likely to agree. The Tetons are "mountains without foothills"; they rise almost vertically from the floor of Jackson Hole. Love and Reed's *Creation of the Teton Landscape* (2nd edition, 1971) is recommended to those interested in Grand Teton National Park. With excellent photographs, including aerial views and block-diagrams, it is a good book for both the layman and the geologist.

The Tetons

Grand Teton rises 13,770 feet above sea level; the rocks in the Tetons are ancient metamorphics and igneous rocks that were formed deep in the earth's crust. Clearly, large uplifts took place in order for these deep-seated rocks to be where they are now. It is also clear that erosion, almost unbelievable erosion, had to take place in order to first expose the ancient rocks and then to dissect them deeply to form the jagged peaks of today. In the days before Hutton's Uniformitarian Principle was accepted, the "catastrophists" attributed mountain ranges to "colossal upheavals" which took place overnight. Such explanations are perhaps more exciting than the one now generally accepted, which follows. (The sequence as outlined in detail for the Tetons by Love and Reed (1971) is summarized here without including the supporting evidence they submit.) The Laramide Orogeny resulted in mountains in the Teton area, but they were very different from the Tetons of today. Instead of being a north-south range, these mountains trended northwest-southeast. They were anticlinal in structure, whereas today's Tetons are fault-block mountains.

During early Tertiary time the original mountains were destroyed by erosion. Thrust faulting then lifted the Teton block several thousand feet and was followed by normal faulting along the east side. The great Teton Fault lifted the Tetons still higher and de-

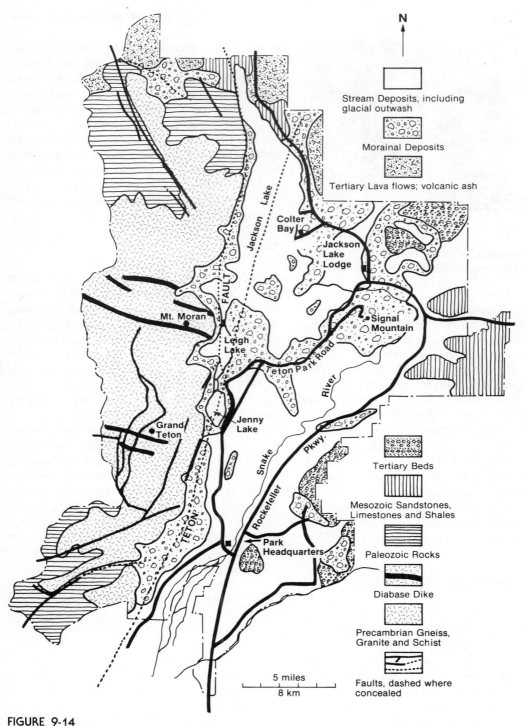

N

Stream Deposits, including
glacial outwash

Morainal Deposits

Tertiary Lava flows; volcanic ash

Colter
Bay

Jackson
Lake
Lodge

Jackson Lake

FAULT

Signal
Mountain

Mt. Moran

Leigh
Lake

Teton Park Road

River

Grand
Teton

Jenny Lake

Snake

Rockefeller

Pkwy.

Tertiary Beds

Mesozoic Sandstones,
Limestones and Shales

Paleozoic Rocks

Diabase Dike

Precambrian Gneiss,
Granite and Schist

Faults, dashed where
concealed

TETON

Park
Headquarters

5 miles

8 km

FIGURE 9-14

Generalized geologic map. (Adapted from Love and Reed, 1971, U.S. Geological Survey and Grand Teton
Natural History Association.)

pressed the Jackson Hole block located on the east. Faulting continued periodically, tilting the Jackson Hole block more steeply westward. Great thicknesses of Tertiary and younger deposits filled the depressed basin.

During the Tertiary, volcanoes erupted, mainly in the surrounding areas — in Yellowstone and in the Absaroka Mountains. Some of the Miocene and Pliocene lavas and pyroclastics were laid down in the northern section of the park.

At the beginning of the Pleistocene, the Tetons still looked very different from those of today. In height and in ground plan they were much the same as now, but they were much less jagged and spectacular. It was glacial action that enhanced the beauty and created the excitement of the Tetons.

Thus we see the Tetons as a mountain range that began to take shape in about mid-Tertiary time. Late Tertiary faulting along the east side was the process that gives the range its distinctive character, much the same as in the Sierra Nevada. The displacement on the Teton Fault, like that of the Sierra Nevada, is large; total vertical movement has been determined to be at least 20,000 feet. The mountains are still rising, and the Jackson Hole block is being tilted still more today. As a result of the continued tilting of the valley area, flood-control engineers have a real battle keeping the Snake River from shifting westward to a lower route at the base of the mountains (Love and Reed, 1971).

Love and Reed (1971) picture a small fault scarp that offsets an alluvial fan at the base of Rockchuck Peak. Inasmuch as the scarp is essentially unmodified by erosion, it is clear that the faulting occurred within the past few hundred years. Additional evidence of recent tectonic adjustments along the Teton Fault came to light when, in 1983, submerged trees were found standing upright in Jenny Lake, at the foot of the Tetons. One tree, a 70-foot-tall Engelmann spruce, was determined to be about 600 years old. It is possible to imagine a large landslide, with trees upstanding, moving en masse down into the lake, or perhaps major faulting taking place, in order to explain this unusual situation. However, Dr. Dave Love of the U.S. Geological Survey urges caution in accepting any specific interpretation until all of the evidence can be considered, including the seemingly anomalous features of certain of the moraines that impound the water in Jenny Lake and other nearby morainal lakes (personal communication, November 1984). In time, we will have an answer to this intriguing problem.

The Precambrian of the Tetons

The Precambrian rocks make the Tetons what they are. These extremely resistant rocks, once they were finally exposed in the upraised mountain block, maintain the jagged features imposed on them by the erosion of water and ice. Lesser rocks would probably have been largely worn away by this time, forming a subdued topography.

The oldest rocks are banded gneisses and schists, both of which were metamorphosed under high temperatures and pressure, therefore at great depths. For some time, these rocks have been known to be older than the 2.5-billion-year-old granites that had intruded into the gneisses. Now, as a result of work by Reed and Zartman (1973), we know that some of the gneisses are as much as 2.8 billion years old.

FIGURE 9-15

Generalized west-east cross-section of the Tetons.

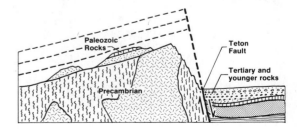

One of the more conspicuous features on the east side of the Tetons is a black band extending from the top to the base of Mt. Moran. This is a dike about 150 feet wide composed of diabase, a rock similar to basalt. Geologists recognized immediately that this dike is younger than the rocks that it cuts through, but how much younger was not known. There is an unconformity near the top of Mt. Moran; a small remnant of Cambrian sandstone rests on the unconformity, below which gneiss and the dike are clearly exposed. As seen in Figure 9-18, the dike does not cut through the sandstone and is therefore older than Cambrian. Now a more precise age has been assigned, based on radiometric dating. The dike intruded about 1.4 billion years ago (Reed and Zartman, 1973).

Teton's Glaciers: Past and Present

As in Glacier National Park, it is the scenery produced by glaciers, rather than the glaciers themselves, that Grand Teton is noted for. The glaciers are there, about a dozen according to Love and Reed (1971), but they are small and tucked away in re-entrants protected from the

FIGURE 9-16

Contorted gneiss in Cascade Canyon, Grand Teton National Park.

sun. Falling Ice Glacier on Mt. Moran is in a re-entrant where the black dike has been eroded back of the mountain face. Teton Glacier, about 3500 feet long in 1970, sits in the shade of Grand Teton. Although small, they are true glaciers and not merely icefields. Like those in Glacier National Park, these glaciers were born since the Thermal Maximum, during the Neoglacial.

In many areas, Wisconsin glaciers obliterated older glacial deposits and glacial features. In the Jackson Hole section of Grand Teton National Park, however, the evidence for a great ice advance about 200,000 years ago is unmistakable. According to Love and Reed (1971), ice from the Beartooth and Absaroka Mountains formed a large piedmont glacier that invaded Jackson Hole from the north. Along the way it was joined by tributary glaciers and pushed southward across Jackson Hole, even through Snake River Canyon, into Idaho. Much later, early in Wisconsin time, essentially the same thing occurred except that the glacier extended only about half as far.

Valley glaciers in the Teton Range were active during both advances of the large piedmont glacier, adding ice to the main mass at the base of the mountains. Then, late in the Wisconsin, there was another advance of the Teton glacier ice which, on reaching the flats, spread out and formed small piedmont glaciers. Rainbow-shaped end moraines mark the extent of the advance and impound the waters of Jenny Lake, Leigh Lake, and others.

Meltwater from the glacier carried vast quantities of sand and gravel out over the *outwash plain*. This area — called The Potholes — is pitted with small, steep-sided depressions. As you recall, potholes are depressions in bedrock formed by the swirling action of gravel-laden waters. The bedrock here is far below the bottoms of the depressions. "The Potholes" are actually *kettles* that were formed when large blocks of ice were partially buried by outwash material; when the ice melted, the depressions or kettles were left. A few can be

FIGURE 9-17

View west to Mt. Moran from top of Signal Mountain; Skillet Glacier with black dike located to the left of the skilled handle.

FIGURE 9-18

Close-up of east face of Mt. Moran and black dike. Note cap of sandstone of top of the dike. (Airview by John Shelton.)

FIGURE 9-19

Looking down over Disappointment Peak from top of Grand Teton to Jenny Lake (left) and Jackson Hole. (Photo by P. Vincent.)

FIGURE 9-20

Telephoto of Teton Glacier protected from the sun by Grand Teton. Note moraine of Little Ice Age advance (arrow), perhaps less than a thousand years old. (Photo by Nan H. Daniels.)

seen from the Teton Park Road south of Jackson Lake; all are visible from the top of Signal Mountain, an excellent viewpoint in the park.

There are many hiking trails that give the park visitor the opportunity to examine at close range the handiwork of the glaciers. Cascade Canyon Trail takes you up a classic U-shaped canyon, past Hidden Falls where the water tumbles out of a hanging valley, into the land of spires, horns, and aretes. Along the way, look back from a vantage point and admire little Jenny Lake and the artistically arcuate moraine that holds back the water. Then, be up on top of Signal Mountain a little before sunrise; the sun on the very tip of Grand Teton and then suddenly on the whole face of the Tetons makes this first-light safari worthwhile. And, if you are there in the fall, you will hear the "rusty-gate" bugling of the bull elk in the trees below you!

Grand Teton National Park is small. It has no spouting geysers and it has no Grand Canyon. But as Love and Reed have said, few places "can rival the breath-taking Alpine grandeur of the eastern front of the Tetons."

TIMPANOGOS CAVE NATIONAL MONUMENT

Timpanogos, a 250-acre national monument, was established in 1922. It is located in American Fork Canyon in the Wasatch Mountains southeast of Salt Lake City, Utah. The mile-and-a-half-long trail leads from the visitor center (5665 feet) up the side of the canyon to the cave entrance at 6730 feet above sea level. The cave is open only during the summer. Have a jacket in your pack; the weather in the cave is predictable — a cool 43°F (6°C) or perhaps slightly cooler.

Actually, three caves are involved but they are now connected by man-made tunnels. You go through Hansen Cave into Middle Cave and then into Timpanogos Cave, the largest of the three, investigating the Coral Gardens, Father Time's Jewel Box, the Cavern of Sleep and the Chimes Chamber, among others.

The geologic sequence is long and involved and only the essentials, mere glimpses, are given here. During the Paleozoic Era the area was both below and above the sea; during Mississippian time the marine Deseret and other limestones were deposited. For a long time the Deseret in which the caves were eventually formed remained deeply buried beneath younger sediments — until the end of the Mesozoic. Then, during the Laramide Orogeny, the area was uplifted and eroded. At times during the Tertiary there was additional uplift and faulting; the Wasatch Mountains, a fault-block, were uplifted on the east side of the Wasatch Fault.

Streams eroded deep canyons, exposing the Paleozoic rocks in their walls; in places, as in American Fork Canyon, they cut down into the Precambrian rocks below. You will see some of the Precambrian at the visitor center, before you start up the trail.

Once the river had cut into the Deseret formation, conditions were right for subsurface waters to begin dissolving out the limestone.

Solution took place along major fractures which were formed as the mountains were uplifted. Gradually the openings were enlarged and interconnected, thus forming the tunnels and hallways of the three caverns.

But chemical precipitation soon began to decorate the ceilings and the floors with *speleothems* of several types — stalactites, stalagmites and flowstone, in some places highly colored. (For a more complete discussion of cave-forming processes, refer to the Oregon Caves section of Chapter 4.)

The stalactites are growing longer, for Timpanogos is a "living cave." At some time in the distant future the canyon will be widened and these caverns will be no more. But take heart; likely some other caverns now unknown will probably be exposed. Geologic processes go on forever.

DINOSAUR NATIONAL MONUMENT

Here in the Uinta country, the Park Service salutes the mighty dinosaur, that awesome stupid beast which somehow managed to rule the world during much of the Mesozoic Era. Why the dinosaur became extinct at the end of the Mesozoic may never be positively known, but we can be certain that here at least, it was not man's doing!

Dinosaurs roamed the West for a long period, but it is in rocks of late Jurassic age, particularly the Morrison Formation, that their skeletons are most widely preserved. Many skeletons have been excavated from a large number of places in the Rocky Mountain region, but when in 1909 Earl Douglass of the Carnegie Museum found "Dinosaur Ledge," it was soon recognized as a bone bonanza. Douglass lived among the dinosaurs for several years, digging them out and putting them back together. In 1915, President Wilson designated the quarry area as a national monument, all 80 acres of it. In 1938 it was enlarged to include the scenic canyons of Green River and

of Yampa River, a main tributary. This area of about 326 square miles extends eastward across the Utah-Colorado line. For those who are only momentarily bemused by dinosaurs, the monument includes some fantastic structural and erosional features in the canyons of the Green and Yampa rivers. Ecological relationships are strikingly set forth here. Pinyons and junipers that thrive on the near-bare sandstones are absent from the shale areas where grasses and other lowly plants are well established. In this dry area the soils are directly related to the underlying bedrock, and the different plants have selected the soils best adapted for their growth.

Geologic History

The Laramide Orogeny produced here an east-west range of folded and faulted mountains, the Uintas. The major structure is a large, asymmetrical anticline broken by several major faults that parallel the trend of the range. The Uinta Fault breaks through the steeply dipping beds on the north flank; Willow Creek Fault near the base on the south side, coupled with local folds, resulted in steep dips in the dinosaur quarry area.

Erosion has breached the anticline, exposing Precambrian rocks along the crest of the range. On the flanks are rock strata representing most of the geologic periods from the Cambrian to the Tertiary. These strata are exposed in textbook fashion in the walls of Green River and Yampa River canyons. Major John Wesley Powell was the first geologist to see this outdoor geological museum as he was fighting the rapids in Lodore Canyon of Green River in 1869, on his way to Grand Canyon. He and his party managed to skirt around Upper and Lower Disaster Falls and the worst of the rapids, and finally reached Steamboat Rock, a huge rib of rock that rears up high at the junction of the Yampa and the Green. Much later they emerged from Split Rock Canyon where the Green leaves the mountains. Powell (1875)

FIGURE 9-21

Visitor Center and bone quarry at Dinosaur National Monument.

had seen the inside of an anticline; the same strata which they had left behind as they entered the canyon reappeared as they came out at the lower end, but here the beds were dipping in the opposite direction. He reasoned that the river had been there first and that the anticline had risen up from below athwart the course of the stream. The uplift was gradual, he said, and the river had cut down through the uprising rocks and maintained its course. He coined the term *antecedent* for this type of stream, a term that was immediately accepted. At that time no detailed geologic work had been done in the Rocky Mountain Region and therefore he did not know when the Uinta Mountain anticline had developed. Much later it was determined that it is a Laramide structure and that it had been there *before* Green River began to cut its canyon. Therefore this segment of Green River is *not* an antecedent stream but a *superposed* stream that had cut down through the old structure, the anticline. Even though Powell was wrong about the Green River, he contributed a truly significant concept that has been applicable in other areas.

There are many large and grotesque erosional features in the canyons of the monu-

ment; unfortunately, close-up pictures can be obtained only from a boat. Distant views from a plane are truly exciting, as are those from the rim at such points as the Echo Park Overlook.

The Dinosaurs

Before the Laramide Orogeny produced the Uinta Mountains — in fact, even before sea water invaded this part of the Rocky Mountain Geosyncline — the area was low-lying and there were widespread swamps with abundant vegetation. This was an ideal habitat for dinosaurs, and they were there in large numbers during the latter part of the Jurassic Period.

In Morrison time, streams were depositing muds and sands in the inland basins; in the monument area a thick series of mudstones, shales, and sandstones accumulated before the Cretaceous sea covered the region. In one layer of sandstone, the "Dinosaur Ledge," skeletons of dinosaurs, large and small, occur in numbers yet unknown, because the supply is still good. This baffled the early workers; it appeared that in this one area there had been an unbelievable concentration of dinosaurs during the late Jurassic. No area could have had the carrying capacity to support such a large number of voracious eaters! In time, it

FIGURE 9-22

Dinosaur bones in the Morrison Formation. (Photo by National Park Service.)

became evident that this was a large sand bar, or perhaps a delta, and many dinosaurs, caught in floods, had been rafted in from other areas. This then was the burial ground for the entire drainage area.

Dinosaurs are intriguing to large numbers of people. Gigantic Brontosaurus and terrifying Tyrannosaurus can be relied upon to attract attention in any paleontological museum. And now "Supersaurus" has been discovered — the one that overshadowed all others. The little fellows, however, have not received the notice they deserve. One that was about the size of an overfed turkey ran like an ostrich and was far from slow-footed. The entire menagerie is colorfully presented by Ostrom in his article in *National Geographic* (August 1978).

Long believed to have been cold-blooded animals, at least some are now regarded as warm-blooded, as Ostrom and others have reported. Another common misconception is that there were hordes of dinosaurs of all sizes and shapes and then suddenly, at the end of the Cretaceous, none! Actually, certain groups evolved and gained prominence, only to die out even before the beginning of the Cretaceous. As Alexander (1981) and others have made clear, the so-called "Great Dying" may have involved a few million years, not merely a few days or years.

A colder climate, with resulting changes in vegetation, is one of several possible causes of extinction. Or possibly the small mammals decided that dinosaur eggs were their favorite food; these mammals were abundant by that time, and were probably clever and conniving. However, one of the more recent theories is far more spectacular. It involves the splashdown of a giant asteroid or meteor near the end of the Cretaceous Period (Alvarez et al., 1980). The dust and pulverized debris were thrown high into the air, enough to block sunlight from reaching the earth. With the sudden blackness, photosynthesis was quenched and the food supply drastically diminished.

That a large meteor did enter the earth has been established; a thin layer of sediment at the Cretaceous-Tertiary contact contains an abundance of iridium, an element generally rare except in meteorites. Whether the timing of the two events was mere coincidence is debatable. There is an equally significant question: If the dinosaurs were done in by starvation, how were most other animal groups able to survive? For the present, it appears that the meteorite theory, exciting though it is, can be included only as a possible cause of extinction.

To pause and reflect for a moment, dinosaurs have attracted our attention but generally have not commanded our respect. Slow-witted, overgrown brutes — "Terrible Reptiles," even "Terrible Ineptiles" — they surely deserve something better. At very least, we should give them credit for ruling the world for many millions of years longer than we have!

A chilling extension of these ideas was recently suggested at the "World after Nuclear War" conference in Washington, D.C., in October, 1983 (Holden, 1983). According to Carl Sagan and others, a major nuclear attack would generate clouds of dust from the blasts and burning buildings that could reduce temperatures in the mid-latitude areas as low as $-9.4°F$ ($-23°C$) and cause them to remain below freezing for several months. Combined with other factors, this could be the ultimate catastrophe for humans and send them and many other species along the path of the dinosaurs.

Whatever the cause of their disappearance, the variety and adaptations of the dinosaurs were remarkable (Ostrom, 1978). Many different types, from the 84-foot-long Diplodocus down to the 6-foot-long Laosaurus, have been excavated from the quarry in Dinosaur National Monument.

Dinosaur skeletons have gone to several museums — Carnegie Museum in Pittsburgh, National Museum in Washington, American Museum in New York, University of Utah in Salt Lake City, and the Denver Museum of Natural History.

The quarry is now housed in the Visitor

Center where workmen—with hammers and chisels and, on occasion, jack-hammers—demonstrate the techniques used in the quarrying of dinosaur skeletons.

For more details, a good source is Wallace R. Hansen's descriptive material on the back of the U.S. Geological Survey topographic map of Dinosaur National Monument, published in 1966. It includes geologic cross-sections and pictures of the monument and a discussion of the dinosaurs by G. Edward Lewis.

FIGURE 9-23

Steamboat Rock, a monolith at the confluence of the Yampa and Green rivers, Dinosaur National Monument. (Photo by National Park Service.)

● National Park
○ National Monument
▲ National Seashore/Lakeshore

PLATE 10
The Southern Rockies.

CHAPTER 10

SOUTHERN ROCKY MOUNTAIN PROVINCE

The Southern Rockies consist of a series of ranges, generally parallel, that extend from southern Wyoming across Colorado into northern New Mexico. The early explorers came from the east; hence the Front Range was the first major range they encountered. It extends from near Canon City, in southern Colorado, northward into southern Wyoming, where the name Laramie Range is also used. Had the explorers come from the west, the Sawatch Range might have been called the Front Range. The Sangre de Cristo Range is the easternmost range in southern Colorado and in New Mexico.

FIGURE 10-1

Looking south along the Front Range to Pikes Peak, with Red Rocks Park west of Denver in foreground. The lower area to the left, including the hogbacks, is the Colorado Piedmont which is composed of sedimentary rocks, less resistant to erosion than the Precambrian metamorphic and igneous rocks of the Front Range. (Photo by John Shelton.)

Prominent geomorphic features of the Southern Rockies are the "parks"—broad open areas partially or completely enclosed by mountains. The Laramie Basin in Wyoming, and North, Middle, and South parks, and the San Luis Valley in Colorado are the principal ones. These are structural basins, downfolded and/or downfaulted blocks of the earth's crust. Their topographic position, however, is the result of more rapid erosion of rocks softer than those of the surrounding mountains.

West of the parks the main ranges are the Medicine Bow, Park, Mosquito, and Sawatch ranges in Wyoming and Colorado and the Jemez-Nacimiento Mountains in New Mexico. In southwestern Colorado, west of the San Luis Valley, the San Juan Mountains, composed mainly of Tertiary volcanics, occupy an irregularly shaped area.

The Southern Rockies are higher than either

FIGURE 10-3

Gravel pit in esker remnant between Twin Lakes and Independence Pass, south of Leadville, Colorado.

FIGURE 10-2

Ancient Precambrian algae in the metamorphosed dolomite in the Medicine Bow Mountains, southern Wyoming. (Photo by M. E. McCallum.)

of the other Rocky Mountain provinces. Fifty-four peaks are more than 14,000 feet (4270 m) high, Mt. Elbert (14,431 feet; 4400 m) in the Sawatch Range being the highest. There are several prominent peaks in the Front Range; among them, from north to south, are Longs Peak (14,255 feet; 4346 m) in Rocky Mountain National Park, Mt. Evans (14,264 feet; 4348 m) west of Denver, and Pikes Peak (14,110 feet; 4300 m) west of Colorado Springs.

There are numerous vantage points from which to view Front Range scenery and geologic features—one at the top of the well-worn Longs Peak Trail. The trail was blazed by Major John Wesley Powell and his party in 1868, according to early records. Major Long, early explorer for whom the peak was named, merely saw the peak and was satisfied, without climbing it. For others inspired by less climbing, the view is superb from the top of Mt. Evans at the end of the highest paved road in the United States. Go early in the morning before the clouds roll in and engulf the peak. A five-minute trail takes you from the parking lot up to the top. If you ignored the sign warning against going to the top during an electrical

FIGURE 10-4

Wolford Mountain near Kremmling, Colorado. Precambrian rocks (tree covered) were thrust up and over Cretaceous shales during the Laramide Orogeny.

storm, and you hear "z-z-z-z" all around you and your hairs are standing on end like tiny lightning rods, it is time to come down! On your way down from the parking area, stop at Summit Lake, which is in one of the cirques flanking the mountain. If you are there in July or August, search among the wildflowers for the tiny blue Alpine Forget-Me-Not, truly the most beautiful flower in the world!

Bedrock and Structural Geology

In the Front Range and in the Medicine Bow Mountains, Precambrian rocks are widely exposed — rocks that range in age from about 1.0 to about 2.5 billion years. Rocks which from field relations are known to be older have not been accurately dated. Metamorphics such as mica and chlorite schists, gneisses, and quartzites are abundant, with lesser amounts of marble. Granitic magmas intruded into the metamorphics at different times during the Precambrian, forming the Pikes Peak and other batholiths. Diabase and basalt dikes cut through all of the Precambrian rocks here, as in the Teton Range. *Pegmatites* — extremely

coarse-grained granites — cut across older Precambrian granites and metamorphics. In places huge quartz and potassium feldspar crystals are associated with large sheets of muscovite mica; less commonly, beryl and other rare minerals are found in the pegmatites.

Certain of the ranges, notably the Mosquito Range west of South Park and the Sangre de Cristo Range, are principally upturned sedimentary rocks, largely Paleozoics. Still others, the San Juans and Jemez-Nacimiento Mountains, are dominantly volcanics — mainly Tertiary volcanic breccias and lava flows. Of special interest is the Valle Grande Caldera, also known as the Valles Caldera, on the western flank of the Jemez Mountains in northern New Mexico. Here, early in the Pleistocene about 50 cubic miles of rhyolitic pyroclastics were blown out, causing immediate subsidence of the walls and forming the caldera that is about 15 miles (24 km) in diameter (Smith et al.,

FIGURE 10-5

Tops of the "Eggs" in Wheeler Geologic Area (formerly a national monument), about 13 miles east of Creede, Colorado. Here, volcanic tuff weathers into unusually symmetrical geometric forms.

1961). The Sawatch, the highest range in the Southern Rockies, is composed of Precambrian metamorphics extensively intruded by early Tertiary magmas; one section of the range, including Mt. Princeton (14,197 feet; 4328 m), is composed of Tertiary "granite" of the Princeton batholith.

The Laramide structures of the Southern Rockies are complex; apparently they are mainly the results of compressional forces that caused the rocks to fold and later fault. A large thrust fault dips westward beneath the Front Range in the Pikes Peak area; others on the west side dip eastward under the mountains. There are also strike-slip faults, some of which date back to the Precambrian (Badgley, 1960). Although difficult to prove, it appears that major Precambrian faults determined that the Southern Rockies are not farther west, more in line with the general trend of the Middle and Northern Rockies.

Later faulting, mainly normal faulting, occurred during the mid-Tertiary. This faulting

FIGURE 10-6

The Gangplank, an extensive remnant of Pliocene pediplain, along the Colorado-Wyoming border.

was recognized early, particularly in the mines where Laramide ore deposits were offset by the faults (Lovering and Goddard, 1950). Recent work in other areas, largely by the U.S. Geological Survey, indicates that some of these high-angle faults have large displacements.

Volcanism was associated with this tectonic activity, especially in the San Juan region of southwestern Colorado and to the south in New Mexico. Volcanic activity occurred also in the Never Summer Range along the western border of Rocky Mountain National Park and in the South Park area, where there are extensive lava flows and welded tuffs. The prevailing westerlies carried fine particles of volcanic ash eastward over the plains; in the southern Colorado Front Range, ash buried many plants and animals in Florissant Fossil Beds National Monument.

Geomorphic Sequence

After the Laramide Orogeny ended late in the Eocene, erosion continued, and by the end of the Eocene an erosion surface of low relief — a peneplain — had been formed in most of the Rocky Mountains. The peaks and ridges that stood prominently above the peneplain surface are called *monadnocks,* after Mt. Monadnock in New England. The peneplain, the end product of a fluvial cycle, was only a few thousand feet above sea level; remnants of the peneplain are now as much as 12,000 feet (3660 m) above sea level, with monadnocks like Longs Peak rising still higher.

Uplift, mainly broad up-arching, occurred periodically during the Oligocene and Miocene epochs. A second fluvial cycle began with uplift and continued well into the Pliocene, and a second erosion surface was formed, leaving remnants of the first peneplain standing above, as high benches. The interpretation of Rich (1938) and others that this younger surface was a pediplain — a dry-climate peneplain — is accepted here. In granite areas, notably in southern Wyoming and northernmost Colorado and in the Pikes Peak area, a thick *regolith* of physically weathered granite is indicative of a dry climate.

Uplift near the end of the Pliocene brought the mountains up to their present height and initiated a third fluvial cycle, the one that is in progress today. The streams cut canyons — as much as half a mile deep — below the pediplain surface which in many places was narrowed to mere ridges of accordant heights.

It was not until this latest uplift occurred that the boundary between the Southern Rockies and the Great Plains was delineated. When uplifting began, the relatively smooth pediplain surface extended eastward from within the mountains across the area that was to become the Great Plains Province. The evidence is clear; in southernmost Wyoming there is a narrow remnant of the pediplain surface — the Gangplank — which slopes gradually up from the plains to the crest of the Front Range. To the south, the resistant-rock area became mountains, and the area of weak sedimentary rocks was lowered by erosion to form the Col-

orado Piedmont Section of the Great Plains Province. Thus, it was primarily a lithologic break that determined the boundary between the two geomorphic provinces. Unlike the abrupt eastern boundaries of the Tetons and the Sierra Nevada, where faulting was responsible, the eastern boundary of the Front Range is primarily the result of differential erosion.

In the Southern Rocky Mountains there are, at canyon-top level, ridges and flattish areas at 9500 to 10,000 feet — remnants of the late Pliocene pediplain. High above are benches from 11,500 to 12,000 feet that are older — remnants of the Eocene Summit peneplain, above which rise the many monadnocks. The steep-walled youthful canyons were cut after the latest uplift, near the end of the Pliocene. The uplift elevated the mountains to their present heights — high enough for glaciers to form.

Glaciation

The first glaciers were ice caps that formed along the high divides and extended in lobes down the then-undissected pediplain surface. During the long interglacials, canyons were cut by the streams, and valley glaciers became the dominant type. The evidence of the early glaciers is fragmentary, essentially erased by later advances; almost all of the prominent glacial features of today were formed during the several Wisconsin advances within the last 70,000 to 80,000 years. About 11,000 years ago these great glaciers began their last retreat with the warming of the climate which culminated in the Thermal Maximum about 5000 years ago. The cirques which were enlarged during Wisconsin time were occupied only by cirque lakes; a few of the cirques were reoccupied by glaciers during the Neoglacial. Ten small glaciers or glacierets are known to occupy the more sheltered cirques; icefields — potential glaciers — occupy others.

Rock glaciers have already been discussed — those in the Alaska Range. Rock glaciers are

FIGURE 10-7

Sill about two miles southwest of Lyons, Colorado. Columnar jointing is similar to that in Devils Postpile, California, and Devils Tower, Wyoming.

FIGURE 10-8

Rock glacier near Cameron Pass, northern Colorado; Nokhu Crags in upper left.

FIGURE 10-9

Lake Marie, with cliffs of the Snowy Range, southern Wyoming; note large taluses covering lower sections.

found in many mountainous areas, but concentrations are confined to only those localities where optimum conditions exist. A near-glacial climate is the primary requirement. Also essential is a rugged topography where large talus deposits form at the base of cliffs. Rocks that weather readily into large and durable fragments also contribute to the process. All of these requirements are met in the San Juan Mountains of southwestern Colorado. Here, at the heads of many glaciated valleys, rock glaciers have developed in the cirques and have enlarged by extending themselves down-valley.

The rock glacier shown in Figure 10-8 and many others are active today, as determined by examining their frontal sections. Forward movement renders the frontal section highly unstable. (Try if you must to climb up the front of this rock glacier!) Perhaps more convincing, lichens have not had time to grow on newly exposed surfaces of the boulders. Those rock glaciers which are now inactive are the ones in which the ice core has been so reduced by melting as to make it ineffective in moving its load of rocks.

Mountains of the Past: Vanished Mountains

The present mountains have been there for a long time, so long in fact that we might forget to ponder the question of whether there were other mountains before them, mountains completely destroyed long before the Laramide Orogeny.

The ancient faults referred to above were major structures of ancient mountains — Precambrian mountains — which were erased by erosion by the end of the Precambrian. In this way, the deep-seated granites and older deep-seated metamorphics were exposed, to be deposited upon during the Paleozoic. Recalling that the Precambrian encompasses about seven-eighths of geologic time, there may have been more than one generation of mountains. As we will see later, in the Grand Canyon area two major ranges came and went, one after the other, before the Paleozoic rocks were laid down. No suggestion is made, however, that mountain-building in these two widely separated areas was synchronized.

More recently, yet long before the Laramide Orogeny or the Rocky Mountain Geosyncline which preceded the orogeny, a large part of what is now the Southern Rockies was mountainous — during mid-Paleozoic time. These Paleozoic mountains, also referred to as the Ancestral Rockies, resulted from uplifts during the mid-Paleozoic. As compared to the present Rockies, they were never spectacular; erosion prevented their becoming majestic. And they, like their predecessors, were erased; by mid-Mesozoic time the Rocky Mountain area was a lowland, with many swamps and

many dinosaurs, particularly during the Jurassic, just before the time of the Rocky Mountain Geosyncline. Thus, we may regard the Rockies as merely the latest mountain range to occupy this site.

The Southern Rockies and the adjacent Great Plains have been regarded as "relatively stable" provinces. Statements have appeared indicating that no significant earthquakes have occurred in this area during this century. And so it was until October 18, 1984.

This earthquake, which measured 5.5 on the Richter scale, shook parts of six states: Wyoming, Colorado, Nebraska, Montana, South Dakota, and Utah. Its epicenter was about 40 miles (66 km) southeast of Casper, Wyoming. Although many people along the Colorado

FIGURE 10-10

Longs Peak, landmark of Rocky Mountain National Park, rises to 14,255 feet (4346 m) to dominate the northern Front Range. (Photo by Gregory C. Nelson)

Front Range felt the jolt and recognized immediately that it was an earthquake, others were completely unaware of it. The damage was generally light, although a number of buildings in certain areas sustained moderate to severe damage. No injuries were reported.

Localized earthquakes have been reported from several places in this general area, from the upper Laramie River Valley in northern Colorado and from the San Luis Valley in the southern part of the state. Many people remember the earthquakes that jolted the Denver area during the 1960s. Studies indicated that minor adjustments took place along a fault—sudden slips caused by the injection of large quantities of toxic waste into a deep disposal well. The liquid that was pressured into the fault zone provided sufficient lubrication for movement to take place. A few of the quakes were of moderate intensity, but the area affected was very limited, unlike that of the 1984 earthquake.

Will this single, medium-size earthquake discourage any of the thousands of people who were about to move westward into the Front Range area—an already seriously impacted section of the West? "Go west, young man" may not be the best maxim here. Also, will this quake cause architects to consider designing large buildings to withstand severe earthquakes? On both counts, it is probable that at least one more (perhaps devastating) quake will be needed.

The Southern Rockies have been subjected to intensive geologic study since the region was first explored, and there is no dearth of published material. For an overview of what was accomplished and for the latest interpretations of the rocks and geologic sequence, refer to the *Geologic Atlas of the Rocky Mountain Region.* And for the general reader, John and Halka Chronic's *Prairie, Peak and Plateau: A Guide to the Geology of Colorado* is an easy-to-read, colorfully illustrated booklet (1972).

ROCKY MOUNTAIN NATIONAL PARK

High mountain scenery, mainly the work of Alpine glaciers, is preserved in Rocky Mountain National Park in the northern Front Range of Colorado. Enos Mills was to Rocky Mountain what John Muir was to Yosemite; Mills' efforts finally bore fruit, and in 1915 about 410 square miles were set aside as a national park. Stanley Steamers sprang into action to carry visitors around the park and back to the famous Stanley Hotel or other lodging in or around the town of Estes Park. At that time, cars did not go over the top because Fall River Road was not built until 1920. Now, visitors can see representative features of the park by driving from Estes Park over Fall River and Milner passes, 11,796 (3600 m) and 10,758 feet (3280 m) high, respectively, down the Colorado River to Grand Lake, a beautiful gla-

FIGURE 10-11

Huge boulder, transported by a large valley glacier and deposited in the moraine at Bear Lake Parking Area, Rocky Mountain National Park, Colorado.

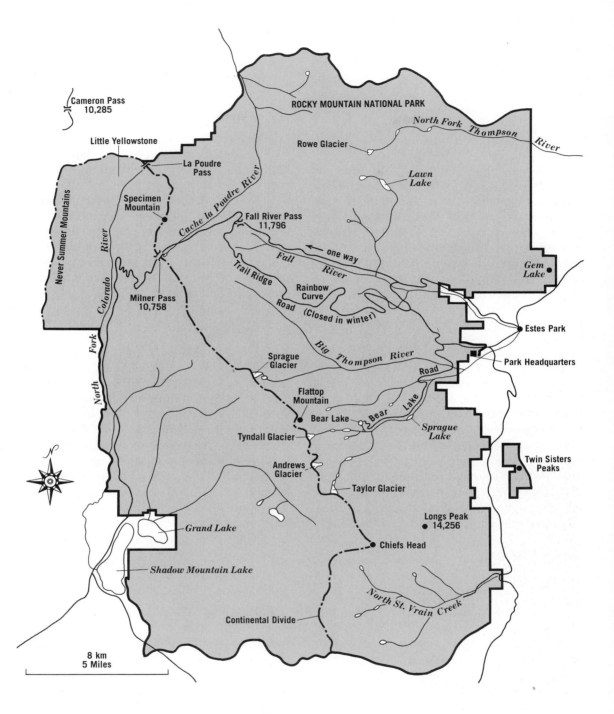

Cameron Pass
10,285

Little Yellowstone

La Poudre Pass

ROCKY MOUNTAIN NATIONAL PARK

North Fork Thompson River

Rowe Glacier

Lawn Lake

Specimen Mountain

Never Summer Mountains

Colorado River

Cache la Poudre River

Fall River Pass
11,796

one way

Gem Lake

Trail Ridge

Fall River

Milner Pass
10,758

Rainbow Curve

Road (Closed in winter)

North Fork

Big Thompson River

Estes Park

Park Headquarters

Sprague Glacier

Bear Lake Road

Flattop Mountain

Bear Lake

Bear Lake

Sprague Lake

Tyndall Glacier

Andrews Glacier

Twin Sisters Peaks

Taylor Glacier

Longs Peak
14,256

Grand Lake

Chiefs Head

Shadow Mountain Lake

North St. Vrain Creek

Continental Divide

N

8 km
5 Miles

PLATE 10A

Rocky Mountain National Park.

cial lake hemmed in by a terminal moraine. From the several overlooks, especially Forest Canyon Overlook and the visitor center at the top, a good overview of the park is readily available. Other roads lead into the Fall River, Moraine Park, and Bear Lake areas, and numerous trails take the more ambitious into the remote sections — the real park.

With elevations ranging from 7800 to 14,256 feet (2378 to 4346 m) atop Longs Peak, several life zones are represented in the park. Elk, deer, and marmots (rock chucks) are likely to be seen by sharp-eyed visitors. Of the smaller animals, the cony, or pika, is most unusual. An unlikely looking little guinea pig – rabbit, the cony is more often heard than seen because it blends in with the gray boulders of its home. Listen for the cony's raucous shriek at Rock Cut on Trail Ridge Road. It does not hibernate but spends its summers storing "hay" for the winter; it is the "haymaker."

Several general references, notably Alberts' 1963 booklet, *Rocky Mountain National Park* and the National Geographic's 1966 book *America's Wonderlands* are available. For more specific geological information, a 1974 publication by Richmond of the U.S. Geological Survey is recommended. Both the title, *Raising the Roof of the Rockies,* and the material presented are intriguing; particularly significant are his discussions of glaciers and glacial features in Rocky Mountain National Park.

General Geology

The park is located along the crest of the Front Range where Precambrian rocks dominate. The rocks within the park are almost all Precambrian granites, granitic gneisses, and schists; the Never Summer Mountains along the western side, however, are mid- and late-Tertiary volcanics, mainly pyroclastics. Ashflows and wind-deposited ash probably covered much of the park; a remnant of one of the ashflows, believed until recently to be a rhyolite flow, is readily seen from Trail Ridge Road

at Iceberg Lake. Specimen Mountain, a few miles east of the Never Summers, was believed to be a volcano (Wahlstrom, 1944). However, recent studies by R. B. Taylor, reported by Richmond (1974), indicate that Specimen Mountain is a craterlike erosion remnant of an ashflow derived from one of the Never Summer volcanoes. It may be some time before a consensus is reached; pending that decision, both interpretations are given.

In Rocky Mountain, the rocks are very old and very young; rocks intermediate in age, Paleozoic and Mesozoic, covered the area at one time but were stripped off as the mountains were uplifted above the sea, early in the Tertiary. Some of these rocks "lean" against the mountains in the hogback area some 20 miles (32 km) east of the park.

The Little Yellowstone area, in the northwestern part of the park, was so named because it is generally similar to Yellowstone Canyon. Here, small tributaries of the Colorado River have cut through rhyolitic volcanics which are hydrothermally altered and highly colored, but mainly yellow. This scenic area of pillars and spires is located north of Lulu City, a ghost town, and is reached only by trail.

Glacial features and small live glaciers, together with the above-timberline scenery, contributed support to the concept that this area should be a national park. Today's glaciers are tiny, called glacierets by some, but the work of earlier glaciers — U-shaped valleys, hanging valleys, cirques, cirque lakes and huge lateral moraines — make the park a veritable museum of glacial geology.

Geomorphic History

Within the park there are extensive remnants of the late Eocene peneplain, the origin of which was discussed earlier. The slightly undulating top of Flattop Mountain was so recognized by Lee (1923), who first used the term *Flattop Peneplain*. The term has been applied by many workers, including Van Tuyl and

FIGURE 10-13

Vertical jointing in granite near Longs Peak produces unusual landforms. Note emergency cabin for climbers. (Photo by E. Kiver.)

Lovering (1935), to the entire Southern Rockies area; Harris (1963) and others prefer to use the term *Summit Peneplain* for the older surface and *Subsummit* for the lower, younger erosion surface, following the lead of Bevan (1924).

The Tundra Curves area along Trail Ridge Road, like the surface of Flattop Mountain, was originally interpreted to be a remnant of the late Eocene peneplain. In recent years, other interpretations have been suggested. Richmond (1974) believes that the erosion surface

FIGURE 10-14

Summit (Flattop) Peneplain surface along Tundra Curves section of Trail Ridge Road. Telephoto from Toll Memorial Nature Trail.

extends across the 28-million-year-old ash-flow and is therefore much younger. Conceivably, however, the ashflow merely filled a broad valley that extended across the old surface. Should a *paleosol* be found immediately below the ashflow, the older interpretation would be supported.

Another possibility has also been considered, namely that the surface was not formed by erosion but by *altiplanation* processes, mainly *solifluction*. According to this theory, the surface was formed in recent times and at this high elevation. The term *solifluction* means "soil flow," the gradual downslope movement of unstable, almost fluid soil material. Clearly, solifluction has been active in recent time; small terraces were formed on slopes where, due to oversaturation, loose materials were rendered mobile. The thrust of the question is whether, by such processes, sharp peaks can be dulled and flattened.

There is other pertinent evidence which should be considered. A seven-foot-deep soil pit on the top of Flattop Mountain revealed a mature soil profile that could have been developed only under a warm and humid climate. This soil was formed when the peneplain was still intact at perhaps 3000 feet; it is a paleosol which has been in "cold storage" since the area was uplifted. Boulders on the surface and throughout the profile prevented the soil from being eroded away. It appears therefore that altiplanation processes were effective in modifying the surface but not in forming it. Until there is more convincing evidence to the contrary, the surface is still Late Eocene in age.

When uplift began late in Eocene time, the rejuvenated streams began to dissect the mountains and destroy the peneplain surface. After repeated interruptions by uplifts, the second and lower subsummit surface was formed, leaving the remnants of the Summit Peneplain high above.

In like manner the subsummit surface was largely destroyed after the latest major uplift near the end of the Pliocene. Examination of the remnants of the subsummit surface led to the conclusion that it was formed under relatively dry climatic conditions and is therefore, strictly speaking, a pediplain. The deep canyons were cut below the pediplain surface in the ongoing fluvial cycle, sometimes referred to as the "canyon-cutting cycle."

Glaciers and Glacial Features

The scenery of the higher mountains is the work of glaciers, primarily the glaciers that advanced several times during the Wisconsin. These were valley glaciers that at their maximum moved down to about 8000 feet on the east side but were melted at a slightly higher elevation on the warmer west side. Headward erosion in the cirques at the heads of the glaciers narrowed the flat peneplain area; glaciers in the cirques around Longs Peak steepened the sides of the old monadnock, thus forming sheer walls epitomized in the famous East Face. Chasm Lake at the base is one of many cirque lakes in the park.

Chains of rock-basin lakes at different elevations are on the treads of the "Giant's Stairways" or Cyclopean Stairs in certain of the gorges, Fern Canyon for one. Roches moutonnées are found in a number of the valleys; a small one in Moraine Park is readily identified as it stands above the flat floor of the valley. Sharp peaks and pinnacles are found in considerable numbers; two with names are the Sharkstooth and the Little Matterhorn.

A few of the cirques are occupied by tiny glaciers — such as Andrews, Tyndall, and Taylor — which were born during the Little Ice Age. They are all ensconced in deep Wisconsin cirques on the east side of the Continental Divide; even so, by the late 1930s, at the end of the warming trend, they had dwindled to the point where they appeared doomed. They have responded to the cooling of the past 40 years and have increased significantly in volume, although they are still confined well within their cirques.

End moraines are found in a number of places, but most of them are not well pre-

served. A notable exception is the end moraine which, tied onto a lateral or side moraine,
impounds water in Grand Lake just west of the
park; this earth-fill dam has functioned well for
more than 11,000 years. There are several
well-formed lateral moraines, the largest and
best-formed being South Lateral Moraine on
the south side of Moraine Park. This long ridge
extends for more than two miles out beyond
the canyon mouth. Somewhat smaller but perhaps more intriguing is the lateral moraine that
formed Hidden Valley. As is readily seen from
Rainbow Curve Overlook, the lateral moraine
blocks off Hidden Valley. The Fall River glacier
moved down the main valley past the mouth of
Hidden Valley and deposited the material in
the morainal ridge. This required the stream in
Hidden Valley to find a new and longer route in
order to join Fall River. Hidden Valley is hidden behind the lateral moraine (Figure 10-17a
and b).

Other Geologic Features

Periglacial features are found high in the tundra
areas of the park. The term *periglacial* means
"around the glaciers," but it is often used to
denote a cold environment where "frost action"—freezing and thawing—has formed
unusual but incompletely understood features. Frost riving of layered rocks in the park
—mainly gneisses and schists—produces flat
slabs, and by frost heaving many of the slabs
are left standing on edge. Stone-stripes extending downslope are apparently formed in
this way. In some cases the pattern is roughly
circular, and "stone-rings" are seen on more
gradual slopes.

Where fine-textured material becomes saturated, mass-wasting in the form of solifluction
takes place and small terraces or *terracettes* are
formed, even on very gradual slopes. In limited
areas, solifluction may be related to *permafrost*
which is present at depths of about four feet
where willow thickets provide sufficient protection.

Tors are prominent features that rise

abruptly above the high slopes in the park.
They were formed by differential weathering
along fractures; eventually irregular towers of
rock were formed, some of which resemble
sea stacks. Excellent examples can be seen on
the nature walk above the Rock Cut Stop on
Trail Ridge Road.

FIGURE 10-15

Telephoto looking west to Diamond Face (East
Face) of Longs Peak.

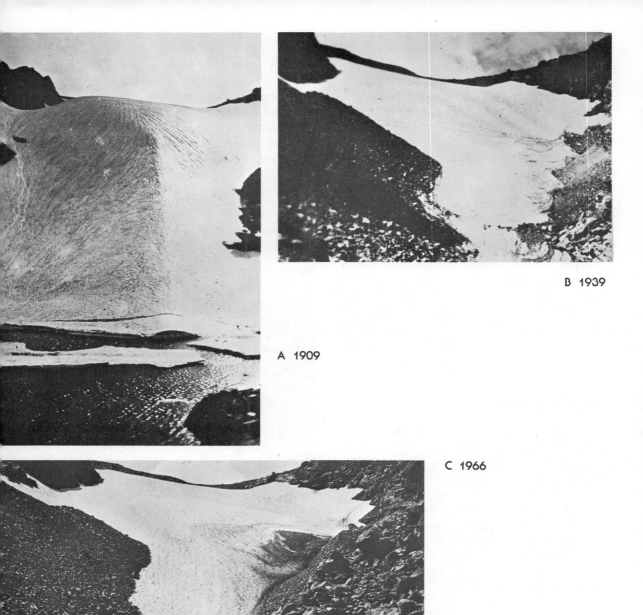

A 1909

B 1939

C 1966

FIGURE 10-16   Andrews Glacier in 1909 (A) when it was retreating; (B) when it was at low ebb, probably a stagnant icefield; and (C) in 1966, after expanding modestly during the latest cooling trend. (Photo A by R. G. Coffin; Photo B by National Park Service.)

FIGURE 10-17*a*   Looking down from Rainbow Curve Overlook, over Horseshoe Park (middle ground) and Hidden Valley (lower right).

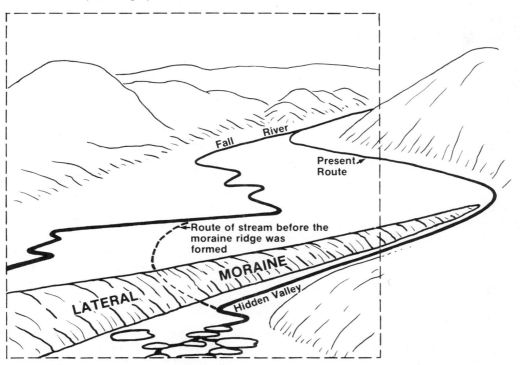

FIGURE 10-17*b*   Sketch showing the origin of Hidden Valley as pictured in Figure 10-17*a*.

Waterfalls are common in the park. Some are at the mouths of hanging valleys; others are formed along main streams. In many cases those of the latter type have resulted from stream erosion at cross-fractures or cross-faults, in some cases merely accentuating the stairstep-like profile of the valley formed originally by glacial abrasion and quarrying. Alberta Falls and Chasm Falls are examples.

As indicated earlier, most of the lakes in the park were formed by the glaciers, either by erosion or by deposition. A few exceptions may be of particular interest: Gem Lake, about 3 miles (5 km) northeast of Estes Park, was out of reach of the glaciers. Its location — perched up in a low saddle — and its origin are both unusual. Apparently the saddle is there because of badly fractured granite that weathered and eroded more rapidly than the rock on either side. The lake basin is also related to the fracturing; the feldspars in the granite weathered chemically to form clays along the fractured surfaces. The tiny clay particles were readily carried away by the winds that frequently whistle through the saddle. The tiny depression — a deflation basin — was gradually deepened and enlarged to form a tiny lake — Gem Lake.

While there, take note of the sand piled up by the wind on the leeward side of the lake. And across the lake notice that the granite is layered. *Sheet-jointing* is the term used to describe these layered structures. As the granite was raised from far beneath the surface and the overlying rocks were eroded away, the upper part of the granite expanded and the horizontal fractures were formed; once exposed, the fractures were widened by weathering, making them easier to see, particularly from a distance. Those in the cliffs north of Estes Park can be readily seen from the main highway.

The trail to Gem Lake is gentle except at the upper end; however, the features along the way and the views from various points make the struggle well worthwhile. Watch for the

FIGURE 10-18

Gem Lake in northeastern part of Rocky Mountain National Park. Note the Twin Sisters in the distance, about ten miles to the south.

Hole-in-the-Rock on your left as you go up, and don't miss the view of Twin Sisters (Figure 10-18).

Sprague Lake, along the road to Bear Lake, has special interest. The Nature Trail around the lake is a "Five-Senses-Trail" designed to accommodate handicapped visitors, including those in wheelchairs. Sprague Lake is not a glacial lake; in fact, it is artificial. Innkeeper Abner Sprague needed a fishing spot for his guests, so he built a low dam across the creek. The Park Service purchased the property years ago, and in 1957 they razed the old hotel. Those interested in the early history and the origin of local names will find *High Country Names* (Arps and Kingery, 1977) a joy.

Lawn Lake (10,987 feet; 3330 m), high in the mountains north of Fall River, is partially artificial. At the end of a 6-mile (9.6 km) trail, it has long been a favorite of ambitious hikers. One of the early-day hikers had an inspiration; if a dam were built on top of the morainal dam, a very substantial amount of water could be stored for irrigation down on the flat-lands.

This was done in 1902, several years before Rocky Mountain National Park was established. The building of a dam on morainal material of unknown composition has proved to be poor engineering practice, here and elsewhere. Early on July 15, 1982, the aging Lawn Lake dam failed, and the floodwater torrented down the steep channel of Roaring River, carrying everything with it, including huge boulders. At the mouth of the canyon, the water spread out and deposited an enormous boulder fan on the floor of Fall River Canyon. The floodwaters rushed down Fall River Canyon and through the business district of Estes Park. Property damage ran into the millions of dollars, but miraculously only three lives were lost. Fall River Road in the park was buried beneath many gigantic boulders; recently, the road was reopened after being built up and over the boulder fan (see Figure 10-19).

Lawn Lake is still there, but it is even smaller than before the dam was built. According to Chief Park Naturalist Glen Kaye (personal communication, October 1982), the lake level is five or six feet lower than it was before 1902. The disaster had one beneficial result: The program of inspection of dams in Colorado, including old dams, has high priority.

This was the second flood to disrupt normal visitation in the park in recent years. On July 31, 1976, the worst flood in Colorado's history devastated Big Thompson Canyon between the towns of Estes Park and Loveland. Although the cloudburst storm, and consequently the flood, was downstream from the park, large sections of U.S. Highway 34 were ripped out, and a major access to the park was closed for an extended period. More important by far, more than 100 lives were lost during the flood.

If possible, visit Rocky Mountain National Park soon after the main tourist season is over. If you should hike up the Bierstadt Trail you can look down on the valley and see an ocean of golden aspen leaves quaking in the breeze. Perhaps you will be high on the trail at one of those one-in-a-million moments when whirlwinds updraft showers of gold leaves high into the air, then release them to float gently down.

When you climb Longs Peak, you will be up where the air is pure and the views are superb —to the south, west, and north. But do not look to the east; on many days, far too many, you would see a pall of orange-brown smog sitting on top of Denver and other Front Range cities and towns. On occasion, fortunately not often, upslope winds carry the smog high up into the eastern part of the park. It is indeed a sad commentary on modern living when the environment of our national parks is no longer safe.

FIGURE 10-19

Huge boulders deposited by the flood of July 15, 1982, after Lawn Lake Dam failed.

FLORISSANT FOSSIL BEDS NATIONAL MONUMENT

The name Florissant, derived from the French, was given to a little valley and a town about 35 miles west of Colorado Springs because of the profusion of wild flowers. The name was fortuitous because a few years later, in 1874, Dr. A. C. Peale discovered that about 35 million years ago another valley, also with abundant flora, had occupied the area. Fossil flora, exceptionally well preserved, easily justified the creation of the national monument, after almost a century—in 1969. In addition to plant fossils, the locality is world-famous for insect fossils; more than 1100 species have been collected and identified over the years. But there is still more! Several species of fish, birds, and small mammals were also entombed and fossilized.

FIGURE 10-20

One of several fossilized sequoia stumps in the Florissant Fossil Beds National Monument, Colorado. Stump was exposed by excavation of volcanic mudflow material.

What combination of circumstances could have been responsible for the burial and preservation of such a wide variety of life forms? It was not one coincidence but several that made it possible; it was not the labors of one geologist but of several that put the pieces of the puzzle together.

Our story begins with a broad Oligocene valley, perhaps about 3000 feet above sea level. Volcanoes several miles to the west started erupting, and lava flowed northeastward from Mt. Guffey to block off the valley and form a 12-mile-long lake—Lake Florissant. Remnants of the basaltic lava flow can be seen from the road a short distance south of the monument.

Swampy areas developed around the lake and soon there was lush vegetation, an ideal habitat for insects. Trees familiar to us included maples, beeches, birches, hickories, and willows. But there were also Sequoias, which today are confined to restricted environments. Palm trees were there, indicating that a subtropical climate prevailed at that time.

The volcanoes erupted again and this time belched forth large amounts of tephra. The coarser materials fell near the vents, but the westerly winds carried the fine ash and dust eastward and spread it over a large area, including the lake and the associated swamps. Where there had been life in abundance, there was death and desolation. Trapped in the suffocating volcanic cloud, the insects and other animals were no more able to escape than the plants.

Volcanic mudflows 10 to 20 feet thick surrounded the giant Sequoia trees encasing the lower sections of their trunks. Later, lava flows and flow breccias covered the entire area, thus forming a protective caprock. Petrifaction of the tree stumps took place slowly under essentially optimum conditions. Sealed in by the mudflows and buried by lava, there was very little decomposition of the wood. Silica from the volcanic ash filled the cells of the wood, rendering it almost indestructible.

The final chapter was written when the Southern Rockies were uplifted late in the Pliocene. Streams eroded valleys in the volcanics and in this specific valley exposed but did not destroy the fossil beds. The tops of some of the silicified sequoia stumps were exposed.

The preservation of leaves and insects was phenomenal, because burial was under water in extremely fine-textured volcanic ash. In certain instances even the most delicate veinlets in the wings of insects were perfectly preserved.

Because of the quality and abundance of the fossils, scientists from many parts of the world collected specimens which are now on display in a number of university and public museums, including the Denver Museum of Natural History. S. H. Scudder was an avid collector before the turn of the century; Scudder's Diggings are now open to the public — to see but not to take the fossils. T. D. A. Cockerell, who collected for the University of Colorado Museum, referred to the deposit as an "ancient Pompeii," as he collected dragonflies and a myriad of other fossils, shortly after 1900.

FIGURE 10-21

Well-preserved moth, one of more than 1100 species of insects found in Oligocene ash, Florissant Fossil Beds National Monument. (Photo by E. Kiver.)

The Florissant area is beautiful rolling country about 8300 feet above sea level. Land developers were looking hungrily at it prior to final settlement by the government. At just the right moment, a group of concerned citizens fought for and saved the monument for posterity. Florissant is open to the public throughout the year; however, summer is the best time for a visit, because snow may cover everything except the higher Sequoia stumps.

At Florissant, there are no awe-inspiring canyons, no breath-taking waterfalls, and no beautiful glaciers. Instead, there is merely the record of a series of geological coincidences without which this brief moment in a recent chapter of geologic history would have been lost.

GREAT SAND DUNES NATIONAL MONUMENT

The Great Sand Dunes National Monument, an area of about 58 square miles, was created as a national monument in 1932. It is located in the only true desert in the Southern Rockies — the San Luis Valley in southern Colorado. Here, at the base of the Sangre de Cristo Mountains which lie to the east, are some of the world's tallest sand dunes.

Geomorphic Development

The San Luis Valley is the southernmost of the "parks" in the Southern Rockies. It lies between the San Juan and La Garita mountains on the west and the Sangre de Cristos on the east. The depressed valley block has had a great thickness of lake sediments deposited upon it, according to Powell (1958), who investigated the valley's groundwater supplies. Many of the alluvial fans that extend out from the mountains are important sources of water. Particularly significant in the economy of the valley are the artesian wells that produce large quantities of irrigation water from the huge

FIGURE 10-22

Air view north over Great Sand Dunes to the Sangre de Cristo Mountains. (Photo by R. L. Burroughs.)

fans along the western side, at the base of the San Juans and La Garitas. A large lake once covered most of the valley as indicated by widespread lake sediments; now, a relict lake, San Luis Lake, occupies the bottom of the enclosed basin in the northern part of the valley.

The sand dunes are located at the base of the Sangre de Cristos west of Mosca Pass. At one time, it was assumed that most of the sand was derived from deposits of Medano and other creeks that flow out of the Sangre de Cristos; however, mineralogical studies by Johnson (1971) and others indicated that essentially all of the sand was derived originally from the rocks in the San Juans about 50 miles to the west. Streams carry weathered materials out onto the valley floor where the winds pick them up and transport them across the valley to the dune area. The prevailing westerlies, actually southwesterlies, sweep across the poorly vegetated valley, pick up sand and dust, and carry it eastward; when the winds begin to rise to funnel through the pass, they drop the sand in the dune area. Quartz and feldspar are the main minerals in the dunes, but there is an unusually wide variety of accessory minerals, including garnet and magnetite.

Magnetite is not a common mineral in most sand dunes, yet black bands of magnetite sand are common on some of the dunes in the monument. With a specific gravity almost twice that of quartz and feldspar, magnetite is not transported as readily by the wind. The magnetite was not derived from the San Juans but from the Sangre de Cristos on the east. Medano Creek, which flows along the eastern side of the dunes, deposits the heavy magnetite in its broad streambed at the base of the dunes. But how does it get up on the crests of the dunes? Although the westerlies are the prevailing winds, for short periods early in the year violent winds roar down out of the canyons to the east and lift the sand, including the magnetite, high up on the dunes. Concentration of the magnetite is accomplished by gentler winds winnowing out the lighter minerals, and leaving the black bands of magnetite along the dune crests.

These same violent winds produce some unusual dune shapes, reversals of the asymmetry produced by the prevailing winds. Once the violent winds subside, the prevailing winds set about to reshape the dunes as they should be, with their slip-faces facing toward the mountains.

The dunes area appears to be essentially stationary. At times, dunes encroach on and even block the stream, thus forming temporary lakes. The next flood out of the mountains removes the dune front; thus, there is a battle for occupation of this boundary area — one that appears to be a standoff at the present time.

The Great Sand Dunes are of interest to botanists and hydrologists, as well as to geologists. The struggle for survival of the shrubs and trees in the tension zone along the edge of the dunes — where both the winds and the floods are powerful agents — makes this a particularly fascinating area.

Although there are good views from the valley or — after a frustrating climb up the shifting sands — from the top of the 700-foot-high dunes, it is also impressive to look out over the top of this mountain of sand from the Montville Trail leading up to Mosca Pass.

BANDELIER NATIONAL MONUMENT

Bandelier National Monument, located in the Jemez Mountains about 45 miles west of Santa Fe, was set aside in 1916 in order to preserve several archeological sites. Because an unusual geologic situation made it possible for the early inhabitants to make their homes here, this monument is included with those that are more specifically geological.

The monument consists of plateaus with deep, steep-walled canyons cut in volcanic tuff. The main canyon, where the visitor center and the Tyuonyi Ruin are located, was cut by Rito de los Frijoles, the "Little River of the Beans." Only part of Frijoles Canyon is accessible by car; the remainder of the monument is explored on foot. Some of the trails are steep and visitors from low areas may find the going rough at about a mile and a half (2.5 km) above sea level. The trail up to Tsankawi Ruin in the

Detached Section of the monument is in part over an old trail cut deep into the tuff by the feet of countless ancient Indians. Other trails lead downcanyon from the Visitor Center to the Upper and Lower Falls near the Rio Grande, and to the Stone Lions, the Painted Cave and the Kiva House.

The geologic story is exciting, with volcanic explosions periodically breaking the stillness of the Jemez Mountains through much of Tertiary and Quaternary time. Basaltic lavas poured out over large areas, later to be buried beneath thick volcanic tuffs.

West of Bandelier, in the Valle Grande country, a huge caldera was formed by explosion and collapse during the early Pleistocene. The sequence, including the development of the 15-mile-wide Valles Caldera, is detailed by Smith, Bailey and Ross (1961).

As the latest main geologic event, streams cut deep canyons through the tuff, dissecting the Pajarito Plateau. One good geologic section is exposed a short distance downcanyon from the visitor center. Here, the basalt flow and the tuff are in full view.

FIGURE 10-24

Excavated ruins near visitor center, Bandelier National Monument.

FIGURE 10-25

North Cliff, Frijoles Canyon. Note weathering pits and excavations in tuff.

In places the tuff is poorly consolidated and weathering pits were soon formed. These pits were enlarged by the ancient Indians, in some cases for homes, in others for granaries in which to store grain, beans, and other foods.

A few of the ruins date back to the twelfth century; it is not known whether habitation was continuous, but the canyons were abandoned about 1550 A.D. Life in the good days at Frijoles Canyon is depicted in *The Delight Makers* by Adolph Bandelier, for whom the monument is named. Bandelier made extensive investigations of the prehistoric ruins in the late 1800s. One puzzle the archeologists have yet to solve — why did the Indians leave this delightful spot?

● National Park
○ National Monument

PLATE 11-1 Map of Colorado Plateaus.

CHAPTER 11

COLORADO PLATEAUS PROVINCE

The Four Corners, where Arizona, Utah, Colorado, and New Mexico meet, is near the center of the vast Colorado Plateaus Province. The province was so named because much of the region is drained by the Colorado River and its main tributaries — the Green, the Little Colorado, and the San Juan rivers. (See Figure 11-1 in color section.)

It is a land of lonesome beauty where, from many vantage points, the mesas, plateaus, and canyons seem to extend to infinity without interruption. It is our most colorful province; rocks of all colors — brilliant reds, salmon pinks, yellows, browns, grays and white — are exposed, layer-cake fashion, in the high cliffs. Most of the region is dry and essentially treeless except along the streams in the canyons and valleys; the high plateaus, like the Kaibab north of Grand Canyon and the High Plateaus section of Utah, have higher precipitation and are covered with pine, spruce, and fir forests.

But the scene is changing. Now we see strip mines and coal-fired power plants springing up in various places; the Four Corners Power Plant near Shiprock, New Mexico, is fired with coal from what is reportedly the world's largest strip mine. The air that was once clear is now polluted in large areas, including many of our national parklands. How can this be possible? The Clean Air Act, passed in 1977, specifies that the air over the parklands will be kept clean, restoring where necessary the visibility that has been impaired. Because essentially all of the pollution originates outside park boundaries, this law requires a general cleanup of the atmosphere. Although most Americans are strongly in favor of clean air, certain industrial interests are much less so. Under pressures from lobbyists, the House Energy and Commerce Committee considered an amendment to the Clean Air Act, one designed to render the law ineffective. Although the assault failed, it did so by only a small margin. This action serves as fair warning that the future of our parks is not assured. In the meantime, the Environmental Protection Agency apparently considers the enforcement of this law to have relatively low priority.

General Geology

In most of the geomorphic provinces already discussed, Cenozoic mountain-building and/or volcanic activity were basically responsible for today's landscapes. As the adjacent Basin and Range Province was similarly affected, the Colorado Plateaus Province appears to have been an "island of tranquility" in the sea of tectonic upheaval. G. K. Gilbert and other early geologists were at a loss to explain how this vast area escaped the folding, faulting, and igneous intrusion that had profoundly changed the remainder of western United States.

The problem is less burdensome now that we know more about the region. There are numerous structures, some of major proportions, as shown in Plates 11-A,B,C,D. The folds are generally broad flexures rather than the close folds in the Rocky Mountains, and the major faults are gravity faults. Therefore, it appears that vertical movements were dominant in the Colorado Plateaus, from the Paleozoic to the present. The interpretation may become evident when the plate tectonics of the continent are more clearly understood.

Structures less obvious than folds and faults are nonetheless important in the development of many landforms. In a number of places in the Colorado Plateaus country, parallel fractures have resulted in the formation of parallel

GEOLOGIC CROSS SECTION OF THE GRAND CANYON –

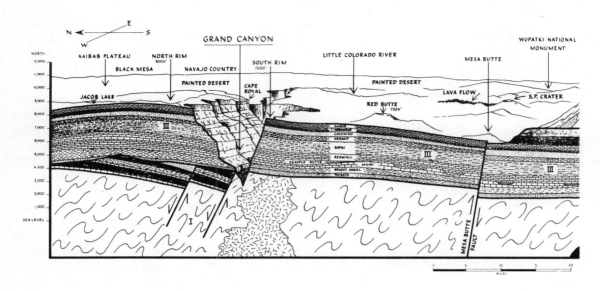

GRAND CANYON

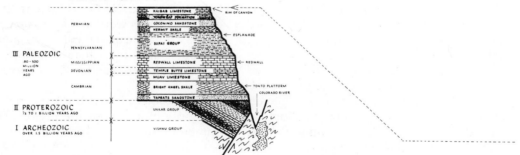

In the region between the Grand Canyon and the Verde Valley, earth history is revealed on a gigantic scale. As revealed in the bottom of Grand Canyon, two episodes of mountain building and erosion occurred during the Archeozoic and Proterozoic Eras, from two billion to one-half billion years ago. From the beginnings of the Paleozoic Era to the end of the Mesozoic Era – five hundred million to sixty million years ago – this area was essentially a low-lying plain, sometimes submerged under the sea, at other times a flood plain crossed by sluggish rivers and on occasion a desert with blowing sand dunes. During this time period over 10,000 feet of sediment accumulated – rocks present today

PLATE 11-A

SAN FRANCISCO PEAKS ~ VERDE VALLEY REGION

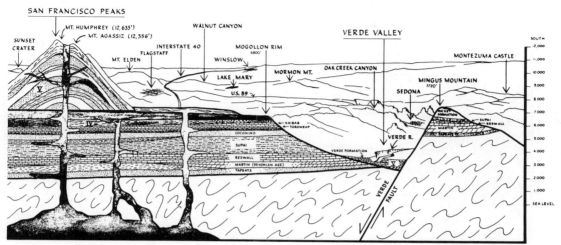

SAN FRANCISCO PEAKS

SUNSET CRATER

MT. HUMPHREY (12,633')
MT. AGASSIZ (12,356')

WALNUT CANYON

VERDE VALLEY

SOUTH
12,000
11,000

INTERSTATE 40
FLAGSTAFF
MT. ELDEN
WINSLOW

MOGOLLON RIM
6800'

MONTEZUMA CASTLE
10,000

LAKE MARY

MORMON MT.
OAK CREEK CANYON

MINGUS MOUNTAIN
7720'
9,000

U.S. 89

SEDONA
8,000
7,000

V

KAIBAB
TOROWEAP

VERDE R.

SUPAI
REDWALL
6,000

IV
COCONINO
5,000

SUPAI
VERDE FORMATION
4,000

REDWALL
MARTIN (DEVONIAN AGE)
TAPEATS
3,000

VERDE
FAULT
2,000
1,000
SEA LEVEL

Artwork by Dick Beasley. Geology and cross-section by William J. Breed, Museum of Northern Arizona,
with assistance from: Barton Wright, Ben Foster, and Pam Lunge.

NAVAJO RESERVATION ~ BLACK MESA

SAN FRANCISCO PEAKS ~ VERDE VALLEY

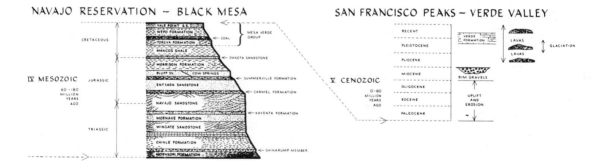

IV MESOZOIC

CRETACEOUS

YALE POINT S.S.
WEPO FORMATION
TOREVA FORMATION
MANCOS SHALE
MORRISON FORMATION
BLUFF SS.   COW SPRINGS
ENTRADA SANDSTONE

MESA VERDE GROUP
COAL
DAKOTA SANDSTONE
SUMMERVILLE FORMATION

JURASSIC

60 - 180
MILLION
YEARS
AGO

CARMEL FORMATION

NAVAJO SANDSTONE

MOENAVE FORMATION

TRIASSIC

WINGATE SANDSTONE

KAYENTA FORMATION

CHINLE FORMATION

MOENKOPI FORMATION
SHINARUMP MEMBER

V CENOZOIC

RECENT
PLEISTOCENE
PLIOCENE
MIOCENE
OLIGOCENE
EOCENE
PALEOCENE

VERDE FORMATION
LAVAS
GLACIATION
LAVAS

RIM GRAVELS

0 - 60
MILLION
YEARS
AGO

UPLIFT
AND
EROSION

at Grand Canyon and in the Zion Canyon region to the north and Black Mesa and the Navajo Reservation to the east. In Cenozoic times from sixty million years ago to the present this region was uplifted, the Mesozoic rocks were removed by erosion and eventually canyons such as Grand Canyon and Oak Creek Canyon were formed. During the latter part of the Cenozoic, outpourings of lava built the volcanic field near Flagstaff including the San Francisco Peaks. These peaks were later modified by glaciation. This area of the Colorado Plateau is only slightly disturbed by faulting and folding and is mainly underlain by horizontal beds - in great contrast to the Basin and Range regions south of the Mogollon Rim.

PLATE 11-B

GEOLOGIC CROSS SECTION OF THE CEDAR

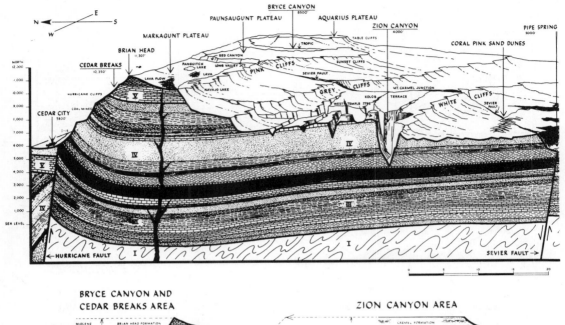

BRYCE CANYON AND CEDAR BREAKS AREA

V CENOZOIC 0-60 MILLION YEARS AGO	MIOCENE	BRIAN HEAD FORMATION
	EOCENE	*Pink Cliffs* WASATCH FORMATION
IV MESOZOIC 60-180 MILLION YEARS AGO	CRETACEOUS	*Grey Cliffs* STRAIGHT CLIFFS - WAHWEAP SANDSTONE / TROPIC FORMATION / DAKOTA SANDSTONE / COAL

ZION CANYON AREA

	JURASSIC	CARMEL FORMATION / NAVAJO SANDSTONE — *White Cliffs*
IV MESOZOIC 60-80 MILLION YEARS AGO	TRIASSIC	KAYENTA FORMATION / MOENAVE FORMATION / CHINLE FORMATION — *Vermilion Cliffs* / SHINARUMP MEMBER / MOENKOPI FORMATION

In this region the forces of erosion have laid bare 1 billion 500 million years of earth history. The oldest rocks, those of the Archeozoic, Proterozoic and Paleozoic, are found in the walls of the Grand Canyon. The Mesozoic forms the temples and towers of Zion. The most recent, the Cenozoic, is exposed at Cedar Breaks and Bryce. Presumably all the layers of the Cenozoic and Mesozoic at Cedar Breaks and Zion once extended over the region of the Grand Canyon. The relentless wearing of the waters has stripped the layers back to the north forming the celebrated "Great Rock Stairway" of the Vermilion Cliffs, the White Cliffs, the Grey Cliffs and the Pink Cliffs.

PLATE 11-C

BREAKS — ZION — GRAND CANYON REGION

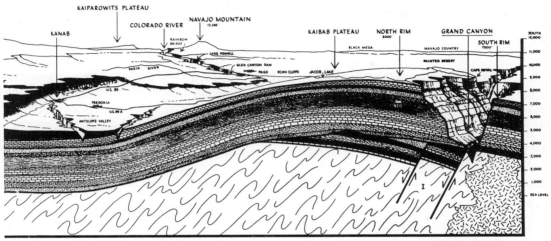

KAIPAROWITS PLATEAU

KANAB   COLORADO RIVER   NAVAJO MOUNTAIN 10,388'   KAIBAB PLATEAU 8000'   NORTH RIM 8000'   GRAND CANYON   SOUTH RIM 7000'   SOUTH 12,000

RAINBOW BRIDGE   BLACK MESA   NAVAJO COUNTRY   11,000

LAKE POWELL   PAINTED DESERT   10,000

GLEN CANYON DAM   PAGE ECHO CLIFFS JACOB LAKE   CAPE ROYAL   9,000

PARIA RIVER   8,000

U.S. 89   7,000

FREDONIA   6,000

U.S. 89 A   5,000

ANTELOPE VALLEY   4,000

3,000

2,000

1,000

I   SEA LEVEL

Artwork by Dick Beasley. From an original by Peter Coney, Middlebury College.
Revised by William J. Breed, Museum of Northern Arizona.

GRAND CANYON

KAIBAB LIMESTONE ← RIM OF CANYON

PERMIAN   COCONINO SANDSTONE

HERMIT SHALE ← ESPLANADE

SUPAI GROUP

PENNSYLVANIAN

III PALEOZOIC
180 - 500
MILLION
YEARS AGO   MISSISSIPPIAN   REDWALL LIMESTONE ← REDWALL

DEVONIAN   TEMPLE BUTTE LIMESTONE

MUAV LIMESTONE

CAMBRIAN   BRIGHT ANGEL SHALE ← TONTO PLATFORM

TAPEATS SANDSTONE   COLORADO RIVER

II PROTEROZOIC
½ TO 1 BILLION YEARS AGO   UNKAR GROUP

I ARCHEOZOIC
OVER 1.5 BILLION YEARS AGO   VISHNU GROUP

Published by the Zion Natural History Association, Zion National Park, Springdale, Utah 84767,
in cooperation with the National Park Service, U.S. Department of the Interior.
ONE DOLLAR. © 1975 Zion Natural History Association.
Lithography by Northland Press, Flagstaff, Arizona.
ISBN 0-915630-02-8

PLATE 11-D

drainage patterns which, in turn, have led to the creation of high rock walls, or *fins*. Later, as the fins are destroyed, such features as arches, standing rocks, columns, and natural bridges come into being.

The main fractures are parallel to the trends of the anticlines and are developed along the axes and on the flanks. In one large area, in southeastern Utah and southwestern Colorado, some of the anticlines are of unusual origin: They were formed by the intrusion of salt! This process, detailed by Hite (1972) and Lohman (1975), is briefly given. During part of the Pennsylvania Period, a thick layer of salt was deposited in a large inland basin. The resulting salt formation, appropriately named the Paradox, lies beneath younger rocks in the Paradox

Basin. Salt is a peculiar rock because under pressure it will flow laterally from one place to another, to areas where the pressure is less. There, it will dome up the overlying rocks; or, if the area of less pressure is elongate, a long anticline will be formed. As the overlying rocks are being slowly arched up, parallel fractures develop (Figure 11-2).

Later, when the rocks overlying the salt are largely eroded off of the salt anticline, groundwater leaches out the salt, particularly along the crest of the fold, leaving the overlying rocks without support. They collapse in a manner not unlike the collapse of rocks in a caldera. Here, however, the process is much slower and the collapsed block is distinctly elongate. We will see in more detail how this unusual

FIGURE 11-2

Parallel fractures, common in parts of the Colorado Plateaus, are well developed in the Arches National Park area; view westward over fins and arches to Salt Valley, a graben. (Photo by National Park Service.)

sequence of events predetermined many of the features in Arches and Canyonlands national parks.

As noted, a large part of the province is underlain by sedimentary rocks that are generally flat-lying; consequently, in many places where one plateau rises to a higher one, the latter is capped by younger strata. In a number of places, however, vertical faults separate plateaus of different elevations, particularly in the western part of the province.

Rocks representing all of the geologic eras are exposed in the Colorado Plateaus Province; all are represented within the Park System, in one park or monument or another. In your travels you will see certain formations again and again. You will soon learn to recog-

nize the more prominent formations by the landforms. Some are cliff-formers; others form flatter slopes, perhaps dissected into *badlands*. Ordinarily each formation has its characteristic color or range of colors; however, several of the Mesozoic formations are of continental origin rather than marine, and the color may be frustratingly different from place to place. One example is the Entrada which is red in Arches National Park, but in Colorado National Monument, only 60 miles (100 km) away, it is buff colored to salmon-pink. The Navajo Sandstone, with its eolian (dune-type) cross-bedding, is a widespread formation readily recognized in a number of the parks in the Colorado Plateaus. You may be surprised to learn when you go into Colorado National Monument that

FIGURE 11-3

Goosenecks of the San Juan River, near Mexican Hat, southeastern Utah. (Photo by J. A. Campbell.)

the Navajo is missing; there, erosion removed the Navajo from that area prior to the deposition of younger strata. Such unconformities are common in the Colorado Plateaus, but by piecing together the geologic sections from the several areas, the missing formations are found and fitted into place.

Elevations range from about 5000 feet (1524 m) in the lower plateaus to more than 11,000 feet (3354 m) in the Aquarius Plateau of the High Plateaus Section of the province, along the western border. There the plateau stands high above the low-lying Basin and Range Province, which lies to the west of the Hurricane and Sevier faults.

Several areas are covered by Miocene and younger volcanics. The Mt. Taylor area in New Mexico, San Francisco Peaks in Arizona, a large area in southwestern Utah, and Grand Mesa in western Colorado are some of the more important volcanic areas. Locally, as in the Henry Mountains in southeastern Utah, intrusives such as stocks and laccoliths are exposed.

The landforms are dominantly the products of stream erosion under arid and semiarid conditions. Generally, the plateaus are capped by thick and resistant strata, either well-cemented sandstones or limestone. In humid areas, limestones weather readily and form lowlands; in the Colorado Plateaus, limestones such as the Kaibab — the rimrock at Grand Canyon — are highly resistant.

Abrupt topography is the rule, with escarpments, or "scarps," rising almost vertically from one plateau level to a higher one. Easily eroded shale generally forms gradual slopes that extend out from the base of the scarps or cliffs.

The Colorado Plateaus Province is gradually getting smaller — by scarp retreat along the western and southern borders — as the adjacent Basin and Range Province is enlarged. Individual plateaus are also getting smaller by scarp retreat. Blocks of rock fall from the cliffs, either pried loose by frost-wedging or by the removal of the shale support by erosion. In either case, it would seem that there should be an abundance of fallen blocks — taluses perhaps — along the base of the cliffs. Talus deposits, however, are essentially nonexistent and even scattered boulders are rare. The scarcity of boulders prompted the assumption that active scarp retreat had ceased, perhaps when there was a change in climate. This assumption appears to be invalid, however, because Schumm and Chorley (1966) found that (1) many sandstone boulders essentially disintegrate by impact and (2) others weather rapidly once they are partially buried in material that tends to hold moisture. Therefore, scarp retreat continues today although the boulders disappear about as rapidly as they form.

Deep, steep-walled canyons are characteristic of the province. In places where canyons are closely spaced, the flat plateau surface has been destroyed, thus forming isolated peaks such as Shiva Temple in Grand Canyon.

Where there are large areas of shale, extensive badlands have been developed by the dissection of the easily eroded materials by many small streams. Badlands are found in Petrified Forest National Park, Capitol Reef National Park, along the Colorado River in the Grand Junction, Colorado, area, and in many other places.

The geomorphic history of the Colorado River is too involved for detailed discussion. Early "greats" such as Powell (1875), G. K. Gilbert (1877), Dutton (1880), Emmons (1879), and Wm. M. Davis (1901) delved into the problem but did not solve it completely. Recently, Hunt (1956) attempted to resolve the problem as to whether the Colorado is a superposed stream or an antecedent stream by proposing that it is both, in at least certain of its segments. Those wishing to make an in-depth study of this complex problem should refer to Breed's summary article in the *Four Corners Geological Society Guidebook* (1969). More recently, Baars, (1972) published his account in his book *Red Rock Country*, also recommended.

Canyon de Chelly   **Monument Valley**

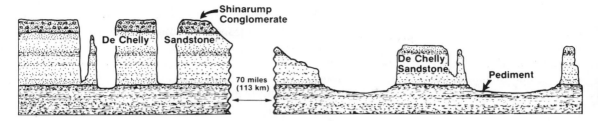

FIGURE 11-4

Generalized cross sections, showing early stage (left) of the arid cycle in Canyon de Chelly area, Arizona, and an advanced stage (right) in Monument Valley, Utah. For details on the stratigraphy, readers should refer to published geological reports.

FIGURE 11-5

Canyon de Chelly with Spider Rock, a spire that rises about 800 feet (244 m) above the canyon floor.

FIGURE 11-6

Monument Valley, southern Utah, with mesas and buttes standing above the pediment surface.

The evolution of landforms in dry regions is well illustrated in the Colorado Plateaus where flat-lying, massive sedimentary formations are common. Unlike humid regions where the steep sides of youthful valleys are gradually flattened, in the plateau country scarp retreat is the dominant process. As a result of the Pliocene uplift, the streams cut narrow, steep-walled canyons; later, they shifted back and forth laterally, undercutting the cliffs and causing large masses of rock to crash down, thus widening the valley floor, as in Canyon de Chelly (Figure 11-5). Eventually, when the streams and their tributaries widen their valley floors, the original plateau surface is dissected into a series of *mesas;* when the mesas are reduced to *buttes,* the cycle is nearing completion, as in Monument Valley (Figure 11-6). Here, broad pediments extend out from the buttes, thus forming a new surface, one that may be uplifted and dissected sometime in the future. When essentially all of the buttes are erased, the cycle will be completed. Thus, Monument Valley contains much more than colorful scenery, more than a setting for a Western movie!

The Navajo Indians have done their part with the parks concept in setting aside this unique area as a Tribal Park. It is, in essence, a national park.

Pleistocene glaciers covered several areas of the province, among them the San Francisco Peaks, the Aquarius Plateau, the Wasatch Plateau, and Grand Mesa. In summary, although the Colorado Plateaus were shaped mainly by stream erosion, in places volcanism was dominant; in other places glaciers put the final touches on the landscape.

Within the Colorado Plateaus Province, there are eight national parks — more than in

FIGURE 11-7

"The Chessmen", fashioned by weathering and by the wind, in Goblin Valley about 20 miles north of Hanksville, Utah.

any other province. Of the numerous national monuments, those with particular geologic significance are included. Each park and monument has its own assemblage of rock formations, many of which are brilliantly colored, and its own geologic structures and individual landforms. Each contains a chapter of geologic history, and together, much of the 4.5 billion years of geologic time is accounted for.

In general, the sequence will be to travel through the Plateaus in a counterclockwise direction, beginning with the Black Canyon of the Gunnison National Monument near the Southern Rockies boundary. The enlarged map, which shows the location of the parks and monuments, may be of assistance.

BLACK CANYON OF THE GUNNISON NATIONAL MONUMENT

The Black Canyon of the Gunnison River is located in Colorado near the boundary of the Colorado Plateaus with the Southern Rockies. Here the Gunnison River has cut such a deep and steep-walled canyon that at the bottom it is rather dark and gloomy, hence the name Black Canyon. It is deeper, its walls are steeper, and it is more spectacular than the famous Royal Gorge. Clearly, Black Canyon belongs in the Park System, and it was so designated in 1933.

The geologic story revealed in the Black Canyon is, at very least, unique. Where else has a stream that really doesn't belong there cut such a spectacular gorge? Gunnison River isn't where it used to be; until recently, it flowed westward several miles to the north of Black Canyon — where the West Elk Mountains are now. Nothing is more diverting to a stream than to have a volcano burst out in the middle of its valley. That is precisely what happened to the Gunnison River a few million years ago when the first of the West Elk volcanoes began to erupt. Now, the stream flows southward, then westward through Black Can-

yon and northward — around the West Elk Mountains. The full story, including earlier events vital to the formation of Black Canyon, is presented by Hansen (1965).

Black Canyon Rocks

The gorge is about 2700 feet (826 m) deep in one place; at the Narrows, it is about 1700 feet (519 m) deep, but at this point it is only 1100 feet (335 m) across. Clearly, in order to maintain nearly vertical walls of this height, the rocks must be both extremely strong and resistant to weathering.

FIGURE 11-8

View of the Black Canyon and the Gunnison River.

All of the rocks in the canyon walls are Pre-cambrian, and they were formed deep in the earth's crust. Most of them were highly meta-morphosed by strong compressive forces, mainly gneisses and quartz-mica schists, with lesser amounts of black hornblende schist. Granitic magmas intruded to form irregular masses, dikes, and sills, which interlace to but-tress the metamorphics, further strengthening them.

A short distance back from the rim, Meso-zoic sedimentaries rest on the Precambrian, with a major unconformity between them. Far-

ther north, the West Elk volcanics cover the sedimentaries.

Black Canyon History

As mentioned, prior to the eruption of the West Elk volcanoes, the old Gunnison River flowed generally westward; now the Gunnison follows a circumferential course around the south side of the West Elk Mountains. Proba-bly the river was shifted southward again and again as the volcanoes expanded, before being allowed to maintain a permanent course in its present location. Volcanoes were also erupt-ing to the south, in the San Juan Mountains. The Gunnison merely followed the lowest route between the two volcanic areas. If the river had been shifted a few miles farther to the south, beyond a major fault, it would have cut its valley into weak sedimentary rocks — a val-ley like many others, entirely unlike the Black Canyon.

When the Gunnison cut through the vol-canics and sedimentaries, it cut into a buried ridge of hard Precambrian rocks. There was no other way for it to go because it was now con-fined in a deep valley; therefore, it was super-posed across the resistant ridge. The overlying rocks adjacent to the canyon were stripped off, mainly by tributary streams. Spires and sharp peaks such as Curecanti Needle rise from the bottom of the canyon.

Because of the depth of the canyon and the superior resistance of the rocks, it was once assumed that canyon cutting required many millions of years. Now that the sequence of events is known, it is obvious that canyon cut-ting did not begin until after at least most of the paroxysms of the West Elk volcanoes had died down, perhaps only a few million years ago. Hansen suggests that the Gunnison River may have done its work during the last 2 million years.

Considering the hardness of the rocks, the rate of cutting was abnormally rapid. Equipped with an abundance of good cutting tools —

FIGURE 11-9

Granite dikes and sills (light colored) that cut through metamorphics in the Painted Wall. (Photo by Na-tional Park Service.)

hard, abrasive rocks carved out of the Sawatch Range — the Gunnison was equal to its task. High-volume flows were assured as the streams head in the lofty, well-watered mountains; high-velocity flows through the canyon were assured because of an unusually steep gradient, one that was even steeper in earlier times.

The Gunnison is actively downcutting today; the canyon will be deeper tomorrow and Curecanti Needle will rise higher above the floor.

Black Canyon is not a large monument — only about 22 square miles — but not all of the choice park areas are large.

MESA VERDE NATIONAL PARK

Mesa Verde was established in 1906 to preserve extensive and unusual archeological records of the Cliff Dwellers of southwestern Colorado. They were apartment-dwellers; rock and adobe-mud apartment houses high on the cliffs, beneath overhangs, are of many types and sizes.

The Park Service has restored some of the ruins, after a long period of abandonment which began about 1299 A.D. During early occupancy, the Mesa Verde Indians — the Anasazi — farmed the "green plateau" and lived near their fields atop the mesa. About

FIGURE 11-10

Spruce Tree House below park headquarters. Cliff House Sandstone (Cretaceous) forms the shelter above the ruins.

1200 A.D. they decided to move down under the overhangs of the Cliff House Sandstone. Here they were protected from the elements, and intruders could be dealt with—with dispatch. But in 1276 a long drought began and, according to the archeologists, this was the beginning of the end for the Mesa Verdes.

At Mesa Verde, as at Bandelier National Monument, geological developments made it possible for the Anasazi to utilize the areas as they did. The geologic story begins with the deposition of mud late in the Cretaceous Period—mud in vast quantities that became the thick Mancos Shale. Later, in what is referred to as Mesa Verde time, this part of the geosyncline received shallow-water sediments, mainly sand that was cemented to form the Point Lookout Sandstone. Then more mud and sand were deposited in the sea; the presence of coal layers, however, indicates that the area was slightly above the sea for a time. Again below the sea, the massive Cliff House Sand-

stone was laid down as the youngest member of the Mesa Verde Formation.

The Laramide Orogeny, which so deformed the Rockies, also affected the Colorado Plateaus; certain areas were faulted and folded, and others, including the Mesa Verde area, were merely uplifted. As the result of this and later uplifts, Mesa Verde is now about 8000 feet (2439 m) above sea level; the latest uplift, late in the Tertiary, caused the streams to cut deep canyons, leaving the flat intercanyon areas as mesas.

By weathering and mass-wasting of the soft rocks beneath the Cliff House Sandstone, the overhangs were formed. In places, as at Cliff Palace, slabs of the roof spalled off and fell in, leaving broad archways high above the base. In these large niches beneath the overhangs, the Mesa Verdes built their apartments—at Cliff Palace, Spruce Tree House, Balcony House, and many other places. Here they were well situated; adding to their well-being was a reliable water-supply from springs at the base of the Cliff House Sandstone. Their crops did well—until the Big Drought—on the mesa above. Important too, at least to the archeologists of today, they had giant "dispose-alls" that were the canyons, right at their front doors.

Mesa Verde is open throughout the year but many of the cliff dwellings can be visited only in the summer. The Wetherill Mesa section, in the southwest part of the park, was opened for summer visitation in 1973. Private vehicles are not allowed in this section. Buses take visitors part of the way; then they transfer to the mini-train, which winds through the pinyon-juniper forest to the edge of the mesa, above Long House Ruin.

Views of the surrounding lowland area are excellent from a number of points on the mesa. Until recently, the view of Shiprock—a huge, volcanic neck that rises abruptly above the flats—was one long to be remembered. Now, on certain days the smog from the nearby Four Corners power plant essentially blots out this striking feature.

FIGURE 11-11

Air view of Shiprock, northwestern New Mexico, and one of the radiating dikes. (Photo by S. A. Schumm.)

COLORADO NATIONAL MONUMENT

In western Colorado, near Grand Junction, about 28 square miles of the Uncompahgre Plateau was designated as a national monument in 1911. John Otto was responsible for this colorful part of the Colorado Plateaus becoming a monument. Otto was so entranced by the fantastic erosion forms in the canyons that he single-handedly carved out trails up the canyon walls, so that he could share this wild country with others. Finally, with the support of the Grand Junction Chamber of Commerce, he was successful in persuading the Secretary of the Interior that it should be a national monument. To celebrate, Otto cut steps and handholds up the side of a 450-foot-high sandstone monolith and on the 4th of July planted the American flag on top of Independence Monument.

In the early days, if you went into the monument at the south end, you followed the Serpents Trail, a 54-switchback road which took you up out of No Thoroughfare Canyon to the top of the mesa. Now, a paved highway which has replaced the Serpents Trail takes you up out of the canyon and around the rims of the canyons to the visitor center, and then down Fruita Canyon to Interstate 70 near Fruita, Colorado.

Colorado National Monument is a good place to get acquainted with plateau geology. Exposed in textbook fashion in the sheer walls of the canyons are many of the formations which are repeated in the other plateau park areas. Also, the Park Service, under the guiding hand of Stan Lohman of the U.S. Geological Survey, entices you into the geology of the monument in a simple but effective way — by putting up markers identifying each geologic formation as you come to it along the road. To further your education, interpretive plaques at the overlooks outline the evolution of the canyons and explain the origin of the other geological features you see below you. In addition, in a well-illustrated booklet, *The Geologic Story of Colorado National Monument,* Lohman tells the story of the monument area, fitting it neatly into the broader area, the Colorado Plateaus.

The Geologic Story

The monument is an elongate area on the northeastern flank of the Uncompahgre Plateau, also known as the Uncompahgre Uplift. The rocks are mainly Mesozoic sedimentaries that are essentially flat-lying except along the northeastern border of the monument area. Here, they dip steeply toward the northeast. If you look across the broad Colorado River Valley to the Book Cliffs you will see that they are flat-lying or nearly so. But those rocks are Tertiary in age, much younger than those in the monument. With horizontal rocks on both sides of a dipping section, the structure is a monocline. Originally all of the rocks were flat-lying, but along one section they were flexed or bent down, as shown in Figures 11-13 and 11-14.

FIGURE 11-12

Independence Monument rising high above canyon floor, Colorado National Monument.

FIGURE 11-13

Cross section showing flat-lying sedimentary rocks resting on granitic basement rocks.

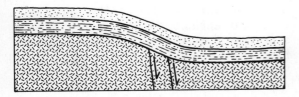

FIGURE 11-14

Schematic cross section of rocks shown in Figure 11-13 after faulting in basement and folding of younger rocks — a monocline.

Another main structure is an unconformity below the sedimentary rocks. Exposed in the bottoms of the canyons are black rocks — ancient Precambrian rocks — below the unconformity. You will see a good exposure soon after you enter the monument from the north, in Fruita Canyon. As elsewhere in the monument, the bright red Chinle (Triassic) rests on these black Precambrian rocks. These are the same Precambrian rocks that form the walls of the Black Canyon of the Gunnison and this is the same unconformity. Because the Precambrian rocks are extremely resistant and the old erosion surface was fairly smooth, the period of erosion must have been extremely long; it is called the "Great Unconformity."

There are also two younger unconformities, as shown in Figure 11-15. Above the red shales of the Chinle are the bold cliffs of Wingate Sandstone, capped by the Late Triassic Kayenta formation, mainly sandstone and conglomerate. Immediately above should be the Navajo; instead, the Entrada rests on an eroded surface in the Kayenta, above the unconformity. The Navajo and Carmel formations were removed by erosion during early Jurassic time, while the lower Entrada was being deposited in surrounding areas. Then the varicolored mudstones and siltstones of the Summerville and the Morrison were laid down. During the Morrison time, dinosaurs slogged about in the shallow lakes and swamps. Dinosaur bones have been found in the Morrison outside the monument; probably more bones are resting peacefully, deep in one of the hills inside the boundary — to be resurrected some day by erosion.

Three Cretaceous formations, the Burro Canyon, the Dakota, and the Mancos, are exposed in the higher hills. After the shales and sandstones of the Burro Canyon were laid down there was a period of erosion prior to the deposition of the gravels which became the basal member of the Dakota. The interval represented by this unconformity was relatively short.

Tertiary formations were probably laid down and then entirely eroded away when the Uncompahgre Uplift occurred late in the Tertiary. At this time the beds were up-arched, forming the plateau and the monocline, which was in places broken by faults. Streams adjusted their courses to the new structures and carved out the exciting features of today.

Today's Landscapes

Canyons — deep, steep-walled canyons — and flat-topped mesas are the two major landforms in the monument. Some of the canyons are U-shaped in cross-section, suggesting that they might have been carved out by valley glaciers. In the Colorado Plateaus, however, only limited areas — areas much higher than any in the monument — were glaciated. And no glacial deposits are found here. The canyons, therefore, must have been formed by stream erosion. Where are the streams? You are likely to drive from entrance to entrance without

GEOLOGIC AGE GEOLOGIC FORMATIONS

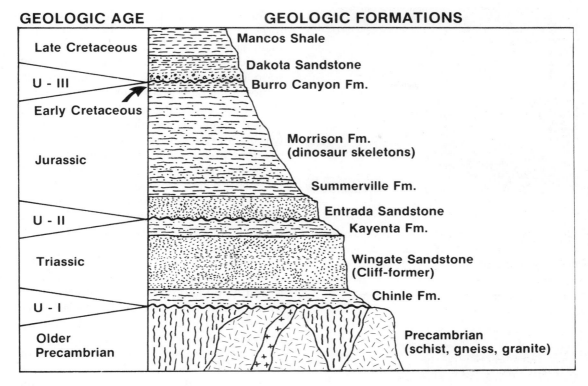

GEOLOGIC AGE	GEOLOGIC FORMATIONS
Late Cretaceous	Mancos Shale
	Dakota Sandstone
U - III	Burro Canyon Fm.
Early Cretaceous	
Jurassic	Morrison Fm. (dinosaur skeletons)
	Summerville Fm.
U - II	Entrada Sandstone
	Kayenta Fm.
Triassic	Wingate Sandstone (Cliff-former)
	Chinle Fm.
U - I	
Older Precambrian	Precambrian (schist, gneiss, granite)

FIGURE 11-15

Columnar section of formations in Colorado National Monument. U-I: major unconformity. U-II and U-III: disconformities. (Adapted from Lohman, 1965 U.S. Geological Survey.)

seeing a drop of running water, only dry streambeds. But if you are there at the right time, during a high-intensity summer storm, you will see erosion taking place, and at a high rate. You may see a wall of water convert the dry streambed into a raceway. The stream will likely resemble a mudflow, as the loose, weathered material that has accumulated since the last flood is flushed out of the watershed.

FIGURE 11-16

View east down U-shaped, unglaciated canyon floored by Precambrian rocks; Wingate Cliffs rise above Chinle Shale slopes.

FIGURE 11-17a

View northwest to the monocline in Colorado National Monument; Redlands Fault is located along the base of the cliff.

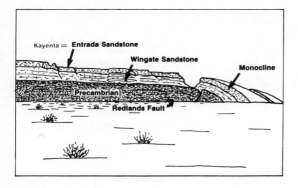

FIGURE 11-17b

Sketch of geologic formations shown in Figure 17a. (For more details, see Lohman, 1981.)

The U-shaped valleys are the result of a coincidence that began when, during the Triassic, the poorly resistant Chinle shales were laid down on the resistant Precambrian metamorphic and igneous rocks, and then the thick, cliff-forming Wingate Sandstone was deposited on top of the Chinle. When the streams have cut down through these and younger sedimentaries, they find rough going in the Precambrian. They shift from side to side and undercut in the soft Chinle, causing huge slabs

of the overlying Wingate to fall. Vertical fractures in the massive Wingate play a part in maintaining near-vertical cliffs as the canyons are widened. The broken slabs of sandstone soon disintegrate, and this sand forms the slopes at the base of the cliff, thus forming the U-shaped valleys or canyons.

As the canyons are widened the mesas between them are narrowed, until only high narrow rock walls remain. Then, when sections of the wall tumble down, only monoliths or monuments are left. Independence Monument, the 450-foot-high monolith in Monument Canyon, is the sole survivor of such a wall, as Lohman (1981) points out. The caprock is the Kayenta, which is much more resistant than the Wingate that forms most of the monolith.

When the Kayenta is removed, the Wingate is weathered and eroded into unusual forms, some of which are similar to huge beehives or coke ovens. The Coke Ovens in the upper part of Monument Canyon are examples.

Archways are common in the walls of the canyons, particularly in the Wingate Sandstone. By weathering and/or undercutting, blocks of the sandstone fall down, eventually forming a broad, graceful inset arch overhead. The archways north of Artists Point are most impressive.

You can see many intriguing features from the rim; however, trails that lead down into the canyons afford opportunities to see at close range not only the geologic features but the plants and wildlife — including, with luck, the bison. Take a full canteen of water and save most of it for the long pull up out of the canyon, particularly in the heat of the summer. Colorado National Monument is open throughout the year.

ARCHES NATIONAL PARK

Any other name would not be appropriate for Arches National Park. There are other interesting features, but arches clearly dominate the landscape. Arches, baby arches, and windows are so numerous that you soon lose count. The park was established as a monument in 1929 and upgraded in 1971. Located north of the Colorado River, about five miles north of Moab, Arches is in the redrocks country of Utah.

An attractive visitor center is located at the park entrance on U.S. Highway 163, and paved roads now lead into several sections of the park. Those of us who were there in the early days are immediately aware of the change from the primitive to the modern. At that time a visit to the monument was an adventure, particularly for the visitors who were marooned by a flooded stream. For a fascinating story of the Arches of yesterday, read Stan Lohman's U.S. Geological Survey Bulletin 1393 (1975). Among those who think back wistfully to the primitive Arches is Edward Abbey who, in his *Desert Solitaire,* speaks eloquently in behalf of wilderness. Fortunately, there remain today several areas of the park which can be reached only on foot.

The Arches

Architects may believe archways, the Gothic Arch and the arch bridges to be products of their own fertile brains but this fallacy should be laid to rest. They have merely copied natural arches — arches that have stood for thousands of years, a true test of design.

Arches and natural bridges are similar in general appearance, but geologists insist on the two designations because they are formed by different processes. A natural bridge spans a valley of erosion and an arch does not; arches therefore cannot be formed by stream erosion. Instead, they are formed by weathering; in Arches National Park they were developed by the "holing through" of thin walls or "fins" of the Entrada Sandstone of Jurassic age. The calcium carbonate cementing material in the Entrada is not uniformly distributed, and chemical weathering is most rapid where the

sandstone is least cemented. Here, a tiny hole or "window" is formed which, when enlarged sufficiently, is called an arch. Windows enlarge slowly by weathering, and rapidly by the collapse of blocks of roof rocks. In 1940, the falling out of a huge block almost doubled the size of the opening beneath Skyline Arch.

When does a window become an arch? Obviously, the dividing line is an arbitrary one, and the number of arches and windows, regardless of specific size, changes with time. Perhaps by now one or two more windows have become arches; conceivably an old arch somewhere may have fallen. Now, Delicate Arch stands lonely on a high ridge where others once stood. (For details on the origin of natural bridges, see Natural Bridges National Monument.)

Natural arches are found in a number of places in the Colorado Plateaus but the concentration in this small area deserves special attention. Before arches can form, fins must be present. These high walls of rock are bounded by parallel, vertical fractures along which weathering takes place. Therefore, the fracturing that occurred in Arches was particularly right for the development of fins (refer to Figure 11-2).

Geologic History

Rocks as old as Pennsylvanian and as young as Cretaceous crop out in the general area of the park, recording a variety of ancient environments. During part of the Pennsylvanian, shallow enclosed basins were the sites of chemical precipitation of salts from entrapped sea water. As evaporation took place, the waters became sufficiently concentrated to cause precipitation of salt (halite), gypsum, and in places potash salts. Later, marine deposition occurred when the sea covered the area that is now the Colorado Plateau. But the area was also above the sea, as indicated by the eolian sands of the Navajo Formation. These "frozen dunes" are particularly well exposed in the

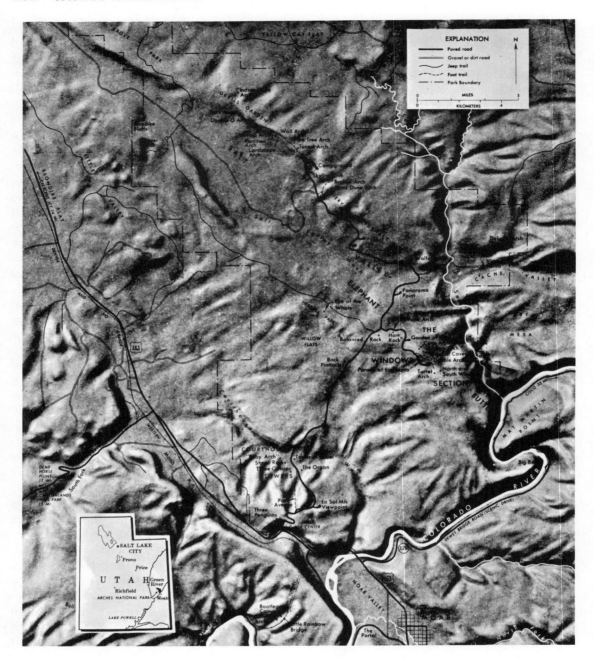

FIGURE 11-19

Map of Arches National Park.

southeastern section of the park.

In Colorado National Monument, the Navajo was absent, having been eroded off entirely; in Arches National Park, only part of it was stripped off. Then, the Entrada Sandstone was laid down on the erosion surface, now an unconformity. The Entrada is mainly red and reddish brown, due to iron oxides that coat the grains of sand. That it is merely a coating can be verified by observing the color of windblown sand. Where it is near its source, a red sandstone cliff, the sand is distinctly red. If it has been carried for several miles, it may be coral pink. Far from the source, the sand will be essentially white, the red coating having been worn off of the quartz grains.

Younger formations that were laid down were later eroded off most of the area; therefore, the Entrada and the Navajo are the dominant formations in the park. However, two of the salt anticlines described earlier lie within the park. The Salt Valley anticline is now a *graben* (German for 'grave'), as a result of collapse along the crest, where the salt was leached out. Within the graben, down-dropped masses of rock as young as the Upper Cretaceous Mancos Formation have been preserved. The Cache Valley graben, formed in the same way, lies to the southeast and can be seen from Panorama Point.

Uplift, folding, and faulting, mainly in Tertiary time, initiated the erosion that eventually exposed the Navajo and the Entrada. One large fault, the Moab Fault, is readily observed from a viewpoint on the park road about a mile from the visitor center. The rocks on the skyline on the left side of the valley are older than those on which the visitor center rests.

But it was the folding that was primarily responsible for the concentration of arches in this one area. The anticlines and synclines trend northwestward, and the main fractures are vertical and parallel to the fold axes. Therefore the fins are also parallel. Where the fractures are closely spaced, weathering is at its maximum, and the freed sand grains are swept away by wind and streams, leaving the fins standing high as walls. The spacing of the fractures, some closely spaced and others widely spaced, is a critical factor in the formation of the fins in which first the windows and then the arches are developed.

The complete sequence of development of arches begins with "Baby Arch," scarcely more than a window, as seen in Figure 11-20 in color section. When enlarged by the downfall of sandstone blocks, it will be similar to Skyline Arch (Figure 11-21). Landscape Arch represents the last stage; because of its length (291 feet; 89 m) and its thinness (11 feet; 3 m), its days are numbered. Visitors are no longer allowed on top, in order to postpone the downfall of the arch, and you. A violent sonic boom could cause it to collapse. Fortunately, as Lohman (1975) relates, the Air Force has been persuaded to route their supersonic aircraft around Arches and other national parks. In many places, only the abutment rocks still stand, representing the completion of the cycle. But there are many windows, soon to become arches, to take their place.

The Fiery Furnace — Devils Garden Section

In this northern section, you will see perhaps more fascinating features than in any other part of the park. At Stop 20 in your Road Guide (available at the Visitor Center) is a maze of sandstone walls and pinnacles called the Fiery Furnace. If the light is right, you will have no difficulty understanding why it was so named. Stop 25 is the end of the road but the beginning of a trail that you really should not miss — up through the Devils Garden with its 25 arches, including the world's largest, Landscape Arch, already mentioned. Wear hiking boots and carry a full canteen of water.

The Old Records

The early inhabitants of the Colorado Plateaus left their marks in many places. Both picto-

FIGURE 11-21
Skyline Arch, much enlarged but still structurally stable.

FIGURE 11-22

Landscape Arch, old and frail.

FIGURE 11-23

Narrow corridor between two high fins in Devils
Garden, Arches National Park.

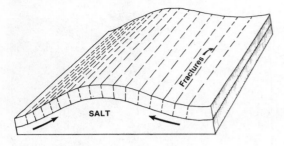

FIGURE 11-24a

Schematic block diagram showing the development of a salt anticline. Salt, plastic due to the weight of overlying rocks, is forced laterally to arch up the rocks where pressures are less.

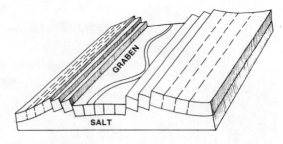

FIGURE 11-24b

Later stage than that shown in Figure 11-24a. After salt has been dissolved and carried away, collapse of roof rocks forms a graben (see Figure 11-2).

graphs (paintings) and petroglyphs (etchings) adorn the sandstone walls in the Arches area. As Lohman points out, the early ones knew how to fashion lasting designs — by pecking loose the thin iron oxide coating and exposing the lighter colored sandstone. Excellent petroglyphs are readily accessible near the old Wolfe Ranch, trailhead for the arduous but rewarding climb up to Delicate Arch, the hallmark of Arches National Park.

CANYONLANDS NATIONAL PARK

About 30 river miles (50 km) downstream from Arches National Park, the Colorado River flows into the northeastern corner of Canyonlands. Green River enters the park at the northwest corner; thus the park includes the canyons of both the Colorado and the Green rivers. In addition, there is a small remote area, Horseshoe Canyon Detached Unit, about ten miles west of Green River. Below Canyonlands, the Glen Canyon National Recreation Area extends downstream to the Glen Canyon Dam, near Page, Arizona. In the park areas, we saw the Colorado River first at its source high in Rocky Mountain National Park. Then it was near Colorado National Monument and

Arches, and it flows through Canyonlands and Glen Canyon (Lake Powell). When we add Grand Canyon, we are impressed with the influence of the Colorado River on Park Service developments.

Canyonlands National Park consists of about 525 square miles of wild plateau country — canyons, mesas, pillars, standing rocks, grabens, and arches. Although several attempts were made to establish a park here earlier, success was not forthcoming until 1964.

Canyonlands is not for the casual tourist with a couple of hours to spare. It is largely undeveloped, and advance preparations are necessary if the visitor is to do more than catch glimpses of what Canyonlands has to offer. The main roads are travelable by highway cars, but 4-wheel-drive vehicles are needed to get into the more spectacular areas. Even so, storms may make any of the jeep trails impassable.

The park brochure contains both warnings and helpful suggestions to make your visit a success. More details on "logistics" are included in Lohman's 1974 *Geologic Story of Canyonlands National Park,* along with its abundance of beautiful color photographs. In addition to the geology, he outlines the early

FIGURE 11-25

Map of Canyonlands National Park.

FIGURE 11-25a

Air view of the confluence of the Green River (left) and the Colorado River, Canyonlands National Park. (Photo by G. Anderson)

occupation of the area by the Anasazi, with their cliff dwellings and granaries. These people left records in the form of pictographs and petroglyphs, particularly in the Horseshoe Canyon and Salt Canyon sections.

Major John Wesley Powell and his party explored the canyons of the Green and the Colorado in 1869 and 1871; his accounts of these two voyages appeared in 1875. Although he was not the first geologist to see Canyonlands, his reports were remarkably complete and of lasting significance. Today, boat trips on the Colorado can be arranged in Moab, Utah, northeast of the park. Views from the river, up the towering canyon walls, are fully as impressive as those looking down from the rim.

Canyonlands is a time-consuming park. It consists of three main sections and each one is reached by a distinctly different route. If the Island in the Sky between the Green and the Colorado is explored first, the others may appear less forbidding. The entrance road leads off from U.S. Highway 163 at a point ten miles (16 km) northwest of Moab. The southern or

Needles district is at the western end of Utah Highway 211, a 38-mile-long paved road leading off from U.S. Highway 163 about 40 miles (65 km) south of Moab. The Maze-Standing Rocks District and the Horseshoe Canyon Detached Unit are west of the junction of the two rivers; these remote areas are reached by a jeep road which leads southeast from Utah Highway 24, north of Hanksville.

Bedrock Geology

As in Arches, the rocks in Canyonlands are all sedimentary; the dominant features, however, are older than those in Arches. The Permian Cedar Mesa Sandstone forms most of the ponderous arches, the Needles, and the other erosion forms in the southern section, with the exception of certain canyons that expose the Pennsylvanian Hermosa Formation. In the northern section, the dominant cliff-formers are the Triassic-Jurassic Wingate, Kayenta, and Navajo sandstones. The highly resistant White Rim Sandstone (Late Permian), however, plays an important role, capping a flattish platform between the high cliffs on one side and the deep gorge on the other. The White Rim Trail, for jeeps and horses, winds around on this platform, below the Island in the Sky. The Entrada, important in Arches, forms the Monitor and Merrimac Buttes, which stand out like battleships near the north entrance.

For truly outstanding panoramas of the northern section of Canyonlands, visit Dead Horse Point State Park a few miles east of the north entrance of Canyonlands. There below you is a broad loop of the Colorado River, the stream responsible for the exposure of all the rock formations pictured in Figure 11-1.

Geologic Structures

At first view, Canyonland rocks appear to be horizontal. Actually, however, broad anticlines and synclines with northwest-southeast trends dominate over a large area of the Colorado Plateaus, including Canyonlands. With low dips in most places, it is generally necessary to view broad areas in order to visualize these structures. One structure that is recognized immediately is Upheaval Dome in the northwestern corner of the park. It is a structural dome that has been eroded into a topographic basin, an inversion of topography (see Figure 11-26) in color section. The stream, which with its tributaries carved out the topographic basin, flows out through a gap in the west rim down into Green River. The origin of the dome is a puzzle. Is it, as some think, a salt dome? Salt layers are known to occur in the Paradox Formation (Pennsylvanian) elsewhere in Utah, and salt domes are definitely established geologic structures. Perhaps, as Stokes (1948) suggests, the weight of the rocks on top of the salt caused it to flow in toward this center and dome up the overlying rocks. But why should the salt have moved toward one specific center? Another theory is that it is a cryptovolcanic structure — one in which gases accumulated below the surface and lifted the rocks up by explosion. Still another theory involves the intrusion of magma in the form of a laccolith. Possibly doming and erosion occurred, thus removing sufficient weight to permit salt flowage to accentuate the doming.

The grabens near the junction of the two rivers in the southern part of the park are apparently down-faulted elongate blocks of the earth's crust; Devil's Lane is an outstanding example. Another theory that some favor is related to the salt dome origin of Upheaval Dome. This one, advanced in the park's brochure, holds that large quantities of salt were taken into solution by percolating groundwaters; the overlying rocks simply collapsed into the space vacated by the salt. This theory appears to have support because, as Lohman points out, the fault scarps along the sides of the grabens are fairly recent — too recent to fit into any known crustal movements in the Plateau region. The scarps are almost untouched by weathering and erosion; moreover, the

streams have not had sufficient time to completely adjust to the newly formed topography.

Erosional Landforms

Under this general heading, the processes of weathering and mass-wasting are implied and in some cases are fully as important as erosion itself. Within the confines of the park, fins, columns, pillars, pinnacles, needles and monoliths have been formed in untold numbers by this combination of processes.

Special mention must be made of the ponderous arches situated in remote sections of the southern part of the park, Angel Arch and Druid Arch in particular. Neither of these would have been formed in rocks devoid of fractures or joints. A new trail to Druid Arch involves a round-trip hike of 8.5 miles, well

FIGURE 11-28

Paul Bunyans Potty, an archbridge in the southern section of Canyonlands National Park. (Photo by J. Gregory.)

worth the effort of those in good physical shape. Paul Bunyans Potty, a so-called "horizontal" arch, can be visited with less effort, in a jeep. Not far from the Potty, perched high on the cliff, is Keyhole Ruin, a granary built by the Anasazi, probably during the twelfth century.

The geologic work of erosion continues today. In this dry country where the annual rainfall is less than 10 inches, most of the streams are dry except during "cloudburst" storms that generally occur in the late summer. Flash-floods remove large quantities of accumulated weathered material, deepening the canyons and arroyos.

No matter how you see the park — from a jeep, a horse, or your own two legs — Canyonlands is worth it.

FIGURE 11-29

The natural bridge of the future, Natural Bridges National Monument. When the stream erodes through the base of the "peninsula" of rock, a new natural bridge will be formed.

NATURAL BRIDGES NATIONAL MONUMENT

Within the National Park System each park area has its own particular role in portraying the total environment. Although natural bridges are found in other park areas — Capitol Reef and Rainbow Bridge for example — nowhere known to us are they as completely displayed as here. In this small area of about 7600 acres are young, mature, and old bridges; in addition, there is at least one natural bridge of the future — one that may be born about the time the oldest one collapses.

Natural Bridges is readily accessible from Blanding, in southern Utah, by way of U.S. Highway 163 and Utah 95. Or, follow U.S. 163 north from Kayenta, in northeastern Arizona, through beautiful Monument Valley to Mexican Hat, Utah. After visiting the spectacular Goosenecks of the San Juan, take Utah 261 north to Utah 95. Be prepared for a steep multi-switchback section as you climb out of the valley of the San Juan up onto the plateau.

Natural bridges are formed by more than one geologic process. The famous Natural Bridge in Virginia is the only section of a cavern roof which did not collapse. In the Colorado Plateau, however, it is almost invariably stream erosion which is responsible. Natural bridges of this type, however, are formed only under special circumstances. First the rock should be thick-bedded — massive — and strong enough to support the bridge floor; the Cedar Mesa Sandstone member of the Permian Cutler Formation admirably fulfills this requirement. A meandering stream must entrench itself into the bedrock; then, by undercutting the rock wall on the two sides, the stream eventually breaks through the wall, as shown in Figure 11-29. When this occurs, the stream abandons its looping channel from point X around to Y and flows from X through its new opening directly to Y. By downcutting its new steep channel, the stream enlarges the opening; it is further enlarged by weathering and by the downfall of blocks of rock from above and from the sides. Figure 11-29 shows a natural bridge in the making.

In the monument, Kachina Bridge represents the youthful stage of development. Flood waters continue to enlarge it, and the bridge floor is very thick and bulky. Sipapu

FIGURE 11-30

Owachomo Bridge, 180 feet long, represents a late stage of bridge enlargement. Natural Bridges National Monument displays the entire sequence.

Bridge is older and has been greatly enlarged; its floor is about 50 feet thick, roughly half as thick as that of Kachina. Owachomo Bridge is only about 9 feet thick and as it is 180 feet long, it is clearly in the last stage prior to collapse. Large blocks have recently fallen at each end, weakening the structure (see Figure 11-30). In a geological sense Owachomo's downfall is near — perhaps within a few centuries, conceivably much sooner.

After stopping at the Visitor Center to look at the exhibits and see the slide show, drive around the 8-mile-long loop. From the overlooks you can view the bridges from afar. But take time to hike down the trails for a closer look. Stand under old Owachomo and look up at the bottom of the bridge floor, more than a hundred feet above you. Then, if you have confidence that the bridge will not fall down, go around to the north end, climb up and walk across it. There are large solution pits on top and if it has rained recently they will contain water.

This section of Utah contains Indian ruins, mainly cliff dwellings, some of which are as much as 2000 years old. On one of the abutments of Kachina Bridge are a number of pictographs, some of which resemble kachinas — Hopi dancers. Should you hike between the bridges you will see one of the cliff dwellings, complete with kivas and granaries.

In this monument, we were able to visualize the sequence of natural bridge development, from the prenatal stage to old age. Next, we go about 65 miles southwest to Rainbow Bridge, perhaps the most perfect natural bridge in the world.

RAINBOW BRIDGE NATIONAL MONUMENT

Rainbow Bridge is a huge arch-bridge spanning Bridge Creek in southernmost Utah. Located in remote country, it was not widely known in 1910 when it was set aside as a national monument. Visitation was limited for a number of years because it was far from the beaten path. Long trails lead in from Navajo Mountain Trading Post and from Rainbow Lodge, which are in turn at the ends of long, dusty one-track roads. Now that the Colorado River is dammed at Glen Canyon and an arm of Lake Powell extends far up Bridge Canyon into the edge of the monument, tourist boats take visitors from Wahweap Marina, near Page, into the monument.

Architecturally, Rainbow Bridge is inspiring; it is huge — 278 feet long and 309 feet high — and it is almost symmetrical. By what series of geological coincidences was this masterpiece fashioned?

One important incident was the formation of the thick, massive, and homogeneous Navajo Sandstone. The Navajo represents an interval early in the Jurassic, and perhaps including the latest Triassic, when the Colorado Plateaus area was even more arid than now. It was essentially devoid of vegetation in most places and the wind went about its work without obstruction. Wind erosion and deposition were widespread; dune deposits over almost all of the Plateaus area varied in thickness but reached a maximum of about 2200 feet (671 m) in the Zion National Park area (Lohman, 1965).

Another significant incident was the deposition of the Kayenta Sandstone, underlying the Navajo and forming the abutments of Rainbow Bridge. These sands were deposited by water during the latter part of the Triassic. Cementation in the Kayenta is in most places, including the monument area, more complete than in the overlying Navajo. Therefore, the Bridge arch is composed of sandstone with adequate cementation, and the abutments are of firm, well-cemented sandstone.

A series of events resulted in the formation of the bridge. Eons before the creation of Lake Powell, when the Colorado River's channel was much higher up — even above the present surface of Lake Powell — meandering tributaries, including Bridge Creek, emptied into the river. When the Colorado Plateau was uplifted late in the Tertiary, the Colorado River entrenched its channel, causing the tributaries in turn to entrench theirs. Bridge Creek was flowing then at perhaps the level of the present top of the bridge, and one of its horseshoe-shaped meanders encircled the end of a high wall of Navajo Sandstone. The stream, undercutting at the ends of the horseshoe, finally broke through the rock wall; abandoning the horseshoe for a shorter course, from then on Bridge Creek flowed through the hole in the wall. As Bridge Creek deepened its channel beneath the new bridge, the walls adjacent to the stream became unstable and fell in, block by block, widening the opening and lengthening the bridge. Blocks also fell from above until finally a stable symmetrical form was attained. Fortunately, in this dry region chemical weathering is very slow, and Rainbow Bridge is changing at an extremely slow rate.

In recent years a controversy has developed regarding Rainbow Bridge. When Lake Powell is brought up to full height, Bridge Canyon will

be flooded and the water will extend up under the bridge. It will not, however, rise high enough to reach the bridge abutments; instead it will partially fill the 75-foot-deep gorge beneath Rainbow Bridge. It has been questioned whether the water will penetrate into the surrounding rocks and dissolve out the cementing material. If so, would there be sufficient local settling to cause fracturing and ultimately the collapse of the bridge?

Geological investigations directed specifically to this problem apparently failed to reveal anything alarming. It should be pointed out, however, that the weight of Lake Powell water will, over a period of years, cause subsidence — differential subsidence — of the entire area, just as Lake Mead caused significant subsidence of the area around it a short distance downstream.

Well aware of the possibility that this subsidence could cause fractures to develop in the sandstone of the bridge, three agencies, the Bureau of Reclamation, the National Park Service, and the U.S. Geological Survey, are cooperating in the recording of all data pertinent to the problem. A carefully designed monitoring program is now in operation, and should new fractures develop or existing fractures widen, the information will be immediately available so that appropriate remedial measures can be undertaken promptly. The monitoring installations are located inconspicuously and monument visitors will likely not be aware that they exist.

Rainbow Bridge is the largest and many believe the most beautiful natural bridge in the park system, perhaps in the world. It is reassuring to know that it is being treated with such respect.

CAPITOL REEF NATIONAL PARK

Capital Reef National Park was established as a national monument in 1937 and with considerable enlargement became a 241,865-acre national park in 1971. Of Utah's many colorful parks, Capitol Reef is outstanding for its combination of geology, archeology, and ecology. It is a highly photogenic park with rock formations of many colors — brilliant reds, yellows, purples, greens, and grays. The "Reef" is a bold cliff of sandstone which, emblazoned in the setting sun, is one of the many entrancing features of the park. Capitol Reef is also colorful in another way — its recent history. The exploration and settlement of the area, along with its questionable but immortal characters such as Butch Cassidy and his Wild Bunch, are included in Virgil Olson's comprehensive *Capitol Reef: The Story Behind the Scenery,* brilliantly illustrated by David Muench's photographs (1972).

FIGURE 11-32

Capitol Reef. The term "reef," as used here, describes a high cliff or escarpment. (Photo by Gordon Anderson.)

General Geology

Within the park, almost all of the rocks are sedimentary and range in age from Permian through all of the Mesozoic. The park is essentially Mesozoic, because the Permian rocks are confined to small exposures, mainly in the canyon bottoms. Olson includes a geologic column, in color, with the outstanding erosional features characteristic of particular geologic formations — an effective and attractive presentation.

Igneous activity within the park took place in the northern sections, mainly in Cathedral Valley and South Desert areas where dikes, sills and plugs are prominently exposed. In the area west of the park, lava flows cap the mountains; during the Pleistocene, glaciers on Thousand Lake and Boulder mountains carried basalt boulders downvalley, and meltwater streams transported many of them into the park.

The Mesozoic stratigraphic section is relatively complete in the Capitol Reef area. As in other areas of the Colorado Plateaus, deposition was by no means a continuous process.

Many of the rocks are of continental origin and, with local uplifts, areas that had been receiving sediment were eroded. As Gregory (1939) shows in his stratigraphic sections, there are several unconformities — for one, between the Moenkopi and the overlying Shinarump Conglomerate. Inasmuch as there was very little tilting involved in the movements upward and downward, the beds above and below the unconformities are essentially parallel; these unconformities are therefore *disconformities* rather than *angular unconformities.*

Structure: The Waterpocket Monocline

Although there are several monoclines in the Colorado Plateaus, none is as large or as much on display as the Waterpocket Monocline. It is about 50 miles long and extends in a general north-south direction. (The concept of the monocline is presented in the discussion of

FIGURE 11-33

Looking southwest along Waterpocket Monocline. Note flat-lying beds on horizon and dipping beds in foreground.

Colorado National Monument earlier in this chapter. To recall the faulting that results in the folding of the overlying rocks, refer to the cross-sections included with that discussion.) As the folding progressed, the rocks in the Waterpocket Monocline were fractured, each in its own way. Weathering and erosion therefore affected the various formations differently and distinctly different erosional forms were produced.

Erosional Landforms

Differential erosion has produced intriguing landscapes in many places in the Colorado Plateaus, but, as geologist Fred Goodsell has pointed out, it is particularly true in Capitol Reef where the huge Waterpocket Monocline forms the framework. The superior resistance of the Navajo-Kayenta-Wingate sandstone series contrasts strikingly with the weak Entrada that forms the valley below. This scal-

loped ridge can be seen for miles from such vantage points as Halls Creek Overlook, and for tens of miles when viewed from the air. With its precipitous, twisting canyons, it formed a formidable barrier to east-west travel, except where the Fremont River cut its canyon through it. Finally, during the uranium boom of the 1950s, the Burr stock trail was widened into a road; this road is best traveled in dry weather, as Olson warns. The old Waterpocket Fold road follows the strike valley northward for many miles from near Eggnog, a mere dot on the map east of the park. Wet weather and flash floods on steep side streams make this road hazardous at times.

Stream patterns are structurally controlled in much of the park area. For example, in the southern part Halls Creek flows southward for many miles in the strike valley that it has cut in the soft rocks. This broad valley with the high ridge to the west is one of the truly spectacular features in the park.

In Capitol Reef, the Navajo Sandstone displays a wide variety of unusual landforms. In places, as in the Golden Throne, the Navajo is protected by a hard cap of Carmel Sandstone and is flattopped. Where the protecting cap-

FIGURE 11-34

Capitol Dome, as viewed through Hickman Natural Bridge.

FIGURE 11-35

Temple of the Moon (right) and Temple of the Sun, formed in Entrada Sandstone by weathering and stream erosion in Cathedral Valley. (Photo by Fred Goodsell.)

rock has been removed, large rounded domes like Capitol Dome were formed. Elsewhere, massive arches and natural bridges form large "picture-windows" framing exciting and colorful landscapes. Cassidy Arch and the Hickman Natural Bridge are examples. From Halls Creek Overlook of the Waterpocket Fold, if you look through the binoculars long enough you will find a double arch below you on the far side of the broad valley.

The Entrada Sandstone here is stratified rather than massive as it is in Arches National Park; consequently it does not form arches. In Cathedral Valley in the northern area, high steep-sided sharp peaks or "cathedrals" are formed. Where a hard cap of Curtis Sandstone protects the Entrada, impressive monoliths are formed. (Those who wish to see the Entrada Goblins on parade should visit Goblin Valley State Reserve which is between Capitol Reef and Green River, Utah.)

The Morrison Formation, of many colors in-cluding deep purple, forms low, beautifully rounded hills and ridges, particularly along the road east of the monocline. The drab Mancos Shale is in many places completely dissected by streams and rivulets, thus forming classic badlands.

In this arid climate the vegetative cover is generally sparse except locally along the streams. The weathered materials from the Morrison and the Mancos shales are inherently sterile, and the outcrops of these two formations are almost completely barren. Therefore, on the Morrison and the Mancos areas the rate of erosion is particularly high.

Small and sometimes hidden are the "water-pockets." They were important, even vital, in the lives of early explorers. In this dry area most of the streams flow only on rare occasions and water is extremely scarce. The water-pockets are potholes carved out of bedrock by stream abrasion, depressions likely to hold water for extended periods following summer storms.

We have merely sampled the multitude of geologic features in this multi-colored national park. Its eye-catching appeal is known only to those who visit it.

ZION NATIONAL PARK

Established in 1919, Zion National Park includes 230 square miles of canyon country in the southwestern corner of Utah. The early settlers saw that the high monoliths resembled temples and cathedrals, and named the region Zion. It is an all-year park, each season having its own attractions. Utah Highway 15 through the southern section of the park and a branch road up Zion Canyon enable the motorist to view many of the wonders of Zion. There are trails for all, ranging from half a mile (0.8 km) to 36 miles (58 km) for the round trip.

Bedrock Geology

Zion is essentially a Mesozoic park; all of the sedimentary rocks are Mesozoic, but Tertiary and Quaternary lava flows and pyroclastics cover several areas. The Permian Kaibab Limestone is exposed in the bottom of Tempoweap Canyon a short distance outside the park boundary (Eardley and Schaack, 1976). (For details on the geologic section, refer to Plate 11-A,B.)

The oldest Mesozoic rock is the Moenkopi, a thick Triassic formation of interbedded sandstones and shales; separated by an unconformity, about 100 feet (30 m) of Shinarump conglomerate rests on the eroded Moenkopi. Next, the Chinle, mainly shale about 1000 feet (303 m) thick, is bright red and easily identified by its color. As in Petrified Forest National Park, the Chinle in Zion contains silicified logs.

On top of the Chinle is a formation, the Moenave, which we haven't seen before. It is a reddish sandstone, about 350 feet (107 m) thick which in its lower part contains Triassic fossil fish. Resting on the Moenave is about 200 feet (60 m) of Kayenta Sandstone. Dinosaurs were in Zion during this part of the Triassic, as their tracks testify. Rising high above the Kayenta are the bold cliffs of Zion, cliffs of Triassic-Jurassic Navajo Sandstone.

Late in the Triassic and early in Jurassic time, the Colorado Plateaus area was distinctly arid, and wind erosion and deposition were widespread. In some places sediments were deposited by both wind and water; in the Zion area, the Navajo Sandstone was, with minor exceptions, wind deposited. Whether we have in our deserts today anything which will match that of the old Navajo desert appears doubtful. The area covered was extremely large, and in places the Navajo is as much as 2200 feet (670 m) thick. Eolian cross-bedding—curved and wedged cross-bedding—and the frosted grains of quartz sand testify to its being of eolian rather than fluvial origin.

FIGURE 11-36

Eolian (wind) cross-bedding in Navajo Sandstone, the main formation in Zion National Park.

Later in the Jurassic, the wind gave way to the water as the seas encroached over the land, as the Rocky Mountain Geosyncline developed. First the Carmel Formation, sandstone and limestone, was deposited and then Cretaceous shales and sandstones. The Tertiary Wasatch Formation, which dominates the scene in Bryce Canyon, probably covered the Zion area also. However, much of the Carmel and all of the younger formations were stripped from the area as the spectacular canyons of Zion were being carved out.

Uplift during the Tertiary and especially in early Pliocene time initiated a new cycle of erosion, one of such widespread consequences that it was called "The Great Denudation." We will soon get to Grand Canyon where, because uplift was at its maximum and because the mighty Colorado River was there, canyon cutting took place on a colossal scale.

Volcanoes erupted at various times after the uplift and spread lava flows and pyroclastics over parts of the Plateaus, including Zion National Park. Here, some of the cinder cones are relatively unaltered, indicating that they are of recent origin.

Geomorphic Development

The Plateaus were uplifted differentially, causing fracturing and faulting. One large block, of which the Zion area is a small part, was uplifted to about 10,000 feet (3050 m) above sea level. The uplift caused the Virgin River to begin cutting its canyon — Zion Canyon. Tributaries to the Virgin started to downcut and new tributaries also began to develop — along the fractures. Since the main fractures are parallel, many of the tributary streams are parallel. As these tributary streams developed subsequently, under the control of structures in the rocks, they are subsequent streams.

Fractures played another role in the formation of Zion's topography. The fractures are near vertical and so are the cliffs. The streams cut down through the soft sandstones and formed narrow, vertical-walled gorges, some of which remain today. The Narrows of the Virgin River in the northeastern part of the park provide a good example; here, along the trail upstream from the Temple of Sinawava, the gorge is only a few feet wide but the walls rise at least 1500 feet (457 m)! During flood flows, the Virgin is on edge, so deep is the water.

In the downstream areas the canyons are deeper and wider; yet the walls maintain their perpendicularity and the Great White Throne, the hallmark of Zion, rears up more than 2400 feet above the valley floor. In the soft Navajo Sandstone, the streams carry heavy loads of sand; at times, their channels are choked with sand, and the streams shift laterally and undercut the cliffs. Mass-wasting of large slabs occurs, again controlled by the vertical fractures, thus maintaining the vertical walls.

FIGURE 11-37

The Great Arch, an inset arch in cliff of Navajo Sandstone. Zion National Park

Therefore, the cross-section of the canyons is always essentially rectangular, but the width-depth ratio increases downstream. The Great White Throne is varicolored and grades upward from red to pink to the white of the large upper section.

In places, when large slabs fall off, great arches are formed high on the cliff; the Great Arch of Zion is generally similar to the Great Arches in Yosemite. The rocks are very different but both are massive and have vertical fractures; the process of mass-wasting operates uniformly regardless of geographic location. Rockfalls occur from time to time as Alberding (1971) reports. Late in the summer of 1968, a 5000-ton block of rock fell near the end of the Narrows Trail; earlier, in 1958, a much larger mass came crashing down near the Mount Carmel Tunnel on Utah Highway 15.

In the areas where the cross-bedding in the Navajo is best developed, the many bedding planes are planes of weakness and dominate over the vertical fractures in the control of to-

FIGURE 11-38

Checkerboard Mesa, Zion National Park. Weathering along the cross-bedding and vertical fractures produced the checkerboard pattern on the cliff. (Photo by E. Kiver.)

pography. The slopes are fairly steep and, resulting from differential weathering along bedding planes, a stair-step surface is developed—a surface more to the liking of the amateur climber than the near-vertical walls of the Great White Throne. The famed Checkerboard Mesa is a classic example, in the eastern part of the park south of the main highway. Here, the vertical fractures and the bedding planes "checkerboard" the cliff.

Springs are found in a number of places, at Weeping Rock for one. Here, groundwater appears on the surface and drips down from the cliff. The Navajo Sandstone has calcium carbonate as its cementing material; percolating waters charged with carbon dioxide dissolve out the calcium carbonate. When the water reappears on the surface, by evaporation some of the calcium carbonate is precipitated as travertine. Here, the travertine is precipitated from cold water, unlike at Mammoth Hot Springs in Yellowstone.

During dry periods there are a few waterfalls, as at Emerald Pool; there are many waterfalls immediately after a thunderstorm. Some of the tributaries, dry most of the time, have been unable to downcut as rapidly as the Virgin River; consequently, they flow out of hanging valleys along the main canyon.

Zion with its multicolored cliffs is something to see in the daylight; it is a different world in the moonlight. But try it in the blue-blackness just before first light.

BRYCE CANYON NATIONAL PARK

In going from Zion to Bryce Canyon, we travel only about 50 miles northeastward, but we move up the geologic column more than a hundred million years. At Bryce we are in the Eocene Wasatch Formation, which was much more widespread before the Great Denudation began.

The colors at Bryce Canyon are almost unbelievable. The contrast of the deep green of the plateau forest with the reds, oranges, yellows and white of the many grotesque erosional features—all against Utah's deep blue sky—nearly overwhelms some park visitors. Many are awe-struck at the spectacle of the Silent City with its painted labyrinths. Rancher Bryce was more prosaic and pragmatic when he observed that "it's a hell of a place to lose a cow."

Bryce "Canyon" is misleading; there is no canyon, in the usual sense. A canyon is a deep valley cut by a large stream, and there is no large stream near Bryce, which is on the east side of the plateau. Instead, Bryce was formed by weathering and by erosion of myriad tiny streams. See figure 11-39 in color section.

Bedrock Geology

The Eocene Wasatch Formation, consisting of calcareous mudstones and thin lenses of sandstones and conglomerates, is the only prominent formation in the park. Underlying the Wasatch in the southeastern corner of the park, there are exposed drab shales and sandstones of Cretaceous age.

Geologic History

The same uplift that brought Zion up to its present height also elevated the Bryce block, the Paunsaugunt Plateau, to its present position. Here also, fracture systems were developed, but in rocks very different from those in Zion; consequently, the erosional forms are entirely different.

The Wasatch Formation was laid down in near large lakes beginning about 60 million years ago. Muds, mainly limy muds, were deposited in vast quantity. Periodically, sands and gravels were spread in thin layers over different parts of the basins. At the end of Wasatch time, as much as 2000 feet of sediments had accumulated in the deeper basins (Gregory, 1957). The muds were compacted to form mudstones, and the sands and gravels were cemented together in varying degrees. The colorations were probably, in part, inherited from the older formations that supplied the materials. In part, they were formed later by the oxidation of iron-bearing minerals when the sediments were finally exposed. Limonite and hematite appear to be the main coloring agents. Certain layers, particularly the sandy beds, are light-colored — almost white — in some places but not in others, suggesting that the coloring material has been leached out.

The "Canyon" is suggestive of a huge but shallow cirque with large inset partitions extending upward from the floor and outward from the walls. The partitions are the fins, many of which are divided into a series of columns and pinnacles. Weathering, both physical and chemical, was the main process, and the weathered materials were removed by wind and running water. But a definite fracture system had to be there in order for today's features to be developed. The primary direction of fracture is the direction of the fin; the secondary cross-cutting fractures caused the pinnacles to be isolated from the fins.

The sandstone and conglomerate layers form caps and ledges where they are well cemented; where cementation is poor they form constrictions which in time result in the toppling-off of pinnacle tops. Thor's Hammer may lose its head in the not too distant future. But as some are destroyed others are being formed, and Bryce Canyon will be there for a long time — until it runs out of Wasatch.

Hike down the self-guiding trail to Queen's Garden — with your camera — where you see Gulliver's Castle, the Totem Poles and much more. From the trail where two fins rise high, look up at the sky — the *blue* sky. After this, if you need more, the Peekaboo Loop Trail — about five hours of it — takes you to the Hindu Temples, and the Wall of Windows, and other wonders. Do not neglect to drive to Rainbow Point on the Promontory at the south end of the park; the Natural Bridge is on the way.

CEDAR BREAKS NATIONAL MONUMENT

When you go from Bryce Canyon westward to Cedar Breaks, you find that you have neither gained nor lost any part of the geologic column; here is another area of the Wasatch Formation, again at the edge of a plateau. As would be expected, the fantastic features of Bryce are duplicated here — but on a somewhat reduced scale. There is a marked difference in elevation; at Cedar Breaks the rim above the "breaks" is about 10,400 feet high, more than 2000 feet higher than at Bryce. A short distance north of the monument, basalt-capped Brianhead Peak is 11,315 feet high — at or slightly above timberline in this area. A Forest Service shelter is at the top, from which you can see out over the Basin and Range

Province that extends up to the base of the breaks far below.

Because of the elevation, precipitation is higher and temperatures are lower; therefore, many of the plants and animals are different from those at Bryce. The cantankerous little cony, which we saw last in Rocky Mountain National Park, is with us again at Cedar Breaks. Here, in addition to pines, there are fir, spruce, and aspen up on the plateau; down below in the breaks are junipers, which were mistaken for "cedars" by the early explorers. Bristlecone pines grow along the rim; one whose age is known has been there for about 1600 years, according to the Park Service brochure. And in the upland meadows, the array of wildflowers is magnificent.

The geologic features are very similar to those at Bryce — fins, columns, pedestals, and pinnacles. In addition, however, a fault is well exposed, showing only a small displacement but providing definite evidence of the forces that fractured the rocks that gave rise to Cedar Breaks. It is a beautiful area, not too crowded and therefore highly desirable, particularly when most of the plateau areas are being baked in the hot sun. See figure 11-41 of color section.

THE GRAND CANYON

A View From Space

For centuries geologists mapped geologic formations and structures on foot; later they observed them from the air and used aerial photographs. Now, space and computer technologies provide long-range views which often reveal structures too large to see from the ground — structures essential in the new plate tectonics interpretations.

On the facing page we observe a section of about 30 by 40 miles (50 by 67 km) of the Grand Canyon, as it appeared from the Landsat satellite 570 miles (920 km) out in space. We see the Colorado River flowing southward in Marble Canyon (top), to be joined by the Little Colorado (right). As it bends to the west the river plunges into the narrow and steepwalled Inner Gorge. Here, as we look deeper and deeper into the earth, we are traveling farther and farther back into the ancient history of our planet.

Grand Canyon Village on the South Rim is only barely visible, but Bright Angel and Vishnu faults are seen distinctly at 1 and 2 (see inset), cutting across the canyon and adjacent plateaus.

The unnatural colors result from the use of both infrared and visible light; the images, which are electronically recorded, are radioed back to earth for processing by special computers. The reds are forests of pine and fir on the high plateaus; lighter colors define rocks and desert soils with only a sparse vegetative cover.

This view is a small part of a full Landsat frame taken June 19, 1976, and designated Scene Number 2514-17200. It was processed at the Environmental Research Institute of Michigan from U.S. Geological Survey data tapes.

Made available by Anthony Morse of Synoptic Views, Box 193, Fort Collins, Colorado 80522.

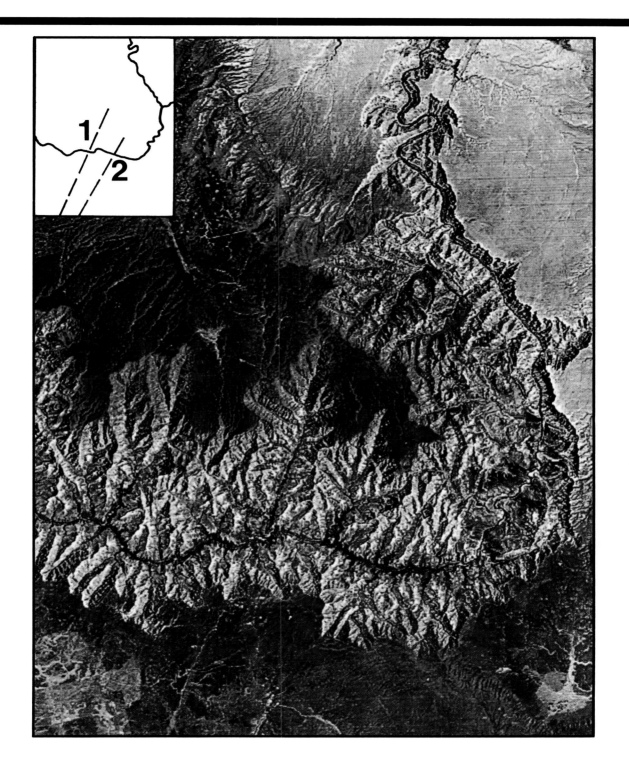

FIGURE 1-4

Aa flow blocks highway in Hawaii Volcanoes National Park. (Photo by National Park Service.)

FIGURE I-4

Smoky quartz surrounded by green Amazon stone feldspar. Pikes Peak District, Colorado.

FIGURE 1-12

Lava toes protruding downslope from a pahoehoe flow of Mauna Ulu. Hawaiian Volcanoes National Park. (Photo by W. H. Parsons.)

FIGURE 1-6
Lava cascading down sea cliff into Pacific Ocean from Mauna Ulu eruption in 1972. Pillow lavas were formed beneath the sea. Hawaii Volcanoes National Park. (Photo by W. H. Parsons.)

FIGURE 1-14
Haleakala "Crater" with cinder cones, including Puu O Maui, rising from the floor. (Photo by National Park Service.)

FIGURE 2-4
Airview east over Aniakchak Caldera, six miles wide from rim to rim, western end of Alaska Peninsula. (Photo by M. W. Williams, National Park Service.)

FIGURE 3-3
Yentna Glacier with its multicolored moraines, on the south flank of the Alaska Range. (Photo by Joan B. Perkins.)

FIGURE 4-1a
Sea arches near Santa Cruz, California, 1964.
(Photo by R. S. Creely.)

FIGURE 4-11
Bold front of Hubbard Glacier in Disenchantment Bay, southeastern corner of Wrangell-St. Elias National Park. (Photo by Charles G. Mull.)

FIGURE 4-1b
The fierce storm in January of 1980 destroyed one of the arches in Figure 4-1 (1984). (Photo by R. S. Creely.)

FIGURE 5-4
Lake Chelan occupies a long and deep glacial trough on the east side of the North Cascades in Washington.

FIGURE 5-15
Mt. Rainier rising above the Tatoosh Range. (Photo by E. Kiver.)

FIGURE 5-19
Crater Lake, a caldera lake, with Wizard Island cinder cone rising high above the water.

FIGURE 5-24

Spires, columns, and pinnacles in southeastern part of Crater Lake National Park—The Pinnacles. (Photo by National Park Service.)

FIGURE 7-1

The High Sierra. (Photo by Peter Barth.)

FIGURE 7-11
Half Dome as seen from Stoneman Bridge on Merced River, Yosemite National Park.

FIGURE 7-18
Devils Postpile.

FIGURE 8-6
Running Eagle Falls (also called Trick Falls), near Two Medicine Lakes, Glacier National Park.

FIGURE 8-12

Horns, knife ridges, and glaciers in the Arrigetch Peaks area, Brooks Range, Alaska. (Photo by Charles G. Mull.)

FIGURE 9-10

Lower Falls, plunging 308 feet (94m) into Yellowstone Canyon. (Photo by E. Kiver.)

FIGURE 9-13
Above Snake River Valley, The Tetons rise to 13,770 feet (4198 m) Grand Teton National Park.
(Photo by Antoinette Lueck.)

FIGURE 10-12

Little Yellowstone, in the headwaters area of the Colorado River, northwestern part of Rocky Mountain National Park. Streams have dissected hydrothermally altered rhyolites, some that are highly and brilliantly colored.

FIGURE 10-23

Reversing sand dunes, with Sangre de Cristo Mountains in background, Great Sand Dunes National Monument. (Photo by E. Kiver.)

FIGURE 11-1

Typical Colorado Plateaus country—canyons, mesas, plateaus—looking westward from Dead Horse Point State Park, Utah. Note giant loop in the Colorado River, below.

FIGURE 11-18

Delicate Arch, Arches National Park. (Photo by National Park Service.)

FIGURE 11-20

Arches that have fallen; note window or Baby Arch in the fin to the left.

FIGURE 11-26

Air view down into Upheaval Dome, with loops of the Green River in the distance. (Photo by J. A. Campbell.)

FIGURE 11-27

Wingate Cliff with White Rim sandstone in distance above Green River Canyon. Photographed from near Grandview Point at south end of Island in the Sky, Canyonlands National Park.

FIGURE 11-31
Rainbow Bridge, perhaps the most nearly perfect natural bridge in the world. The bridge is carved out of Navajo sandstone which rests on Kayenta sandstone abutments. (Photo by Steve Nelson.)

FIGURE 11-39

Complex erosional forms in Bryce Canyon National Park. (Photo by E. Kiver.)

FIGURE 11-40

Bryce's pinnacles with Thor's Hammer (center). (Photo by J. M. Harris.)

FIGURE 11-41

Erosion forms in Cedar Breaks National Monument are similar in type and coloration to those in Bryce Canyon. (Photo by L. Hymans.)

FIGURE 11-44
View north across Grand Canyon, John Wesley Powell's "Book of Geology." Kaibab limestone caps the Kaibab Plateau on the north rim.

FIGURE 11-50

View of Colorado River, down 2,200 feet (670 m) from Toroweap Point, western Grand Canyon. (Photo by E. Kiver.)

FIGURE 11-53

Badlands, Painted Desert section of Petrified Forest National Park. (Photo by National Park Service.)

FIGURE 11-58
Pothole bridges, Coyote Gulch in Escalante Country, Utah. (Photo by Liz Hymans.)

FIGURE 12-21   Sunrise, White Sands National Monument.

FIGURE 13-11

Boxwork, a structure formed when calcite is precipitated in the cracks of fractured limestone, Wind Cave National Park. (Photo by National Park Service.)

FIGURE 13-13
Crystals of calcite, colored with compounds of iron and manganese, cover the walls in Jewel Cave National Monument. (Photo by Ron Rothschadl.)

FIGURE 14-5
Little Niagara, a waterfall built by precipitation of a travertine dam across Travertine Creek, Chickasaw National Recreation Area, near Sulphur, Oklahoma.

FIGURE 15-1

Vapors rising above Display Spring, Hot Springs
National Park. (Photo by National Park Service.)

FIGURE 16-2

Corrugated topography on Isle Royale. Tilted layers of resistant basalt form the ridges; poorly
resistant layers of amygdaloidal basalt were more deeply eroded, forming the valleys. (Photo by
National Park Service.)

FIGURE 16-4
Wave-cut arch on Amygdaloid Island, formed when the lake was at a higher level. (Photo by R. J. Larson.)

FIGURE 18-3
View downstream to Delaware Water Gap, with New Jersey's Mt. Tammany beyond (left), Delaware Water Gap National Recreation Area, near Stroudsburg, Pennsylvania. A long geologic history is recorded here: Paleozoic deposition in the Appalachian Geosyncline culminated in folding, faulting, and mountain building and then was followed by prolonged erosion and periodic uplift. The latest major event occurred when the Delaware River cut the gap through the highly resistant Tuscarora Quartzite ridge. (Photo by National Park Service.)

FIGURE 18-6
Several ridges of the Blue Ridge accentuated by "scenic haze," Great Smoky Mountains National Park. (Photo by National Park Service.)

FIGURE 18-7
Deeply weathered soil exposed in road cut. The red coloration is hematite, an iron oxide formed by prolonged chemical weathering of iron-bearing minerals.

FIGURE 19-13
Arch formed by waves along a vertical fracture at Sand Point near Bar Harbor, Maine.

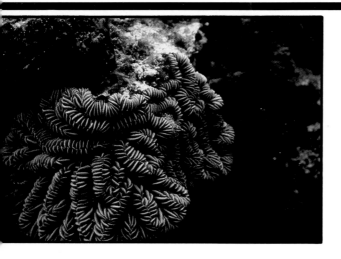

GRAND CANYON NATIONAL PARK

"Do nothing to mar its grandeur," President Theodore Roosevelt said, adding that Grand Canyon was "the one great sight which every American should see." In 1908 he set it aside as a national monument to protect it from exploitation until the time Congress could act. Grand Canyon, an area of about 1100 square miles, was upgraded to national park status in 1919, primarily because both the scenery and the geologic history exposed in the canyon are unparalleled. Certain geologic events are recorded only in the area downstream from the park, and in 1932 this adjacent area of almost 310 square miles was established as Grand Canyon National Monument. Later, in 1969, about 41 square miles upstream from the park became Marble Canyon National Monument. In 1975, these two monument areas were added to the park; thus, Grand Canyon National Park consists of the canyon country along a 280-mile (452 km) stretch of the Colorado River, from Glen Canyon National Recreation Area down to Lake Mead National Recreation Area.

Major John Wesley Powell was the first geologist to explore the Grand Canyon region. In 1869, he led a party of hardy explorers down the Colorado River through the canyon. Why anyone would even consider plunging into such a forbidding canyon, without any maps or other information — in fragile wooden boats — is beyond comprehension. Powell was a Civil War veteran and he had lost his right arm in the battle of Shiloh. What would have meant the end of an active career for many apparently spurred him on, and he became the country's most daring and tireless exploration geologist. Powell saw the Grand Canyon as the "book of geology." In his diary he wrote, "All about me are interesting geologic records, the book is open and I read as I run." He and his men ran the rapids and survived the ordeal; in fact, Powell was so intrigued and so involved that he made a second voyage in 1871. For a fasci-

FIGURE 11-42

Major John Wesley Powell, first geologist to explore and interpret the Grand Canyon region. (Photo by National Park Service.)

nating summary of Powell's expeditions, read Baars' 1969 article in *Geology and Natural History of the Grand Canyon Region.*

Grand Canyon was "discovered" by one of Spanish explorer Coronado's lieutenants, Don Lopez de Cardenas, who was directed there by the Hopi Indians in 1540. The *real* discoverers will never be known; the wind has long since covered their tracks. It was at least 4000 years ago, long before the Anasazi arrived, that these people came into the canyon and made a home in a cave or perhaps beneath one of the huge overhangs. Were they the ones who fashioned the intriguing split-willow animals found in Stanton's Cave, as reported by Wallace (1973)?

Indians still have a hold on the canyon. The Havasu Village, in the western part of the enlarged park, is probably one of the oldest settlements of continuous occupation in conterminous United States. Its inhabitants live in much the same way that their ancestors did for centuries.

The unfolding of the history of the Grand Canyon region is presented in intriguing and understandable fashion by Shelton (1966); his book in *Geology Illustrated,* replete with pictures and reconstructions, is well worth reading. McKee's 24th edition (1966) of his 1931 booklet, *Ancient Landscapes of the Grand Canyon Region,* is also an excellent presentation. For color, and for a collection from Powell's diaries, Porter's (1969) *Down the Colorado* is suggested. With apologies to this remarkable region, only a brief geological summary can be included here.

The Mountains of Grand Canyon

The mountains here are of two general types — the Vanished Mountains and the Mountains of the Future. Both the Archeozoic Vanished Mountains and the much younger Proterozoic Vanished Mountains have long since disappeared, wiped out by erosion. The Mountains of the Future are beginning to be formed — down in Grand Canyon.

The Archeozoic Mountains were high and extensive, but their precise dimensions may never be known. They vanished as topographic mountains before the end of the Archeozoic; but their roots are there, with unmistakable mountain structures, exposed in the Vishnu Schist of the Inner Gorge. They were compressional mountains, like the Rockies and many other major mountain ranges.

During the Proterozoic, about 12,000 feet of sediment and lava flows were laid down on the old erosion surface — rocks of the Grand Canyon Series (Ford and Breed, 1969). Shales, some brilliant red, are the dominant rocks, but sandstones and conglomerates are present. More significant perhaps are the limestones because they contain "heads" of colonial algae — similar to those in the Siyeh Limestone

FIGURE 11-43

Kaibab Trail switchbacks. (Photo by Helen McClure.)

in Glacier National Park. Although these Precambrian rocks have been combed carefully by Walcott (1895) and many others, no other unquestioned fossils have been found. These rocks can best be seen from the North Rim; they are well exposed in the upstream end of the park, forming the Inner Gorge. Eroded fault-blocks, tilted at steep angles, are the roots of the Proterozoic Mountains — fault-block mountains. The erosion that destroyed the mountains near the end of the Proterozoic formed a peneplain that cut across the Grand Canyon Series, the Vishnu Schist, and other Archeozoic rocks. Here and there monadnocks of hard rocks extend into the overlying Paleozoics; otherwise the peneplain was a strikingly smooth surface, as we saw it in the Black Canyon of the Gunnison and in Colorado National Monument.

The stair-step-like character of Grand Canyon is of interest to many people. Why aren't the sideslopes relatively uniform from the rim down to the river? The steepness of the slopes is a direct reflection of rock resistance. In this dry climate, chemical weathering is slow, and limestones such as the Kaibab and the Redwall maintain near-vertical cliffs; later, we will see that in humid regions, limestones weather rapidly and tend to form valleys and other lowlands.

Shales weather readily in any climate, forming the slopes that extend out from the base of limestone and sandstone cliffs. The thick Bright Angel shale (Cambrian) forms the broad, flattish slopes known as the Tonto Platform above the Inner Gorge (see the cross-section in Plates 11-A and B).

The Precambrian metamorphic rocks and granite of the Inner Gorge are much more resistant to weathering and erosion than any of the sedimentary rocks. As the river cut deeper and deeper, the walls became progressively higher, but these rocks, being much stronger, were better able to maintain steep, in places near-vertical, cliffs.

Sequence of Events — Paleozoic

To detail the depositional history of the Grand Canyon rocks would fill a book such as Breed and Roat's 1976 *Geology of the Grand Canyon.* Here, only time-lapse glimpses are given.

As the inland sea encroached over the late Precambrian erosion surface, sands and

FIGURE 11-45

Sedimentary strata exposed in the Grand Canyon of the Colorado River record a span of a billion years of earth history. Photo by U.S. Geological Survey, reprinted from John Wiley & Sons, Inc.

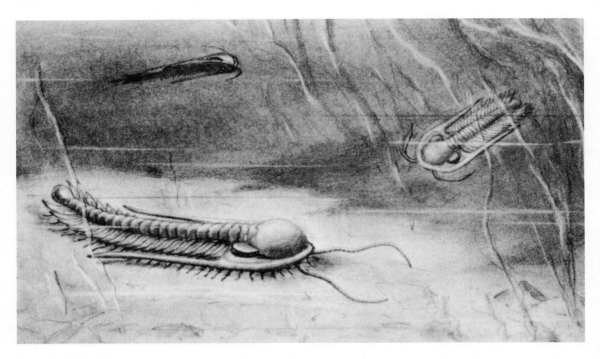

FIGURE 11-46

Restoration of the Cambrian trilobite, *Paradoxides harlani,* one of the giants of its time. About ¼ natural size. Reprinted from Dunbar, 1966, by permission of John Wiley & Sons, Inc.

gravels were laid down which, when cemented, became the Cambrian Tapeats Sandstone. With the deepening of the water in the newly formed geosyncline (depositional basin), the muds that became the Bright Angel Shale were deposited. The sea was teeming with invertebrate animals, such as the trilobite, during this time about a half-billion years ago (Figure 11-46). After making remarkable strides up the evolutionary ladder during the Paleozoic, the trilobites suddenly disappeared—extinct at the end of the Paleozoic, for reasons as yet unknown.

Also during the Cambrian period, the Muav Limestone was deposited, resting on the Bright Angel Shale. Next, there should be Ordovician rocks, but they are not there; nor are there any

Silurian rocks. Obviously, Powell's "book of geology" had several pages torn out. Just what happened during these two geologic periods will never be known; it is likely that some deposition took place, but whatever was deposited was eroded away, prior to the deposition of the Devonian Temple Butte Limestone that rests on the eroded surface of Muav Limestone. This gap in the geologic record, represented by the two formations of distinctly different geologic age and separated by an erosion surface, is an unconformity.

The Grand Canyon area was lifted above the sea and eroded, producing another unconformity, this one between the Temple Butte and the Mississippian Redwall Limestone. In fact, in places *all* of the Temple Butte was eroded away. During Redwall time, the seas were warm and shallow, and marine life was abundant, with brachiopods, clams, and corals of many kinds. By this time, the trilobites had increased in size and complexity, at that time unlikely candidates for the extinction that came later. The Redwall is thick and generally massive, and it forms bold cliffs, mainly red. Actually, the limestone is gray; only the surface is red, colored with iron oxide (hematite, the pigment in red barn paint) that streamed down the cliff from the red Supai Formation.

The Supai is a nonmarine formation that was laid down during the Permian period, on the eroded Redwall Limestone. Thus, the Pennsylvanian period, like the Ordovician and Silurian, is not represented. Although the Supai is largely shale, there are sandstone layers that contain tracks of amphibians, or perhaps primitive reptiles.

The Hermit shale that lies on the Supai is also red and is nonmarine. Some "new" fossils appear in the Hermit-worm trails, insects, vertebrate tracks, ferns of many kinds, and primitive coniferous plants.

The white cliffs that shine out from a distance are made of Coconino sandstone. Eolian (wind) cross-bedding is a distinctive feature. During this part of the Permian, the Plateau country was a desert and the wind piled up dunes over a wide area. Then once again the sea prevailed, and the marine Toroweap limestones and sandstones were laid down.

Above the Toroweap, the bold, generally unscalable cliffs of Kaibab limestone rise vertically to the top of the canyon. Because of its massive character and high resistance to erosion, the Kaibab is the surface rock over a broad area, the Kaibab Plateau to the north and the Coconino Plateau on the south side of the canyon. Marine invertebrate fossils are abundant in certain layers of the Kaibab; also, a few shark teeth have been found.

Thus, we come to the end of the Paleozoic Era during which the Grand Canyon region was both above and below the sea, at times being eroded for long periods. Only invertebrate animals were there at the beginning; vertebrates, some large and well advanced, were present at the end. Land plants appeared sometime during the middle of the era. It was the best of times, it was the worst of times; many species didn't make it through to the end, including the trilobites.

Mesozoic-Cenozoic Events

During the Mesozoic Era, deposition began with the laying down of reddish sands and silts by the sluggish streams that wandered over the Grand Canyon area. In places the Triassic Moenkopi was as much as 1000 feet (303 m) thick. Probably the Shinarump Conglomerate and other younger beds were also laid down; however, all but a few small remnants of Mesozoic rocks have been stripped off. Cedar Mountain, a mesa near the southeastern boundary of the park, provides proof that at least the Moenkopi covered the Kaibab Limestone.

The uplift that rejuvenated the streams that stripped off the Mesozoic rocks occurred sometime early in the Cenozoic Era. The development of a so-called "stripped plain" is probably uncommon, at least on a widespread basis. Here, it was the superior resistance of the Kaibab Limestone that made it possible. The streams shifted back and forth on top of the Kaibab, removing the poorly resistant overlying materials. Remember that the Grand Canyon wasn't there at that time.

Late in the Tertiary Period, no more than 10 million years ago, the region was uplifted again — actually uparched — and then the Colorado River began to cut its canyon (McKee and McKee, 1972). It is estimated that the beginning of the cutting of the Grand Canyon occurred 7 million years ago; thus, the rocks in the canyon are old but the Grand Canyon itself is young. [As you see on Plates 11A-D, the rocks are tilted toward the south;

FIGURE 11-47

View down canyon from Toroweap Point in western part of Grand Canyon. Note large mass of lava (arrow) which formed a dam across the lower section of the canyon. (Photo by Steve Nelson.)

therefore, the North Rim is higher than the South Rim by more than 1000 feet (303 m).]

Downcutting through the sedimentary rocks required a considerable period, and concurrently mass-wasting widened the canyon. Tributaries developed their valleys, in places along the fault zones and major fractures. Bright Angel Creek developed its valley in the Bright Angel fault zone which angles across Grand Canyon; on the south side, Garden Creek follows the same fault and flows into the Colorado River against the current. These are both subsequent streams; Garden Creek is a *barbed tributary* because downstream from the junction, the angle formed by the two streams is distinctly acute.

When the Colorado began to cut into the

hard Precambrian rocks of the old mountain ranges, the rate of downcutting slowed considerably, in fact to a fraction of the earlier rate. Even so, some of the shorter tributaries could no longer keep pace and began to flow out of hanging valleys; the waterfalls became higher as the main canyon was deepened. Mass-wasting took place but at a much slower pace in these resistant rocks. In this way, the Grand Canyon was formed; tomorrow, geologically, it will be deeper, and it will be wider at the top; although these processes may slow down, they never stop.

When the canyon was cut almost to its present depth, volcanoes began to erupt from vents high on the north flank in the Toroweap Valley area. One volcano, called Vulcans Throne, spewed lava down into the canyon, forming a dam about 550 feet high. This lava dam retarded the cutting of this section of the canyon; finally the dam was cut through, and the river has since cut the canyon another 50 feet deeper. A radiometric age of about 1.2 million years has been determined for the lava dam (McKee et al., 1968). This dam and others are in the monument area downstream from the old park boundary. The spine of black rock sticking up out of the middle of the river is

FIGURE 11-48

Vulcans Forge. (Photo by L. Hymans.)

FIGURE 11-49

Colorado River, looking down from Toroweap Point, western Grand Canyon. (Photo by J. K. Hiller, U.S. Geological Survey.)

Vulcans Forge, a basalt plug that has withstood the forces of the river, flood after flood. In addition, a 3000-foot (913-m) near-vertical drop from Toroweap Point makes this a breathtaking as well as a geologically significant section of the enlarged park.

The Mountains of the Future

As the Colorado River cut down through the rocks, its main tributaries cut many side canyons. Tributaries of the large tributaries also cut canyons. The sedimentary rocks above the Tonto Platform were therefore intricately dissected, and tall "temples" were left standing high above the platform—Shiva Temple, Osiris Temple, Isis, Vishnu, Brahma, and Zoroaster temples are a few of the higher ones. These temples are erosional mountains—the mountains of the future. If they rose above a large, flat surface instead of being lost in the vastness below the canyon rim, they would be immediately recognized as mountains. Several million years from now, when the entire Colorado Plateau area is completely dissected—like the area immediately adjacent to the canyon—erosional mountains by the thousands will be there, with peaks rising up to present plateau heights. Eventually these mountains also will be eroded down, as countless mountains have been in the past.

Grand Canyon—A Natural Barrier

The cutting of the canyon effectively isolated the Kaibab from the Coconino Plateau, insofar as certain life forms are concerned. Birds such as the eagle fly nonstop from rim to rim, but small birds apparently develop acrophobia looking down through a mile of air. They fly down and across the river and up again—or elect to stay home. For many animals the canyon bottom is uninhabitable because of the heat—up to 120°F (49°C)—down along the river. Also, food sources available up on the plateaus are lacking in the desert in the canyon bottom. Squirrels, for example, depend upon the ponderosa pines for their mainstay diet. Thus, the Kaibab squirrel is confined to the Kaibab Plateau and another subspecies, the Abert squirrel, is found only on the south side of the canyon. Apparently there has been no intermingling of the two for a few million years, since the canyon was cut below the ponderosa pine life zone. For more details, read Wallace (1973). For a treatise, both general and specific, on all of Grand Canyon's many facets, Stephen Whitney's 1982 *Field Guide to the Grand Canyon* may provide you with everything you want to know.

Seeing Grand Canyon

Some people must see the canyon without getting very far from their cars, looking down into the beautiful but awesome abyss from the rim; others look up from their boats (when they aren't fighting the rapids); still others go down the trail, on foot or mule back.

For those who can't do better, the views from the various overlooks are superb. Don't forget to go around to the North Rim, even though it is about 200 miles (333 km) by road. Before the automobile, visiting it was almost like visiting a foreign land; actually it is a different world, with different climate, plants and animals.

The North Rim is an escape from the crowds of people on the South Rim to the more wilderness-like atmosphere that our national parks are supposed to have. On the Kaibab Plateau the stately ponderosa pines are at their best, towering high above you, and magnificent spruces and firs grow on the areas that are more than a thousand feet higher than the land on the South Rim. From here, you can see the upper section of the canyon—which is out of range from the South Rim overlooks—perhaps from Angels Window or from Vista Encantadora.

Others take the trail down the canyon on foot—theirs or a mule's. For those in good physical condition, the way to get the full impact of Grand Canyon is on foot. Each downward step takes you farther back into the history of the earth. At the river you walk on rocks more than 2 billion years older than the ones you left behind on the rim. To sleep in the starlight under the tall ponderosa pines on the North Rim, shivering and waiting for first light, then to hike down the trail through one life zone after another before finally sweltering in the desert of the canyon floor — this is an experience never to be forgotten. If, however, you become completely exhausted and have to be taken out on muleback — a "dragout," as they call it — you won't forget that either, nor the expense. In other words, it is well to seek the advice of a park ranger before making the decision. In case you do hike down, with luck you may glimpse the fox-faced cacomistle, distant cousin of the raccoon and coati mundi; with luck you won't make contact with any rattlesnakes, scorpions, or catclaws.

You could take the train — the mule train — if you are not too young or too heavy. If you pay attention to what the wrangler tells you, and don't make the foolish assumption that you know more than your mule, your trip down and back will be great. Make your mule reservation well ahead of time, and try to log several hours of riding before your trip.

Another way to see the canyon is from a river boat. The canyon contains many remarkable sights not seen by most visitors because they are accessible only by boat. Some require a bit of climbing, up the cliffs or side canyons, as Major Powell discovered on his first voyage through the canyon. Some of these are missed by people whose goal is to barge through the canyon in record time, using great rafts with big motors. We salute, instead, the river runners who "paddle their own canoes," taking time to make the most of their canyon experience — people such as Liz Hymans, who provided the following information.

In the Granite Narrows section, at Mile 135 on the river runners' map, there is a 3-foot stalagmite like no other in the world. It looks so much like a Christmas tree that nearby Christmas Tree Cave has that name. A huge slump-block of Redwall-Muav limestone hangs overhead, the source of calcium carbonate in the Christmas tree. Drop by drop, the water deposited calcium carbonate to build a tiny mound that eventually grew up to be a tree.

As would be expected, with thick limestones like the Redwall, limestone caverns are not uncommon in Grand Canyon. One, Stanton's Cave, can be seen from the river, up on the east wall, from near Mile 29. It is likely that hundreds of caverns have been formed, only to be destroyed as the canyon was widened.

At Mile 214, the Pumpkin at Pumpkin Springs was also formed by chemical precipitation, but in a slightly different way. In the beginning, warm water poured over a ledge and carbon dioxide was freed from the calcium bicarbonate, thus freeing calcium carbonate from solution. The travertine built out in bulbous form and was fortuitously colored with algae and iron oxides, mainly orange-colored.

Far downstream, near Beaver Falls in Havasu Canyon, large travertine terraces have been built up; they are highly colored, mainly by blue-green algae. In some of the terraces there are indications that they were formed under water, perhaps in a large pool. Then there is the narrow gorge through the black basalt that formed a dam in the canyon — a canyon within a canyon — far below the volcano, Vulcans Throne, the source of the lava. And when least expected, there is Vulcans Forge, a basalt plug rising out of the middle of the river.

This is merely a beginning; those wanting more should read Powell's original descriptions, written in 1869, which were published by Eliot Porter along with his excellent photographs just 100 years later.

Go into the Colorado Plateaus country. Stand on the rim of Grand Canyon and look down into the past; then go down and stand by

FIGURE 11-51

The "Christmas Tree," a stalagmite that is still growing, Grand Canyon. (Photo by Liz Hymans.)

the river. From the rim, you saw it; now you
feel it—the tremendous power of the river.
The roar of the water and the rumble of the
boulders grinding and pounding against each
other, together with the vibration of the
ground under your feet—this is Grand Canyon. At least, it was so when the Colorado
River was mighty, before the harnessing of the
river by Glen Canyon Dam. Pure air and natural sounds were until recently two of the de-
lights of the canyon. Now, Grand Canyon is
sometimes invaded by pollution from the
coal-fired power plant at Page, Arizona; it is
regularly invaded by big power boats that are
noisy and polluting. Whether the Park Service
will be able to restore the wilderness experi-

ence to the canyon is the topic of Liz Hymans'
thoughtful article, "The Flow of Wilderness,"
in *Not Man Apart* (March 1978).

Grand Canyon has many moods—as the
light changes and as the seasons change.
Goldstein (1977) has captured most of them in
The Magnificent West: Grand Canyon. But you
may want a few of your own. Go out on one of
the overlooks, perhaps Yaki Point, when the
sun is low; record the scene as the sun slowly
surrenders the canyon to the moon.

SUNSET CRATER NATIONAL MONUMENT

On the plateau south of Grand Canyon, the
San Francisco Peaks rise almost a mile above a
large volcanic field. Mt. Humphrey (12,633
feet; 3850 m) and Agassiz Peak (12,356 feet;
3767 m), among others, were high enough to
be extensively glaciated. In this volcanic field,
volcanoes, volcanic domes, lava flows and
cinder cones are main features, well described

FIGURE 11-52

Sunset Crater near Flagstaff, Arizona. The cinder
cone was built by the volcanic eruption in 1064 A.D.
(Photo by S. A. Schumm.)

by Pewe and Updike (1970). Breed's block-diagram, Plates 11-A and B illustrates the main volcanic features of this area.

Sunset Crater, one of several cinder cones, was set aside as a national monument in 1930. In 1064 A.D., Sunset Crater erupted and built a basaltic cinder cone about 1000 feet (303 m) high. Earlier in the volcanic sequence, aa lava flows had covered small areas around the crater. As they are also very recent and are located in a dry area, very little alteration of even minor features has occurred. The original surfaces are essentially untouched and squeeze-ups and tiny spatter cones are well preserved.

The Sinagua Indians withdrew from this general area when Sunset Crater began to erupt. Many of their pithouses were buried in volcanic ash and dust that rained down on an area of several hundred square miles. Sunset Crater is therefore as interesting archeologically as it is geologically. Archeologists collected wood from the ash and, using dendrochronological methods, determined the date of the eruption.

The same eruption that drove the Sinaguas away from their homes was also responsible for their return — to an improved farming area! In this dry climate, farming was possible only where moisture conditions were most favorable. Large areas were blanketed with several inches of fine volcanic ash; this porous material retained sufficient moisture in the soil beneath to support good crops. As a result, these people promptly returned to their home territory and many others followed.

Sunset Crater National Monument is located northeast of Flagstaff, Arizona. In this same general area, there are four other archeological monuments. Walnut Canyon is only a few miles east of Flagstaff, and Wupatki is about 15 miles north of Sunset Crater. Montezuma Castle National Monument, about 50 miles (80 km) south of Flagstaff, is primarily archeological, with apartment houses high on the cliff. Their water supply was secure; a lava flow formed a dam with a lake behind it. And Montezuma Well, in a detached section a few miles away, is a sinkhole lake that provided an abundance of water for those who made this their home. Then, several miles to the west, at Tuzigoot, many visitors view the excavated ruins of a prehistoric town. These national monuments, together with the Museum of Northern Arizona, make the Flagstaff area particularly attractive to archeologists and geologists.

PETRIFIED FOREST NATIONAL PARK

Petrified Forest National Park is located in eastern Arizona about 25 miles east of Holbrook. It was originally a national monument but in 1962 national park status was approved. The park consists of two sections; north of Interstate 40 lies the Painted Desert — a multicolored badlands area; to the south are the petrified forests and ancient Indian ruins.

The Colorado Plateaus contain many "painted deserts." All are in areas in which shale outcroppings are widespread, where barren badlands were formed by stream dissection. Each shale formation has a distinctive color or range of colors. In the park, the Triassic Chinle Formation is made up of many layers of shale, each with its own color — some bold, others delicate and subdued. Bands of gray, blue, purple and red, with their various tints and hues, circle the rounded hills nearby and as far as the eye can reach, when viewed from Kachina Point, Chinle Point and other overlooks in the northern section. Locally, sandstone and conglomerate lenses are sandwiched between the shale layers; being more resistant to weathering and erosion than the shale, they form ledges and overhangs.

You may not see even a drop of running water; yet, streams were responsible for the dissection of the badlands. Should you be there during a torrential rain, however, you will see the erosional process in action, with streams muddy and fast-flowing.

The reason for setting the area aside as a national monument was the large number of petrified logs, thousands of them, exposed in the southern section. Many are exceptionally large; some are beautifully colored — jasper and carnelian — a collector's paradise. Rockhounds carted away unknown tons of the choice materials. In the Crystal Forest, there were beautiful crystals of clear quartz and amethyst in the cracks and vugs in the logs. Avid mineral collectors blasted many of the logs to bits to get these gems. The crowning insult, however, came in the form of a huge rock crusher moved in nearby to crush the logs to manufacture abrasive materials! Fortunately for us and for future generations, concerned Arizonans were able to get the national monument established before the rock crusher went into action. See figure 11-54 in color section.

The logs are Triassic in age, much older than those at Florissant and in Yellowstone. Most of the trees represent a pine-like group which is now extinct in the Northern Hemisphere (McKee, 1966). Some grew in swampy lowland areas, others on the slopes. Small lakes contained fish and, on a part-time basis, amphibians. Primitive reptiles were there, too. For a reconstruction of Triassic life in the area, see the artist's conception in Ash and May's *Petrified Forest: The Story Behind the Scenery,* a colorfully illustrated booklet published in 1969.

Some of the trees were buried where they fell; others were rafted in during floods. Picture a large Triassic logjam, or perhaps several, on a low-lying floodplain. Then, before the logs have time to rot, imagine a volcano erupting and spewing forth volcanic ash, burying the logs, in part by ash fall but mainly by stream deposition. Silica derived from the ash eventually fills the wood cells, faithfully preserving the wood structure. The coloring materials that make this wood so attractive are compounds of several elements, mainly iron and manganese.

Finally, after the logs were deeply buried by younger sediments, the Colorado Plateaus were uplifted and streams eroded off the overlying materials, resurrecting these Triassic burial grounds. The streams cut gullies and ravines into the soft rocks, in places leaving log bridges across them. Agate Bridge easily spans a 40-foot-wide ravine, for the log is considerably more than 100 feet (30 m) long. In a number of places, sections of large logs form the protective caps of pedestals. Weathering will eventually weaken the shale beneath, and the log will fall down the slope, as others have done.

Many of the logs have been broken into sections, leading to speculations that earthquake shocks were responsible. Once silicified, the logs were brittle and susceptible to fracturing. It is likely that earthquakes were not uncommon during the uplifting of the plateaus.

The Petrified Forest area was occupied at times after about 300 A.D., and more than 300 Indian ruins have been found in the park (Ash and May, 1969). Puerco Ruin on the Puerco River is near the main park road a short distance south of the Santa Fe Railroad tracks. Here, about 150 masonry rooms enclosed a large plaza; as many as 250 people may have made the village their home. There appears to be a question as to when the site was first occupied, but it was abandoned around 1400 A.D., about a hundred years after the Mesa Verdes had to leave.

Although the Big Drought drove many Indians from their villages, the Puerco Indians apparently had water sufficient for farming. But during the 1300s Puerco River began to actively erode its banks and in time destroyed most of the farmland in the valley. Thus, after surviving the drought, they had to surrender to the malevolent river.

Newspaper Rock is only a short distance from Puerco Ruin. Here the Indians chipped away the iron-oxide coating (desert varnish) and exposed the light-colored sandstone beneath their designs and drawings of humans and other animals. These petroglyphs are found in a number of places in the park.

The park was established to protect the beautiful petrified logs from the few who would destroy them. Even today, there are a few park visitors who would like to take home a small souvenir or two! As you approach the exit, you will see a large sign warning you that your car will be searched; it also mentions the penalty. This area between the sign and the exit might be called the "repentance area" because several hundred pounds of souvenirs have been retrieved from along this short section of the road.

Many fascinating features, including a petrified log house, are there for you. Take a camera and enough time to catch the various moods of the day; if it should rain, the sun will soon be shining again — on a different world of different colors.

CANYON DE CHELLY NATIONAL MONUMENT

Canyon de Chelly, near the town of Chinle in northeastern Arizona, became a national monument in 1931. As at Mesa Verde, it was set aside primarily to preserve archeological and historical records. Similarly, the particular assemblage of rocks and the geologic sequence made Canyon de Chelly a desirable place of habitation for the ancient peoples. The niches beneath huge overhangs provided protected places for the Anasazi to build their cliff houses — apartment houses.

The name "Canyon de Chelly" is the Spanish version of the Indian word *Tsegi* which means rock canyon; the present-day pronunciation is "de shay." The monument is within the Navajo Indian Reservation and many Navajos make it their home, some as employees of the Park Service.

Monument boundaries enclose the main canyon and a large tributary called Canyon del Muerto (Canyon of the Dead) which joins

Canyon de Chelly from the north. Sheer cliffs of massive Permian De Chelly Sandstone rise as much as 1000 feet (303 m) above flat canyon floors. Rio de Chelly and its tributaries flow bountifully during spring runoff and after heavy rainstorms, when floods are common occurrences. Although hazardous, the floods add much needed moisture to the floodplain soils on which the Indians have raised their crops since about 350 A.D. Although the streams are intermittent in places, their flows are perennial where impervious shale is exposed in their channels. This year-around flow in parts of both of the main canyons made Canyon de Chelly a highly desirable area.

The first known inhabitants lived in circular pithouses on the valley floor; later they built masonry apartments (cliff-dwellings) up on the ledges, well out of reach of floods. But as in a number of other places, the Big Drought late in the thirteenth century drove them out of Canyon de Chelly, to find homes along the Rio Grande and other large rivers of the Southwest.

The Navajos moved into Canyon de Chelly around 1700. They were inherently aggressive and their raids on the Pueblos and the Spanish villages along the Rio Grande brought repri-

FIGURE 11-55

Navajo hogan, Canyon de Chelly.

sals; hence Canyon de Chelly was the scene of many fierce battles. In 1815, Spanish soldiers killed 115 Navajos in a rock-shelter in Canyon del Muerto; it was named Massacre Cave. Canyon del Muerto was named later, when Smithsonian scientists discovered the remains of prehistoric burials.

In 1864, Kit Carson was sent in with a detachment of U.S. Cavalry to subdue the Indians and to relocate them on a reservation in New Mexico. This move was ill-conceived, however, and after four years they were allowed to return to Canyon de Chelly where their descendants now live, some in hogans on the canyon floor.

Park visitors can see the canyon and a few of the ruins from the Rim Drive along the south side of Canyon de Chelly, and they are free to hike down the trail to White House Ruin. However, in order to protect the fragile ruins and also the privacy of the Navajos, all other travel is under the watchful eye of an authorized guide or park ranger. Do not forget to look down on Spider Rock, a sandstone spire which rises about 800 feet (244 m) above the floor of Canyon de Chelly near its junction with Monument Canyon. Colorful pictographs are found at several points; those at Standing Cow Ruins are particularly outstanding.

The geologic sequence which made Canyon de Chelly a desirable home began with the uplift of the plateau late in the Tertiary. Rio de

FIGURE 11-56

Rio de Chelly, flowing a sinuous course through Canyon de Chelly.

Chelly and its tributaries cut down through the Shinarump Conglomerate and through the De Chelly Sandstone, in places into the red beds.

Vertical fractures in the sandstone resulted in sheer, near-vertical cliffs. Some of the buff-colored walls have streamers of red from top to bottom, where iron oxides have been deposited. In places at or near the base of the De Chelly, overhangs were formed by weathering; in these niches the ancient people built many of their cliff-houses, safe from floods and sheltered from wind and rain.

Unlike many of modern man's intrusions, these cliff-dwellings blend in with their surroundings. The building blocks are De Chelly Sandstone and the mortar is mud from along Rio de Chelly; both are natural ingredients.

Canyon de Chelly represents an early stage in the evolution of arid landscapes, as illustrated in Figure 11-4. In time — a few million years — we will see here another Monument Valley; shortly thereafter, we will have to change the name Colorado Plateaus to Colorado Plains.

Canyon de Chelly also provides continuity between the historic past and the present. The ancient ruins record the early history; the Navajos are there to tell us of the more recent past, during evening programs near the visitor center.

FIGURE 11-57

El Morro; note slab of Zuni Sandstone that will eventually crash down.

EL MORRO NATIONAL MONUMENT

Carved in Jurassic Zuni Sandstone are the messages of ancient and not-so-ancient travelers, on the face of a high cliff in western New Mexico, several miles east of the town of Zuni.

The name "El Morro," which means "The Bluff," is appropriate because a 200-foot-high bluff rises to the top of a mesa. On top there are ruins, largely unexcavated, which record Indian occupation of the area long before it was invaded by the Spaniards. Hundreds of petroglyphs carved in the rocks record the life of the times.

Inscription Rock is the bluff north of the visitor center. It was a veritable bulletin board for the "tourists" who stopped by, beginning as early as 1583. For example, in 1605 the explorer-soldier Don Juan de Onate left word that he and his men "passed by" on 16 April. The sandstone was easily carved and apparently very few restrained themselves. After the Spaniards explored the area, American soldiers, emigrants and adventurers camped here — because of the never-failing waterhole in this essentially waterless area. This waterhole, which sustained the ancient Indians and later many a weary traveler, is a pothole. The water from late summer storms and snow-melt poured down from the mesa top and carved a pothole in the bedrock. It is a large pothole, one that holds thousands of gallons of precious water.

Here at El Morro, the carvings on Inscription Rock alone justified the creation of the national monument. Today the Park Service takes an extremely dim view of visitors with carving knives who wish to make history. But they have provided for them by placing a large slab of sandstone at the visitor center—a modern-day Inscription Rock.

THE COLORADO PLATEAUS—PRESENT AND FUTURE

We have spent considerable time in the Colorado Plateaus and we have visited many park areas. But there are still others, mainly archeological, such as Chaco Canyon, Aztec Ruins and Hovenweep. Even so, should there be a need for additional parks and monuments, more could be found here. For example, Escalante Canyon in southern Utah has its Stevens Arch, Escalante Natural Bridge, and a host of other outstanding natural features. And Bandera Crater, not far from El Morro National Monument, is worthy of consideration. In addition to the cinder cone and crater, there are lava tubes and the Ice Caves.

Perhaps Glen Canyon would have served us

better as a national park for everyone than as a reservoir for generating hydroelectric power for Los Angeles. What we know for certain is that, with the construction of Glen Canyon Dam, one of our most beautiful canyons was taken from us, and from all generations to come.

We were asleep when the dam-builders quietly obtained approval to build Glen Canyon Dam. The Indians knew the beauty of Glen Canyon; Major Powell knew and described the Music Temple, Cathedral of the Desert, Dungeon Canyon, and all of the other beautiful and awe-inspiring features. But to most of us Glen Canyon was merely "a place somewhere out there in the desert." It is fortunate for us that Eliot Porter got there in time to record the exquisite beauty of the canyon, in color—as the waters of Lake Powell rose higher and higher, to bury them forever. We can enjoy them, as we leaf through *The Place No One Knew: Glen Canyon,* and recall them to mind as we glide over Lake Powell's waters, careful to get out of the way of the monster power boats snorting and hurtling about—with occupants frantically seeking the wilderness experience! It was both ironic and cruel that the lake was named to honor the man who most appreciated the canyon that the lake destroyed. Let us remember that Escalante Canyon is also a beautiful canyon—and that it is in danger, perhaps to be dammed, perhaps otherwise destroyed.

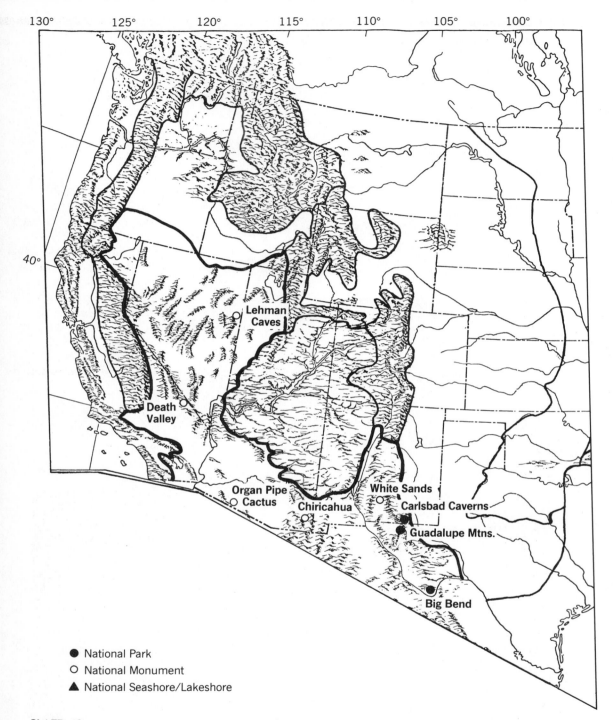

- ● National Park
- ○ National Monument
- ▲ National Seashore/Lakeshore

PLATE 12

Map of Basin and Range Province.

CHAPTER 12

BASIN
AND
RANGE
PROVINCE

The Basin and Range Province is a sprawling area that extends from southern Oregon southward into Mexico and eastward to include the Big Bend country in southwestern Texas. As the name implies, there are mountain ranges — more than 150, if small ranges are included — with basins between them. Many of the basins are enclosed and contain an ever-increasing amount of erosional debris from the surrounding mountains.

How can an area this large be one geomorphic province? In this case, climate is evidently the significant factor; the province is arid, with the exception of small "islands" that rise high enough to intercept moisture. The rocks vary considerably, but in this dry climate the limestones are as resistant to weathering and erosion as many of the sandstones and granites. The ranges therefore are generally jagged with bold faces and cliffs essentially devoid of vegetation. Both the geologic structures and the geologic sequence are similar throughout the

FIGURE 12-1

Basin and Range topography near Ajo, southernmost Arizona. As the pediments enlarge by mountainward extension, the ranges are reduced in size (see Figure 12-9). (Photo by Fred Goodsell.)

province. Thus this large area fits the definition of a geomorphic province reasonably well.

The basins occupy much more of the area than the ranges; in fact, in the distant future the ranges will be largely buried beneath their own waste material, especially in the northern part —the Great Basin—where there is only interior drainage. Even in the remaining area where most streams drain to the ocean, there are local interior basins; the Tularosa Basin in southern New Mexico is one example.

With few exceptions the landscapes are typical of dry regions. The dominant landforms are alluvial fans, *bajadas,* pediments, *bolsons,* and *playas.* Vegetation is sparse and wind erosion is active; large sand dune areas, though less common than one might expect, are found in Death Valley, in southernmost California, and in several other places.

The erosional process is carried on mainly by water, even in this desert region; locally the landforms are clearly the results of wind activity or glaciation. They are, however, local features superimposed on topography formed by water erosion and deposition. Unlike the permanent streams in a humid area, most of these streams are dry much of the time. When intense storms occur, with little or no vegetation to retard runoff, flash floods develop which have tremendous capacity to erode and to transport materials. Therefore, during the brief periods of flood flow, the streams in the Basin and Range Province erode rapidly and quickly alter the landscape.

The Basin and Range Province is being enlarged at the expense of the Colorado Plateaus, as noted earlier. Streams that head on the escarpment that separates the two provinces are cutting headward, enlarging their watersheds. Promontory Butte in east-central Arizona is now within this boundary zone. Actually it is a narrow projection of the Mogollon Plateau; soon, geologically, it will become a true butte within the Basin and Range.

Rocks representing all geologic eras are found in the province. Precambrian granites and metamorphics are the only rocks in some of the ranges; in others, Paleozoic limestones and sandstones are dominant. Mesozoic rocks, which dominate in the Colorado Plateaus, are much less abundant here. Tertiary-Quaternary lava flows and pyroclastics cover a significant number of areas, in places capping tilted fault-blocks. But the basins that occupy most of the province are floored by thick bolson deposits of Quaternary age.

Basin and Range Structures

The structures of the Basin and Range Province are difficult to decipher because they are largely buried beneath *bolson,* or basin, deposits. In many individual ranges the rocks are all dipping in one direction; in some cases lava flows that are the caprocks are now dipping steeply. Early reconnaissance geologists concluded that each of the small ranges was a fault-block that had been tilted; this was the typical "Basin and Range structure." More detailed work revealed that, in addition, there are a considerable number of more complex structures such as overturned folds and large thrust faults. The Keystone Thrust west of Las Vegas, Nevada (Hewett, 1931), and the Muddy Mountain Thrust north of Lake Mead (Longwell, 1928) are classic examples. The conclusion of a number of workers is that the folded and thrust-faulted structures are older; later, when there was relaxation of the compressional forces, normal faulting and the tilting of the fault-blocks took place, probably late in the Tertiary (Mackin, 1960). In the literature, however, when "Basin and Range structure" is used to describe a certain area, it probably means block-faulting and tilting.

Recent studies indicate that the Basin and Range Province is being enlarged in another way—by the spreading of the earth's crust. From detailed investigations in Dixie Valley in western Nevada, Thompson and Burke (1973) suggest that during the past 15 million years the surface area of the province may have

FIGURE 12-2

Fault scarp, near Frenchman in western Nevada. New scarp was formed by the faulting that occurred in 1954 in this tectonically unstable section of the Basin and Range Province.

been increased by 10 percent. The northwestward movement of the Pacific Plate is apparently responsible for the pulling apart of this section of the earth's crust. Again, plate tectonics provides a possible solution of a problem difficult to explain otherwise — namely, "Basin and Range structure."

Pleistocene Glaciers and Lakes

Only a few of the highest areas in the province were glaciated; almost all are in the northern section — East Humboldt Range and the Ruby Mountains in northeastern Nevada are examples. Sierra Blanca in southeastern New Mexico is one exception; it is 12,003 feet (3660 m) high and was extensively glaciated.

Pleistocene lakes, some very large, occupied the interior basins. Lake Bonneville, much larger than its puny descendant, Great Salt Lake; Lake Lahontan with present-day Carson Sink; and Lake Manly in Death Valley are three of the best known. As the climate fluctuated repeatedly from glacial to interglacial, so did the lakes wax and wane, leaving wave-cut notches or terraces — marking stillstands of the lakes — at different levels on the sides of the mountains. In the smaller basins the lakes have long been dry, except for short periods during wet seasons. Some of the water penetrates into the deposits, but in this climate evaporation rates are extremely high.

Caverns

The work of groundwater is recorded in many places in the province. Limestones are widespread, and in those containing well-developed fracture systems, solution is most active. Carlsbad Caverns, in southeastern New Mexico, are world famous. There are also caverns in the limestones in Big Bend and probably in the unexplored sections of Guadalupe Mountains National Park. In the northern part of the province there are a number of caverns; within the park system, there are the Lehman Caves in eastern Nevada.

LEHMAN CAVES NATIONAL MONUMENT

On the hottest day there is a cool spot in the desert in eastern Nevada, near the town of Baker. Lehman Caves are developed at the base of Wheeler Peak, one of the glaciated peaks of the Snake Range. This square-mile national monument, established in 1922, is open to the public during all seasons of the year.

The geologic story began when, during the Cambrian Period, a thick limestone was deposited in this area, along with other sedimentary rocks. Deposition continued throughout the Paleozoic and alternately during the Mesozoic and early Cenozoic. At some later time, perhaps as a result of mid-Tertiary intrusions, the limestone was locally metamorphosed to marble (Drewes, 1958). Faulting and uplift raised the Snake Range to its present position sometime late in the Tertiary. Erosion removed

FIGURE 12-3

Shields (parachutes) in Grand Palace, Lehman Caves, Nevada. (Photo by National Park Service.)

the younger rocks and paved the way for the development of caverns in the mid-Cambrian marble.

Solution by subsurface water along fractures in the marble formed small passageways that were later enlarged to form long, high corridors. Large rooms such as the Grand Palace and the Lodge Room are important units along the underground trail. Stalactites and stalagmites decorate the rooms and passageways. In addition, there are shields, circular plates that extend out from the wall at an angle — like shields sometimes used in indirect lighting. Because many resemble parachutes, that term is also used. These formations, abundant here, are uncommon elsewhere; however, their precise development is not completely understood.

The trip across the desert is long; a visit to Lehman Caves will take you into a different world, one that is well worth investigating.

DEATH VALLEY NATIONAL MONUMENT

In 1933 almost 3000 square miles of southeastern California and westernmost Nevada were set aside as Death Valley National Monument. Death Valley is the hottest, driest, and lowest of our park areas. The maximum temperature recorded — in the shade — was 134°F (57°C). The average annual precipitation is low, only about 2 inches (5 cm). After several years with no precipitation, local storms of high intensity cause floods that in a few hours remodel the landscape significantly.

The topographic relief within the monument is notable. Telescope Peak near the southwestern border is 11,049 feet (3369 m) high; less than 20 miles (32 km) to the east is the lowest point in the Western Hemisphere, 282 feet (86 m) below sea level. This topographic relief of 11,331 feet (3455 m) is less than it was earlier; in this enclosed basin all of the erosional debris carried down out of the mountains is deposited on the floor of the valley. The total thickness of these *bolson* deposits is unknown, but it is great.

The early history of this unique area is colorful yet painful, when the suffering of the early emigrants is recalled. The prospectors came in to find the rich gold deposits they had heard about. It is ironic that it was the lowly borax that made Death Valley famous. Untold tons of this mineral were hauled out of Death Valley in the high wagons drawn by 20-mule teams, as shown in the Wallace Beery movie. Borax and talc are now being mined in Death Valley National Monument; however, mining will be phased out as the existing leases expire.

General Geology

How was Death Valley formed? Is it actually a valley? True valleys are formed by streams, and sea level is the base level of erosion. Death Valley is a structural basin formed by downfaulting of a large block, the "valley," and up-

FIGURE 12-4

Faulted alluvial fan, near Badwater in southern part of Death Valley. Fault scarp is seen on skyline, right of center. (Photo by R. Scott Creely.)

FIGURE 12-5

Myriad streamlets have eroded badlands in poorly consolidated Tertiary sediments near Zabriskie Point, east of Furnace Creek. (Photo by R. Scott Creely.)

faulting of the mountain ranges surrounding it.

The faulting began late in the Tertiary Period, at which time Death Valley and the surrounding mountains began to take shape. At least in large part it was normal faulting, with great blocks being rotated and moved upward or downward. Death Valley was a downward-moving block and is still downward-moving today. Fault scarps across alluvial fans indicate recent faulting, and in one known case the fault scarp is less than a century old.

Volcanism was an active process in the Death Valley area; basaltic lava flows are widespread, particularly in the northern part. Volcanoes erupted in relatively recent times in the northwestern corner, near where "Death Valley" Scotty later built his castle. It was probably not more than 2000 years ago that Ubehebe Volcano built a cinder cone with its half-a-mile-wide, 800-foot-deep crater. In the same area, smaller cones developed during the last 200 or 300 years.

Alluvial fans, most of them very large, are widespread in Death Valley. Conditions are optimum for fan building; large amounts of coarse, weathered materials that accumulate in the mountains are flushed out by periodic torrential floods; in addition, this is a tectonically unstable area where the mountains are being uplifted, and fan building continues unabated. In the stable area, pediments — erosional features — are normally developed at the base of mountains. Denny (1965), who made a study of the landforms of Death Valley, found a few small pediments almost entirely buried by alluvial fans. The logical conclusion is that pediments were formed prior to the most recent period of sustained tectonic activity and that they are now being covered by alluvial fan materials.

So extensive and so closely spaced are the alluvial fans that they overlap each other; a series of coalescing alluvial fans forms a continuous alluvial slope called a *bajada*. Thus, ba-

FIGURE 12-6

Devils Golf Course, Death Valley.

jadas are the dominant landforms in Death Valley; however, where the canyons are more widely spaced, individual fans have maintained their own identities.

Pleistocene climates affected the Death Valley area, albeit in a manner different from that in areas where large glaciers developed. During periods of higher precipitation, the excess water formed large lakes in the enclosed basins of the province. In Death Valley, Lake Manly formed, expanded, and at last waned during the Pleistocene. According to Blackwelder (1933), Lake Manly was as much as 90 miles (145 km) long and 585 feet (191 m) deep at its maximum stage. When the lake level remained constant for an extended period, wave-cut terraces were notched into the sides of the mountains; these ancient shorelines can be clearly seen today. About 11,000 years ago, with the warming and drying that culminated in the Thermal Maximum, Lake Manly disappeared. By evaporation of the lake waters, all of the various salts were precipitated to form the thick playa deposits that floor the valley. Of these salts, borax was one of the most abundant.

The Devils Golf Course, between Furnace Creek and Badwater, consists of the jagged salt ridges and pinnacles that inspired the name. Although the ridges and spires are only a few feet high, the terrain is almost impossible to traverse. For an exhaustive report on the processes involved in the formation of many fascinating features — dessication cracks, salt wedges, salt pans, and salt pools — see Hunt (1975).

Sea level

FIGURE 12-7

The lowest point in Death Valley, 282 feet below sea level. The sign near the top of the photograph is the sea-level marker.

Playa lakes form during storms and then vanish in a few days, mainly by evaporation. The playa deposits consist of silts and clays interbedded with salts that were formed by precipitation as the lake water evaporated.

Sand dunes are found in certain places in this desert area. What is puzzling is why they are not more widespread than they are — where winds are strong and the vegetation is sparse or lacking. The main dune areas are in the vicinity of Stovepipe Wells in the north-central section of the monument (Clements, 1955). Clements points out that most of the dunes are composed mainly of quartz sand, but that an unusual dune area is found about 10 miles (16 km) north of Stovepipe Wells. Here, derived from large travertine deposits, the dunes are made up of travertine sand.

Death Valley has more to offer than its many geologic features; more than 600 species of plants and many animals and birds have adapted to this harsh environment. Plan your trip carefully; the regular season is from October 15 to May 15. The summer sun is merciless and the range in temperature from day to night is great. If you are there during the right season, in a car in good condition, with more than enough drinking water, your trip into Death Valley will likely be one of the best.

ORGAN PIPE CACTUS NATIONAL MONUMENT

In southern Arizona, along the Mexican border, about 516 square miles of desert were set aside as a national monument in 1937. Its primary function is to preserve a typical desert environment, but as an important incidental, classic examples of desert landforms are there in abundance. It is a true desert, receiving an average of about 8 inches of rainfall per year, half of which fall during desert cloudburst storms.

The Ajo Mountains along the eastern border, the Bates Mountains in the western

section, and the Puerto Blanco Mountains in between are composed almost exclusively of Tertiary and Quaternary volcanics, mainly basalt flows and breccias. The mountains rise abruptly above the broad, nearly flat lowlands such as the Valley of the Ajo, which occupies most of the north-northeastern section of the monument.

It might appear that the landforms would be similar to those in Death Valley. Quite the reverse is true; pediments, generally lacking in Death Valley, are well developed and widespread in the general area in and around Organ Pipe Cactus National Monument. Probably this is the result of the difference in degree of tectonic stability. The southern part of the Basin and Range Province is comparatively stable tectonically — much more so than the northern part where Death Valley is located. The streams that flow out of the canyons are seldom loaded to capacity. When they are in flood they have large capabilities to erode, but they have very little loose material to pick up and transport out of the mountains. Consequently, swinging from side to side, the streams plane off the bedrock at the base of the mountains. Coarse materials, the eroding tools, are spread in thin layers over the eroded bedrock, forming the pediment surface. Once this flattish surface is developed, it is then perfected by sheet-flood erosion as McGee vividly described in 1897, after being caught in a desert cloudburst. The Papago country, including the Ajo Mountain area, was recognized by Bryan (1922) as being an ideal place for field research on the problem of the origin of pediments; his studies contributed much to our present knowledge of pediments.

Organ Pipe Cactus National Monument affords unusual opportunities to study plant-moisture–soil-geology relationships. For example, the organ pipe cactus — one of the many species in the monument — shuns the packed soils of the flatter areas but flourishes on the steep, dry, and rocky slopes of the Ajo Mountains. The giant saguaro grows to great

FIGURE 12-8

Cactus-covered pediment with Ajo Mountains in the background (left).

FIGURE 12-9

View south from Diaz Peak along eastern part of Organ Pipe Cactus National Monument, southern Arizona. Pediments have been extended into the mountains, leaving rock hills (inselbergs) rising above the pediment surface. (Photo by Fred Goodsell.)

heights, as much as 50 feet (15 m), on the flats where the soils are finer and hold more of the precious moisture.

This monument has many interesting aspects, some of them colorful, especially the brilliant flame-red blossoms of the ocotillo, usually at their best in April. Wear hiking boots on the trails and carry a good supply of drinking water.

CHIRICAHUA NATIONAL MONUMENT

The Chiricahua Mountains rise above the desert of southeastern Arizona. With peaks as high as 9796 feet (2986 m), they receive both snow and rain sufficient for a dense growth of vegetation in the more favorable sites, particularly in the shaded canyons.

Yucca, cactus, and century plants dot the desert areas; a chaparral of scrub oak and manzanita covers the higher slopes, with the sycamore and Arizona cypress shading the deep canyons. White-tailed deer, peccaries, bobcats, and even the coatimundi are here, along with rodents and birds of many kinds. For the details on the plant and animal life of the area, see Jackson (1970).

The Chiricahuas are composed of upfaulted Precambrian granites and schists flanked by Paleozoics and younger rocks. But within the monument, volcanics carved into weird columns, spires, and balanced rocks are of greatest interest. This maze of rocks, with its myriad columns and labyrinthine passageways, provided an essentially impenetrable stronghold for Cochise and Geronimo, and later for "Bigfoot" Massai. The profile of Cochise can be seen from Interstate 10 near San Simon or looking north from high on the monument road; Cochise lies in stony silence on the mountain north of the monument boundary.

The geologic story might begin early in the Precambrian; it must begin early in the Tertiary or perhaps late in the Cretaceous when volcanoes took charge of the area. As outlined by

FIGURE 12-10

Profile of Cochise, bathed in smog from smelters in the valley.

Yetman (1974), the early volcanics consist of volcanic conglomerates and flows. Later, probably during the Oligocene, sediments were deposited in lava-bound basins before and during the volcanic eruptions of the second series. Then, after a long period of erosion, the area became the scene of highly explosive eruptions that piled up more than 1500 feet (457 m) of rhyolitic tuff and then capped it with a thick lava flow. These explosions immediately preceded the collapse of the volcanic dome that formed the Turkey Creek Caldera a short distance south of the monument. Yetman credits Dr. D. Marjaniemi for the discovery of the caldera and for the interpretation of the tuffs that exploded from that center. Heat welded the tuffaceous materials together and formed rocks that are highly resistant — rocks that were later shaped into the pedestals, columns, and spires for which Chiricahua is famous.

The Pinnacles at Crater Lake were formed by weathering and erosion of poorly welded ash from around columns of highly welded material. In Chiricahua, an entirely different origin is envisioned, one based on intersecting fractures that formed during the most recent faulting and uplifting of the mountains late in

FIGURE 12-11

Tertiary volcanics form a number of mountain ranges in the Basin and Range Province. The Organ Pipes, shown here, are among the many fantastic erosion forms in Chiricahua National Monument, southeastern Arizona. This welded tuff was formed when the Turkey Creek Calders developed in the Chiricahua Mountains.

the Pliocene. Weathering widened the fractures, eventually forming the columns separated by sinuous passageways, some of which were converted into stream courses.

Special mention is made of *accretionary lapilli*, sometimes called *volcanic hailstones*. They are generally uniform in size — about that of marbles — and most are nearly spherical. A good place to see them is along the trail from the Echo Canyon parking area to Massai Point. One theory of origin involves the accretion of volcanic dust around a nucleus during times of extreme air turbulence, when they were repeatedly lifted high and then dropped. If this is the mechanism, the designation as volcanic hailstones is appropriate.

Some of Chiricahua's wonders can be seen from a car. Be sure to see the exhibits in the visitor center and also the slide-tape presentation. Go to the Sugarloaf parking area and hike to the top of Sugarloaf Mountain, where the views are excellent in all directions. Other trails take you to the Natural Bridge and the Big Balanced Rock in the Heart of Rocks section. And you may want to see Duck-On-A-Rock.

BIG BEND NATIONAL PARK

Big Bend National Park, located in the Big Bend of the Rio Grande, contains about 1107 square miles of Texas mountains and desert. For more than a hundred miles the Rio Grande forms the boundary between the park and Mexico; in this section the Rio Grande flows through three separate spectacular canyons or gorges. The Chisos ("ghost") Mountains occupy the central area of the park. The name was well chosen, for from a distance they often appear phantom-like. This is the land of the roadrunner and the javelina.

When Big Bend was no more than just "the Big Bend country" a young geologist became intrigued with this jumbled mass of rocks that no one had been able to make heads or tails of. This was his challenge, more attractive because of its remoteness — wildness, in fact. When Ross Maxwell began the survey there were no roads; the Comanche Trail was the main thoroughfare. A horse named Nugget was his "field assistant"; Nugget carried Ross over many miles of mountain and desert trail during the long days of the survey. This was not the Big Bend of the 1880s as might be assumed, but the Big Bend of the 1930s and 1940s — the last frontier in the 48 states. Slowly and stubbornly the Big Bend surrendered its "past" and the puzzle was put together, minus a few details buried under Chisos volcanics. When the park was established in 1944, Maxwell became its first superintendent. He laid out the roads where they would not be hazardous; yet it was not by coincidence that they pass by, or are within sight of, almost every fascinating geologic feature in the vast area which is Big Bend National Park.

It would be impossible for one man to work out all of the details of such a complex area. In addition to Maxwell's works published from 1939 to 1968, papers by King (1935), Baker (1935), Lonsdale et al (1955), Matthews (1960) and Sperry (1938) are among the significant publications pertaining to Big Bend.

Geographic Setting

Big Bend is a low-lying desert along the Rio Grande, yet the Chisos Mountains rise above it

FIGURE 12-12

The Basin in the Chisos Mountains. (Photo by Ross A. Maxwell.)

to heights as great as 7835 feet (2389 m) at Emory Peak. The Basin is an "amphitheater" in the Chisos, with an opening to the west called "The Window."

The three canyons cut by the Rio Grande are spectacular features. Santa Elena Canyon has sheer walls 1500 feet (457 m) high and is in places only 30 feet (9 m) wide; therefore, in flood the Rio Grande is a "river on edge." Downstream, the Rio Grande has cut deep Mariscal Canyon through Mariscal Mountain. Farther downstream, along the eastern border, Boquillas Canyon is equally striking.

Long slopes covered with mesquite, cactus, and other desert plants extend out from the mountains. These are pediments and remnants of older pediments. Badlands areas are common where streams have dissected easily eroded, poorly consolidated rocks.

Bedrock Geology and Geologic History

Big Bend is composed mainly of Cretaceous limestones which in large areas are buried under Tertiary volcanics. In addition, intrusive stocks, laccoliths, dikes, and sills are extensively exposed. In limited areas, there are Paleozoic limestones and shales, and Cretaceous and Tertiary sandstones and shales.

The oldest rocks in the park are Lower Paleozoics. They were folded and eroded, and on this erosion surface Cretaceous conglomerate was laid down, thus forming an angular unconformity. Cretaceous sediments were almost entirely calcareous muds, however, and several thick and generally massive formations were laid down. The massive and resistant Santa Elena Limestone that forms the rim rock at Santa Elena Canyon is an example.

Folding and faulting during the Laramide Orogeny developed north-south structures that are similar to those in the Rocky Mountains and perhaps were connected to them.

After prolonged erosion and block-faulting in the Big Bend area late in the Tertiary, these older structures became comparatively insignificant; otherwise, Big Bend would be an extension of the Southern Rockies. Here, as in certain other parts of the Basin and Range Province, the newest major structures, with the resulting landforms, are dominant factors in the delineation of the geomorphic province.

Block-faulting of large magnitude formed the mountain ranges of today. The sheer face of limestone at the mouth of Santa Elena Canyon is ample evidence; here, along the Terlingua Fault, a long block of the crust was lifted and rotated, thus forming a mountain range called Mesa de Anguila. Along the eastern side of the park, Sierra del Carmen is another of many fault-blocks in Big Bend.

Igneous activity that continued for a long period began later in the Eocene Epoch; the Chisos Mountain area is an igneous complex of both intrusive and extrusive rocks. Extrusive activity was dominantly explosive, and breccias and tuffs are widespread; lava flows are also common and cap a number of the higher ridges and peaks. Casa Grande, which towers above the Basin, is topped by a lava flow, as is the south rim of the Chisos.

The origin of the Basin has been a matter of controversy. It is in general cirque shaped, leading Jenkins (1958) to the erroneous conclusion that it is a glacial cirque. Its shape led others to believe that it is a volcanic crater, but this too lacks supporting evidence. Rather, it is a basin carved out of volcanic rocks by streams. The complex of rocks was such that, by erosion, a bowl-shaped "amphitheater" was formed.

Volcanic necks and plugs form prominent features, some of which have prompted such names as the Elephant Tusk. One unusual volcanic feature resembles a huge petrified stump and has, in fact, been erroneously identified as such, according to Maxwell (1968). As it is composed of dense rhyolite and rhyolitic pitchstone, it is definitely of igneous origin.

FIGURE 12-13

Santa Elena Canyon as seen from the air. Fault-block of limestones is tilted away from viewer. Mexico is to the left of the canyon; Texas is on the right. (Photo by Hunter's of Alpine, Texas.)

FIGURE 12-14
Map of Big Bend National Park.

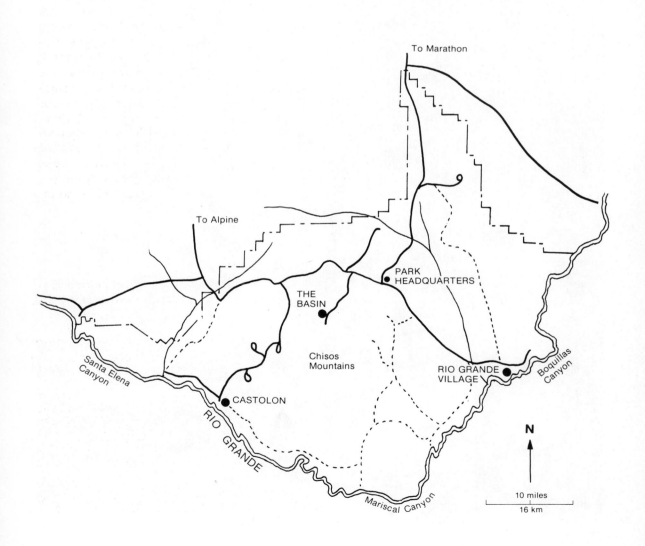

FIGURE 12-15

Volcanic spine exposed by erosion of soft volcanic tuff that encased it.

Geomorphic Features

In addition to the landforms of volcanic origin, those resulting from stream erosion and deposition are widespread and well developed. Pediments are by far the most important of these landforms. These long uniform slopes that extend out from the mountains are graded to the main streams of the area. The bedrock is generally covered with gravelly alluvium from 5 to 20 feet in thickness; locally it is much thicker. Remnants of former pediments are also present, representing earlier cycles of pedimentation. As in the Organ Pipe Cactus area, widespread pediments with few and small alluvial fans indicate that the Big Bend country is relatively stable tectonically.

Although limestones are widespread, caverns are apparently not common features in Big Bend. Maxwell (1968) mentions one — Smugglers Cave — in the wall of Santa Elena Canyon. Apparently, the fractures that are essential in the cavern-forming process are generally lacking.

The canyons are extremely deep and steep-walled. In part, this is due to the massiveness of the limestones; more important perhaps are the climatic factors that retard chemical weathering. The cutting of the canyons began after the enclosed basins, formed by mid-Cenozoic faulting, had been filled to the brim. It was then, according to Maxwell, that the present course of the Rio Grande was established. The Rio Grande flowed across the tops of the mountain ranges and the basin-fill; then, late in the Tertiary, the river began to cut down and became superposed across the resistant mountain ridges where the canyons are now. As down-cutting progressed, the easily eroded basin-fill was removed and the buried mountain ranges were exhumed.

FIGURE 12-16

One of many extensive remnants of pediments in Big Bend. Streams have cut down, leaving the pediment remnants well above present stream level.

FIGURE 12-17

Fossil clams in Cretaceous limestone, Big Bend National Park. (Photo by National Park Service.)

Population Explosions in Texas

During the latter part of the Mesozoic and at times during the Cenozoic, the Big Bend area of Texas was much more thickly populated than now. Most of the Cretaceous limestones are highly fossiliferous, containing both invertebrate and vertebrate remains. Skeletons of marine reptiles such as the mosasaur, ichthyosaur and plesiosaur have been found. At one time late in the Cretaceous the area was slightly above the sea, and the swamps were well populated by dinosaurs, both herbivores and carnivores. Stumps composed of agatized wood occur in the dinosaur-bone beds. Cenozoic rocks contain a large number of different species of mammals, including Hyracotherium, the dawn horse (Maxwell, 1968). And camels and deer were there, too. In fact, from Maxwell and from Matthews' (1960) handbook on Texas fossils, it becomes evident that Big Bend is one of the more highly fossiliferous national parks.

Big Bend is a good park to visit at any time of the year; in summer, if it is hot down along the Rio Grande, it likely will be pleasant high in the Chisos Mountains. A large relief model at park headquarters will help you orient yourself. Do not miss Santa Elena Canyon; a trail leads up into the canyon where you can look up out of this narrow slit in the earth. To get the flavor of the frontier days in Big Bend, go to nearby Castolon Trading Post. Plan to be over in the eastern section just before sunset — when Sierra del Carmen is ablaze. You must take pictures in Big Bend; use a haze filter if there is dust in the air. Also, bring tweezers — for extracting cactus spines from your person.

CARLSBAD CAVERNS AND GUADALUPE MOUNTAINS NATIONAL PARKS

Carlsbad Caverns in southeastern New Mexico and Guadalupe Mountains in Texas are two separate and distinct national parks. Inasmuch as they are separated by only 4.5 airline miles (7 km) and involve the same formations — limestones formed in a huge reef during the Permian — they are included in the same discussion. One main difference is that the rocks that you see in the eerie light of the caverns are the same ones you see high on the face of El Capitan in the Guadalupe Mountains.

In the late 1800s, cowboys in southeastern New Mexico saw thousands of bats apparently explode out of the ground; thus "Bat Cave" — later named Carlsbad Caverns — was discovered. Jim White is given credit for exploring the caverns beginning in 1901. As a result of his accounts, and those of others, Carlsbad was established as a national monument in 1923; in 1930, it was redesignated as a national park.

El Capitan's bold cliff at the south end of the Guadalupes has long been a landmark for travelers in the El Paso extension of Texas. It is of special interest to geologists because exposed in the Guadalupe Mountains is a thick life-reef of Permian age. In addition, with elevations rising from low desert plains to 8751 feet (2668 m) at Guadalupe Peak, several life zones are represented, as Matthews (1968) reports.

For both geological and ecological reasons, Guadalupe Mountains National Park was authorized in 1966. Because of the lack of roads and other facilities, Guadalupe is now a hiker's park. This will not deter those deeply interested in the desert wilderness, hopefully after seeking guidance from a park ranger.

FIGURE 12-18

El Capitan, a majestic landmark that rises abruptly above the plains in Guadalupe Mountains National Park, is composed of Permian Capitan Limestone.

Geologic Development

Late in the Paleozoic, seas covered a large area of New Mexico, Mexico and Texas; in mid-Permian time large *bioherms,* or life-reefs, began to develop, which were generally similar to the Great Barrier Reef off the Australian coast. Many and diverse types of marine life, both plants and animals, thrived in the warm waters. According to Adams and Frenzel (1950), the youngest of the Permian reefs was the huge Capitan Barrier Reef. The resulting Capitan Limestone is the main formation in both Carlsbad Caverns and Guadalupe national parks.

Limestones may be formed by chemical precipitation from sea water. Usually, however, lime-secreting marine animals and plants play an important role in the process; reef limestones are made up largely of the shells and skeletons of organisms. As Press and Sievers (1974) explain, an impressive assemblage of animals — one-celled foraminifera, corals, brachiopods, and molluscs — and plants, mainly algae, build huge reefs in favorable waters, such as clear tropical waters.

A reef is an unstable structure. As the reef grows upward, it expands laterally and forms large overhangs. In time, large blocks break off and slide or roll down the steep front, thus forming a jumbled mass of brecciated blocks which are later cemented together. Such a jumbled mass is excellently exposed in McKittrick Canyon in the northeastern corner of Guadalupe Mountains National Park.

During the Mesozoic, sediments were deposited on top of the Capitan. Uplifts during the Tertiary were accompanied by fracturing and block-faulting. A large fault along the west side of the Guadalupe Mountains separates the mountains from the depressed block, the Salt Basin. The Guadalupe block was tilted toward the east-northeast, exposing the Capitan Limestone in the high cliff known as El Capitan.

Fracturing was particularly significant in the

section that became Carlsbad Caverns, thereby setting the stage for solution by sub-surface water. The Main Corridor as well as smaller corridors are developed parallel to the main direction of fracture—generally east-west in Carlsbad Caverns. As solution continued to enlarge the passageways the walls between them were eaten away and large irregularly shaped rooms, such as the Big Room, were developed. Corridors such as the one which connects the Big Room with the Main Corridor were developed along cross-fractures in the joint system, almost at right angles to the main fractures or joints.

At times, blocks of the ceiling rocks became unstable and fell to the floor. A large block called the Iceberg, weighing about 100,000 tons, is now in the middle of the floor in the Main Corridor. One might assume that a minor earthquake resulted when this huge block crashed down; however, as Park Naturalist Van Cleave pointed out, there are nonvertical stalactites attached to the Iceberg. Since they were obviously formed before the rock fell and were not broken, the Iceberg must have been lowered gently. Van Cleave's interpretation that the corridor was full of water that had a cushioning effect appears logical.

Speleothems—stalactites, stalagmites, columns and other deposited formations—decorate the halls and rooms of Carlsbad. Giant Dome, a huge column or pillar, immediately commands your attention in the Hall of Giants. The King's Draperies are colorful adornments in the King's Palace. The Rock of Ages and the Frozen Waterfall are but two more of the endless array of formations that make Carlsbad Caverns world famous.

Years ago, there were many beautiful pools of water in the Caverns. Between 1936 and 1969, when the senior author re-visited Carlsbad, a definite change took place. By then, there was less water in some of the pools; others were completely dry and chalky in appearance. One of the beauties of Carlsbad had been largely destroyed. Was it the result of a change in climate? Could it be due to overuse or misuse of the caverns? Studies by J. S. McLean of the U.S. Geological Survey, in progress at that time, indicate that humans are the cause. In 1971, McLean reported that approximately 22,000 gallons of water are airlifted out of the caverns each year—through the elevator shafts! The main loss is during the winter when there is rapid exchange of relatively warm, moist cave air for cold, dry air that enters through the natural entrance. As the dry air moves through the caverns toward the elevator shafts it robs the pools of their water.

FIGURE 12-19

One of the huge domes in Carlsbad Caverns, New Mexico. (Photo by Jack Odum.)

We can be certain that the Park Service did not intend to cause deterioration of the cave environment; an impact study would likely have prevented the damage. Remedial measures have now been installed but it will require time for Mirror Lake, Green Lake and other pools to be fully restored to their original beauty.

There is something new at Carlsbad—the New Cave. The Park Service has recently made available to the public the opportunity to experience a cave in its original condition—without paved walkways, elevators, and lighting. An article by Edwards Hay in the June 1974 issue of *National Parks and Conservation Magazine* describes the many features, including a miniature Chinese Wall, that are now available to those who wish to "rough it." Visitors provide their own electric lanterns or flashlight, canteen, hiking boots, and transportation from the visitor center to the New Cave, 23 miles (37 km) distant. Reservations must be made well ahead of time, and only those in good physical condition should apply.

Several caverns have been found in Guadalupe Mountains National Park, but thus far not one of them compares with those at Carlsbad. What remains for the future, when Guadalupe has been completely explored, is of course unknown.

If you visit Carlsbad during the summer months, you will not want to miss the flight of the bats—in the evening, about dusk.

Although other caverns in the park system are well worth seeing, you haven't seen the real one until you have been to Carlsbad. Then go around to Guadalupe Mountains, and even if you do not have the time to hike into the park, the bold front of El Capitan will be there to admire.

WHITE SANDS NATIONAL MONUMENT

Located in the Tularosa Basin in southeastern New Mexico, White Sands National Monument was set aside in 1933 to preserve 230 square miles of the largest gypsum dunes in the world. The Heart of Sands Loop Drive affords an opportunity to see the typical formation of the dunes. Part of the drive is paved, but the shifting dunes have buried certain sections; however, the hard-packed gypsum forms a good road surface.

The origin of the dunes is somewhat complex. Gypsum beds crop out in the San Andres Mountains west of the basin. Gypsum in solution is carried down by streams into Lake Lucero, a playa lake immediately west of the dunes. Here, as a result of evaporation, the gypsum is precipitated. Selenite gypsum crystals that project up from the surface are broken into small fragments by physical weathering and by the wind, and the sand-size particles are blown northeastward to the dune area. The first dune to be formed was probably developed around a shrub that retarded the wind. Once dunes begin to form, they promote further deposition and the dune area enlarges. The dunes are migrating, but at a slow rate; it will probably be some time before the visitor center will be threatened.

FIGURE 12-20

"Ships" in the desert of white gypsum dunes. The "sails" protect visitors from the sun.

Primary structures such as lamination and cross-bedding have been modified by avalanching of the slip-face on the lee side of the dune and by crosswinds eroding at the base. In order to interpret these complex structures, John Douglass of the Park Service made field observations under conditions less than optimum — during sand storms, actually gypsum blizzards. For more information, see McKee (1966), and McKee, Douglass, and Rittenhouse (1971).

The plants in the dune area — such as yucca, iodine-bush, and a few cottonwoods — struggle for survival in the shifting sands, sustained mainly by gypsum water. Natural selection has left only white lizards and white pocket mice, which are unlikely to be seen by predators, as survivors of these two groups, in this world of white.

The white dunes are highly photogenic, particularly when the sun first comes up over the Sacramento Mountains.(See Figure 12-21 in color section.)

FIGURE 12-22

Saguaro cactus, Saguaro National Monument near Tucson, Arizona. The wide range of desert environments in the Basin and Range Province provides habitats for many species of cactus.

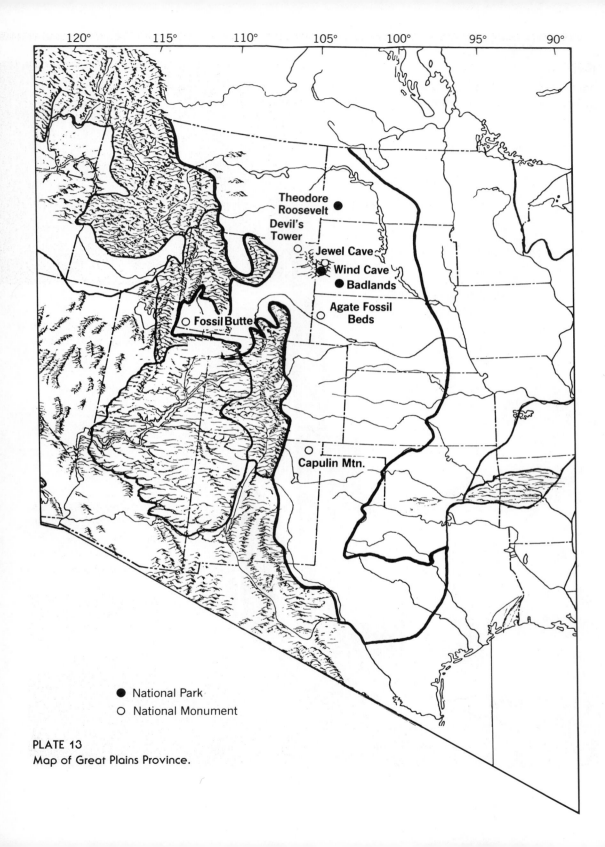

Theodore Roosevelt

Devil's Tower

Jewel Cave

Wind Cave

Badlands

Agate Fossil Beds

Fossil Butte

Capulin Mtn.

● National Park
○ National Monument

PLATE 13
Map of Great Plains Province.

CHAPTER 13

GREAT
PLAINS
PROVINCE

FIGURE 13-1

Garden of the Gods, gateway to the Rockies, near Colorado Springs; vertical walls of Lyons sandstone.

The Great Plains lie east of the Rocky Mountains and extend from south-central Texas northward into Canada. The western boundary is generally a lithologic break, in many places between the Precambrian rocks of the mountains and the sedimentaries of the plains. The eastern boundary with the Central Lowlands Province is a transitional one. Within a hundred-mile-wide (162 km) zone, the dryness of the plains gives way to a humid environment; widely spaced intermittent streams give way to closely spaced perennial streams. Here also, the essentially treeless plains become wooded.

In the Great Plains, almost all the rocks are sedimentary and most are flat-lying or nearly so. However, along the Rocky Mountains the sedimentaries are turned up and truncated, forming hogbacks. Locally, thrust faults form the western boundary, as in Montana where the mountains were thrust out over the Great Plains.

Igneous activity, although not as widespread as in the Rockies, was nevertheless significant in a number of local areas. During the Tertiary, there were intrusions of magma in the Black Hills, in the Devils Tower area of northeastern Wyoming, in western Montana and in the Spanish Peaks area of southern Colorado. Extrusive activity was particularly significant in northeastern New Mexico where lava flows and volcanoes cap the Raton Plateau.

The Great Plains present a monotonous landscape to many people who cross them. If we were able to turn the geologic clock back

FIGURE 13-2

Spanish Peaks, as viewed from the north. The peaks, erosion remnants of a stock that formed by crystallization of magma far beneath the surface, were eventually exposed by stream erosion and shaped into their present form by streams and glaciers.

FIGURE 13-3

Tracks of a bipedal dinosaur which were impressed into the Morrison sands during the late Jurassic, northern Colorado.

FIGURE 13-5

The Four Presidents, Mt. Rushmore National Memorial near Keystone, in the Black Hills of South Dakota. (Photo by Stanley A. Schumm.)

FIGURE 13-4

"Hoodoo Rocks," weathering forms in the cross-bedded Casper sandstone, Laramie Basin, southern Wyoming.

about 5 million years, monotony would be the precise description for the vast fluviatile plains that sloped gradually eastward from the mountains — with only the Black Hills and a few other local mountains rising above them. When, toward the end of the Pliocene, the Rocky Mountains were uplifted, the Great Plains were tilted slightly toward the east. Streams, steepened and rejuvenated by the uplift, cut down and severely dissected certain sections, particularly those adjacent to the mountains. Although large areas remain essentially unaltered, the changes were sufficient to cause Fenneman (1931) to subdivide the Great Plains Province into ten sections, a few of which are mentioned here.

The Raton Section, in northeastern New Mexico, is mainly a lava-capped plateau that extends eastward from the Rockies. Rising above the plateau are a few volcanic cones, including Capulin Mountain. Adjacent on the north is the Colorado Piedmont, a lowland carved out by streams. Here, the Tertiary fluviatile sediments have been largely removed and the underlying Mesozoics and Paleozoics deeply eroded. Along the western border, hogbacks rise prominently above the Piedmont surface (see Figure 10-1). Pediments and pediment remnants, widespread in the Great Plains, are particularly well developed in the Piedmont. Buttes — small erosion remnants — are common, especially near the northern boundary; Round Butte in northern Larimer County and Pawnee Buttes farther east are examples.

The High Plains area to the east of the Piedmont is an extensive remnant of the Pliocene surface that has been little eroded. A narrow remnant of this old surface extends westward in southernmost Wyoming, north of the Colorado Piedmont; it is called the Gangplank.

The Wyoming Basin, often accorded province rank, is merely the western extension of the Great Plains. Like other basins in Wyoming, it is underlain by sedimentary rocks topped in places by late Tertiary stream deposits.

The Black Hills in South Dakota are mountains — domed mountains that were formed during the Laramide Orogeny. As in the Front Range of the Southern Rockies, Precambrian granites and metamorphics are exposed in the Black Hills.

Pleistocene ice sheets pushed southwestward out of Canada and covered the northern part of the Great Plains, the area north of the Missouri River. Prior to glaciation, the Missouri flowed northeastward into Hudson Bay. Blocked off by ice during the Wisconsin advance, the river followed a new course along the front of the glacier. Once established, the Missouri maintained essentially this course after the glacier retreated.

The contrast in landforms on the two sides of the glacial boundary is striking. The glaciated section is noted for its hummocky topography made up of irregular moraines in which kettles, kettle lakes, and morainal lakes are abundant. The streams wander aimlessly about through the glacial deposits, into and out of lakes and swamps. In the area south of the glacial boundary, streams have developed regular drainage patterns and typical fluvial landforms.

Therefore, although broad, monotonous plains still exist, there are many areas where interesting, even spectacular, geologic features have been developed. A number of such areas are now national parks and monuments.

CAPULIN MOUNTAIN NATIONAL MONUMENT

Rising prominently above the volcanic plateau in northeastern New Mexico is a near-perfect volcanic cone, Capulin Mountain. The cinder cone, together with associated lava flows at its base, became a national monument in 1916.

Located about 30 miles (48 km) east of Raton, New Mexico, it is readily accessible, particularly because of a road that encircles the cone and ends at a parking area on the rim of the crater. Many young cinder cones are too unstable for road building; here, the cinders are cemented together by calcium carbonate, thus providing a relatively stable roadbed.

The cone rises more than 1000 feet (303 m) above the lava plateau, to an elevation of 8215 feet (2490 m) above sea level. The crater is about 415 feet (127 m) deep, and the circumference of the crater rim is approximately a mile.

A trail leads down to the bottom of the crater where the vents through which gases were emitted can be observed, essentially unaltered by weathering. Another trail leads around the rim of the crater; from viewpoints, vast areas of the treeless Great Plains can be seen, ending abruptly in the snow-capped Sangre de Cristo Mountains to the west. To the south lies Sierra Grande, a small shield volcano that rises domelike above the surrounding area. At Capulin's base a lava flow is in clear view, so young that the pressure ridges are

FIGURE 13-6

Capulin Volcano, the main feature of Capulin Mountain National Monument, near Raton in northeastern New Mexico.

essentially unweathered. Near the base on the east side is a tiny cinder cone — Capulin in miniature — called Baby Capulin. The stairstep profile of Fishers Peak, about 30 miles (48 km) to the northwest, looms up on the skyline; it is the most prominent erosion remnant of the thick series of basalt flows.

Geologic Sequence

The Laramide Orogeny left the Paleozoic and Mesozoic sedimentary rocks dipping eastward away from the newly formed Sangre de Cristo Mountains. Stream erosion truncated the upturned sedimentaries, thus forming the surface across which the Raton lavas flowed eastward almost to the Oklahoma Panhandle. The age of the oldest flows is not precisely known, but extrusion probably began during the Miocene and continued periodically into the Pliocene.

Late Pliocene uplift rejuvenated the streams, and after prolonged erosion, the surface was formed on which Capulin Volcano was built. Thus Capulin represents the latest stage of volcanism in this region. Although future work may provide us with the precise date of the eruption, it probably occurred about 7000 years ago. Clearly, Baby Capulin erupted after activity in the main vent had ceased.

It is interesting to speculate whether Capulin is an extinct volcano or whether it is merely sleeping.

FOSSIL BUTTE NATIONAL MONUMENT

Fossil Basin, in which Fossil Butte is located, occupies the southwestern part of Wyoming Basin. The monument, located about 11 miles (18 km) west of the town of Kemmerer, was established in 1972 to preserve an unusual assemblage of fossils, both plants and animals, especially fish.

Southwestern Wyoming has been widely known for its fossil fish since 1856 when the discovery was made. Soon after that, when F. V. Hayden was making the first geological survey of the territory, the Union Pacific Railroad made a large cut through a hill 2 miles (3 km) west of the town of Green River, exposing large numbers of fish. This exposure, known as the Petrified Fish Cut, was mentioned in Hayden's 1871 report. In a short time, collectors—both professional and amateur—converged on the area. A dark brown fossil fish on a cream-colored shale background makes a handsome wall plaque, as commercial collectors promptly discovered.

The geologic story of Fossil Basin begins with the Laramide folding and faulting of the rocks in western Wyoming. Once the structural basin was formed, a freshwater lake developed in which huge quantities of sediment were deposited. Fossil Lake, as it is known, was well stocked with the fish that, about 50 million years later, made the area famous.

During early Eocene time, Fossil Lake occupied much of Fossil Basin, and while fish and other animals occupied the lake, forests were widespread in the surrounding swamps and floodplain areas. The sediments that were laid down in the lake—silts and clays with small quantities of fine volcanic ash—became the Green River Formation; the fluvial deposits around the lake became, when consolidated, the Wasatch Formation. In Fossil Butte National Monument, Wasatch beds were laid down before the lake covered that part of the basin; later, as the lake fluctuated, there was intertonguing of the Green River and Wasatch beds in this boundary zone.

Most of the Wasatch consists of varicolored mudstones which erode easily into badlands,

FIGURE 13-7

Diplomystus, an Eocene relative of the herring, 18.25 inches (46 cm) long. (Colorado State University collection.)

FIGURE 13-8

Fossil horse fly from the Green River Formation, Fossil Butte National Monument. (Photo by National Park Service.)

as in Fossil Butte National Monument. Reds, purple, yellow, and gray predominate, contrasting with the whites and light tans of the overlying Green River shale. Here the Wasatch is noticeably different from that in Bryce and Cedar Breaks.

The name "Green River" also designates similar lake sediments laid down in several basins in southwestern Wyoming, northwestern Colorado, and northeastern Utah; at certain times during the Eocene several of the lakes were apparently connected. The organic-rich layers, mudstones and marlstones, are the much-publicized "oil shales."

The fish at Fossil Butte are found mainly in a 14-inch-thick (36 cm) layer which is approximately 155 feet (47 m) above the base of the Green River Formation. This shale layer is thin bedded, with well-defined, parallel bedding planes; therefore, a layer of desired thickness, with a fish mounted on it, can be readily obtained.

Many kinds of fish are found here, including paddlefish, gar, bowfins, herrings, catfish, and perch. Thousands of fish skeletons have been collected, many of them still perfectly articulated. Obviously, they had not been touched by scavengers that in many places left only fragments scattered about. The explanation appears to be that the bottom layer of water was devoid of oxygen but well supplied with toxic gases, including hydrogen sulfide. The fish may have dived into this toxic water and been overcome and buried there, out of reach of scavengers. (Other possible interpretations are discussed by McGrew and Casilliano, 1975.)

Actually, there are many kinds of fossils and some are found in older formations in Fossil Basin, as McGrew and Casilliano report. The Triassic Thaynes Limestone contains a varied assemblage of oysters, clams, and other aquatic forms. The horned Triceratops, believed to be the last dinosaur to become extinct, is found in the Cretaceous part of the Evanston Formation; Paleocene mammal bones are in the upper part of the formation.

In addition to the fish, the Green River Formation contains invertebrates, mainly molluscs and insects, and vertebrates — reptiles and birds. The plant kingdom is well represented — pollen from spruce, fir, and pine, and leaves from oak, elm, maple, beech, and palm. Bats are rarely fossilized because of their fragile skeletons; nevertheless, the Green River Shales yielded a complete skeleton of this flying mammal. Most of the mammal fossils, however, are found in the Eocene Wasatch or in the Pliocene beds. The number of different mammals that made Fossil Basin their homes is more than a little exciting; the mere listing of the names fills a page in McGrew and Casilliano's report. The Dawn Horse, *Hyracotherium,* no larger than a fox terrier, pranced about on his four toes (front) and three toes (hind feet) during the early Eocene. And during the middle Eocene, the lemur-like and tarsier-like primates put in their appearance.

Do not expect to be permitted to "fish" in the monument; instead, merely cast your eyes on the exhibits there. Perhaps you can also visit the museum at the University of Wyoming in Laramie, where outstanding specimens from Fossil Basin and other localities are on display.

DEVILS TOWER NATIONAL MONUMENT

Devils Tower, in northeastern Wyoming, is the most upstanding feature of the Great Plains. It can be seen from points 30 miles distant, rising like a gigantic stump high above the plains. Possibly the term *monument* loomed up in President Theodore Roosevelt's mind; in any case he proclaimed that Devils Tower would be our first national monument, and it was — in 1906. The tower rises 850 feet (261 m) above its base and almost 1300 feet (400 m) above the Belle Fourche River Valley.

After its great size, the most conspicuous feature of the tower is its columnar jointing; columns vary somewhat in size — up to 16 feet (5 m) in diameter. Many are hexagonal but some are 4-, 5-, and 7-sided. They stand almost vertically except at the base where they flare out into a highly fractured mass. Surrounding the tower are large taluses of broken columns.

The tower has naturally aroused the curiosity of all who see it. The Sioux Indians called it Mateo Tepee, meaning Grizzly Bear Lodge. Their explanation of its origin is much simpler

than the one geologists are seeking; they believed that the rock was scored by a giant grizzly that left his claw marks from top to bottom as he tried unsuccessfully to reach some children on top of the big stump.

Mountain climbers are less concerned with its origin than whether they can get up on its top. Many have succeeded, some by using wooden stakes driven into the cracks between the columns. The climb is strenuous and should be attempted only by experienced climbers, after obtaining permission from the monument superintendent. In 1941 a parachutist landed on the tower, unfortunately without any plans to get from the top to the ground. After a few days and cold nights a rescue team brought him down.

The Origin of the Tower

Many geologists have attempted to reconstruct the sequence of events that produced Devils Tower. There is agreement that it is a remnant of an intrusive igneous body composed of porphyritic phonolite, a rock similar to porphyritic andesite. Opinions vary as to size and shape of the original igneous body; however, Jaggar and Howe (1900) concluded that it is a small erosion remnant of an extensive sill. Darton and O'Harra (1907) visualized a broad laccolith that was, with the exception of the highest central section, eroded away. Others believe it to be a plug that was not much larger than it is now. And some have referred to it as an eroded volcanic neck!

When we look at the huge quantity of material in the taluses surrounding the tower, it is clear that it was significantly larger than now. Then, when we find gravel, cobbles, and boulders of phonolite porphyry in the channel of the Belle Fourche River, both inside the monument and downstream, we must increase the original diameter still more. How much of this material was carried completely out of the area will never be determined, but,

FIGURE 13-9

Air view of Devils Tower, Wyoming. Note well-developed columnar jointing. (Photo by Nan H. Daniels.)

considering the amount of time available, it could have been an enormous quantity. The magma intruded in early Tertiary time, probably during the Eocene. Clearly there has been sufficient time for the roof rocks to be eroded away. Are there, then, any reasons to rule out the removal of great quantities of the igneous rock? Probably the size and shape of the original igneous mass will be the subject of enthusiastic discussions for a long time, conceivably until the tower has been eroded away! The block diagram shows one possible interpretation, the one suggested by the early workers.

Devils Tower is huge, but its true size cannot be appreciated from a distance or even with a quick look from its base. Hike around it on the mile-and-a-quarter-long (2 km) Tower Trail and view it from all angles. For those who have more time, there is the Red Beds Trail, almost three miles (4.8 km) long. It is a photo-

FIGURE 13-10

Block diagram showing one explanation of the origin of Devils Tower.

genic monument; a series of pictures at 5-mile (8 km) intervals as you approach the tower will provide a good record of your experience.

WIND CAVE NATIONAL PARK

Wind Cave is located in the southern part of the Black Hills section of the Great Plains Province. The Black Hills are domed mountains that were formed during the Laramide Orogeny. This elongate dome, about 125 miles (202 km) in length, was deeply eroded, thus exposing the Precambrian core flanked by Paleozoic and Mesozoic sedimentaries. In this series of rocks, the Mississippian Pahasapa Limestone is the formation in which the numerous caverns of the Black Hills are developed.

The Pahasapa is a 300-foot-thick (92 m), rel-

FIGURE 13-12

Outcrop of Pahasapa limestone, the "cave former" in Wind Cave National Park. (Photo by National Park Service.)

atively massive limestone that is exposed in a broad zone encircling the Black Hills. During the doming process, stresses developed a distinct fracture system. In addition to the major and minor directions of fracture, local sections of the Pahasapa were essentially shattered.

Wind Cave is unique in that it is a "natural barometer." When a "low" passes over the Wind Cave area, the wind rushes out; when air pressure is higher outside than within the cave, the air rushes into the cave. This barometer was more sensitive when only the natural openings were involved, as was the case when the park was established in 1903.

As in other limestone caverns, groundwater dissolved the calcite along fractures to form the passageways, here for a total length of more than 10 miles (16 km). Outstanding among the many cavern features are the *boxwork* formations. As already indicated, the limestone is shattered or brecciated in places and these tiny, closely spaced fractures were filled with calcite. Later, the calcite in the veins proved to be more resistant than the original rock, which was removed in solution, leaving the thin veins or walls standing in relief. The delicate partitions are pink, brown, and yellow and are translucent. In many places tiny needles of calcite have crystallized on the boxwork surfaces, giving the appearance of frost; therefore it is called frostwork.

In addition to all its other attractions, some

of Wind Cave's calcite is fluorescent and at one point in the tour of the cave, the ultraviolet lamp will probably be used to show the green, red and blue colors.

Once again out in the sunlight, the towering pines of the Black Hills forest provide an inspiring setting. Their deep green color appeared black from a distance; therefore, the early explorers called them the Black Hills.

JEWEL CAVE NATIONAL MONUMENT

Jewel Cave is about 25 miles (42 km) northwest of Wind Cave, almost due west of Custer, South Dakota. Again, the cave is formed in the Mississippian Pahasapa Limestone. Here also, boxwork and frostwork are beautifully developed. Jewel Cave does have its own distinctive hallmark — "jewels." Calcite crystals with well-developed crystal faces — each a scalene triangle — are the jewels. Some are white, others transparent, but most are highly colored; all are extremely attractive.

Certain of the stalactites are unusual. When they begin to develop, they are small tubes that lengthen with time. These are "soda straws." Usually the tubes are gradually filled and the water flows down on the outside, thus increasing the diameter of the stalactite. In Jewel Cave some of the soda straws have continued to lengthen themselves beyond all ex-

FIGURE 13-14

Dogtooth spar calcite crystals. (Photo by National Park Service.)

FIGURE 13-15

White, hydromagnesite "balloons." (Photo by National Park Service.)

pectations, in one case to slightly more than 36 inches (1 m). These soda straws are very fragile and indicate that the area has been subjected to nothing more than minor earth tremors for a considerable period.

In the early years following the establishment of the monument in 1908, lack of supervision permitted vandalism painful to see. The crystals were broken off and carted away, and many of the formations were destroyed. Recently, the Park Service sealed off this "disaster area" and opened up new, unspoiled caverns, this time with adequate protection from despoilation.

BADLANDS NATIONAL PARK

In southwestern South Dakota, about 50 miles (80 km) east of the Black Hills, 380 square miles of badlands were set aside in 1929 as a national monument. The monument was enlarged and upgraded to park status in 1978. The new South Unit contains spectacular badlands features and significant fossil assemblages, notably the Titanothere Beds.

The rocks are flat-lying mudstones, shales and ash beds of the Oligocene White River Formation. The fine texture of the deposits indicates that they were deposited by sluggish, low-gradient streams. In turn, the deposits reflect the geomorphic history of the Rocky Mountains. By the end of the Eocene, when the Summit Peneplain was perfected, the mountains had been essentially destroyed. During much of the Oligocene, the streams were capable of carrying only fine materials, some of which were deposited in what is now the Badlands. Also recorded here are the times when the Absaroka volcanoes erupted; volcanic ash was wind-borne to the area and reworked and redeposited by the streams.

These swampy floodplain areas supported a good growth of vegetation that provided food for many animals. No dinosaurs were there; they had become extinct at the end of the Mesozoic. But other bizarre and ponderous beasts had taken their place; during the Cenozoic the mammals dominated the earth. The huge titanotheres, distant relatives of the modern rhinoceros, were the largest of the lot. They made phenomenal progress from the hog-sized beast of the Eocene to the *Brontotherium* of the Oligocene, and were the largest land animals in America at that time. They were much bulkier than the largest living rhinoceros, perhaps too bulky; they became extinct before the Oligocene ended. *Mesohippus,* the sheep-sized three-toed horse, persevering in the evolutionary process as the ancestor of the modern horse, was on the scene. Many other animals also wandered about on the floodplains, including camels, pigs, dogs and sabertooth cats.

When the animals died, whether from old age, by being mired down in the swamps, or at the hand of the sabertooth cat, many of their skeletons were preserved in the rocks. In the park system, the Badlands, together with Big Bend National Park and Agate Fossil Beds, provide an extraordinarily complete record of vertebrate evolution during the Cenozoic Era. However, without certain later events, we would not know what was buried here beneath the High Plains surface.

The late Pliocene uplift which elevated the Rocky Mountains to their present height also elevated the Great Plains significantly. Streams were rejuvenated and cut rapidly down through the poorly resistant rocks. White River, which is south of the present Badlands,

and the South Fork of Cheyenne River to the northwest were the master streams of the area. Their tributaries cut headward into the highlands adjacent to the entrenched main streams. In these fine-textured materials even small rivulets were able to erode; therefore, an intricate drainage pattern developed and the dissection went on at a rapid rate — a rate that prevented soil formation. Thus, steep and barren slopes developed; these were the original "badlands." But they were not located where the Badlands are now.

As time passed, the tributary streams continued to cut headward, causing the escarpment to "retreat." Now, much of the highland area has been dissected into badlands. During the same period, the tributaries were cutting laterally, planing off the hills along the lower courses of the streams, thus forming a pediment. The enlargement of the pediment area

FIGURE 13-16

Badland erosion of Oligocene beds. (Photo by S. A. Schumm.)

and the headward migration of the badlands zone is continuing today.

Certain sections of the Badlands are somewhat like Bryce Canyon but less highly colored; other sections are more like the Painted Desert section of the Petrified Forest. Elsewhere high gray spires are equally impressive.

Bone-hunting in the Badlands can be frustrating, and particularly so if you find one sticking out of a gully bank. This is a park area, and removal of any type of material is forbidden. Report your find to the naturalist and go on to Rapid City where, at the museum of the South Dakota School of Mines and Technology, you can see many of the inhabitants who roamed the area 35 or 40 million years ago.

The name "badlands" suggests an area that is worthless, but those interested in the wonders of nature will find them well worthwhile.

FIGURE 13-17

Ridges and spires, rising high above the pediment surface in Badlands National Park. (Photo by National Park Service.)

AGATE FOSSIL BEDS NATIONAL MONUMENT

The name *agate* immediately brings gems to mind, or possibly fossils that have been agatized. Instead, the bone beds that inspired the creation of the monument are near Agate, in western Nebraska. Because of the need for preservation of a fabulous assemblage of vertebrate fossils, about 3000 acres were obtained from the Sioux Indians in 1965.

Buried in the Tertiary beds in the Great Plains are thousands of skeletons of the ancestors of our modern mammals. In the Badlands we have a glimpse of the animals that inhabited the area in Oligocene times. Here, near Agate, we see some of their descendants because it is a Miocene graveyard that was exposed when the Niobrara River unceremoniously cut its valley through the middle of the burial ground.

These animals are more advanced than those in the Badlands, but they still lacked about 20 million years of being modern. One attractive fellow, the twin-horned rhinoceros (*Diceratherium*) was there in large numbers. For a fascinating account of the evolution of Tertiary mammals, read Dunbar's chapter entitled "Mammals Inherit the Earth" in his *Historical Geology* (1966).

THEODORE ROOSEVELT NATIONAL PARK

The ruggedness of the North Dakota badlands symbolizes the spirit of President Theodore Roosevelt. An avid big-game hunter in his earlier years, he came to the badlands to hunt buffalo in 1883. He was so enthralled with the area that he bought a ranch and moved onto it the following year. During his ranch years Roosevelt became intensely interested in conservation and later, as president, he designated 18 areas as national monuments and was instrumental in establishing 5 national parks and

the first wildlife refuges. As a tribute to this remarkable man — big-game hunter turned conservationist — in 1947 and 1948, Congress set aside about 70,000 acres in three separate units, one of which was Roosevelt's Elkhorn Ranch. In 1978, the national monument was upgraded to park status. The three units are all north of Medora, park headquarters, in westernmost North Dakota.

Geologic Development

The Paleozoic and Mesozoic rocks; deep beneath the park; record a long period of crustal stability and the thick accumulation of sediments, mainly marine. Minor warping of the beds produced a number of broad arches and large structural basins. The park area occupies a small part of the Williston Basin which contains important accumulations of oil and gas.

The inland sea, in which Cretaceous sediments were laid down, retreated from most of the continent by the end of the period; locally, the Cannonball Sea was the last holdout, as recorded in the Early Paleocene rocks of the area. Later in the Paleocene, after the sea withdrew, streams deposited sediments that were derived from the rising Rocky Mountains to the west. The building of the Great Plains had begun.

Sandwiched between the stream-deposited sediments, beds of *lignite* are locally present. When compacted by the weight of overlying sediments, lush vegetation growing in extensive swamps was converted to lignite, a low grade of soft coal.

As mountain-building continued, volcanoes to the west belched forth large quantities of ash and dust that blanketed large areas of the Great Plains, including the park area. Silica derived from these ash beds played a major role in the petrification of cypress, sequoia, cedar, and oak trees (Schock, 1974).

During the remainder of the Tertiary, streams flowing eastward and northward deposited their sediments over the Great Plains, which extended from the Rockies to the Mississippi River and northward into Canada. Then, only a few million years ago, vertical uplift of the Rockies and Great Plains rejuvenated the streams, causing them to downcut and erode away the poorly consolidated rocks. Thus the younger Tertiary rocks that once covered the Theodore Roosevelt area were stripped off. This was the cycle of stream erosion during which the badlands would ultimately be formed; albeit not without interruption.

When, during the Pleistocene, the continental ice sheets moved down out of Canada and covered the area, stream erosion was obviously impossible; also, the north-flowing streams, blocked off by the ice, were diverted to new courses. The lower sections of the Little Missouri, the Yellowstone, and other streams were relocated in their present southeastward courses (Bluemle and Jacob, 1973).

During the Wisconsin advance, the ice blocked off the Little Missouri River, creating a large lake that covered the North Unit of the park. When the glacier melted away, the lake was drained and badland-forming stream erosion resumed, and it continues today. As the streams continued to downcut in the poorly consolidated sediments, some of the side slopes became oversteepened and unstable. Therefore, landslides are common landforms in the park.

The lignite beds exposed on the side slopes are susceptible to grass fires, lightning strikes, and perhaps spontaneous combustion. Once ignited, the lignite may burn for many years. In places the heat is sufficient to bake the overlying sediment into a hard, reddish-colored material that, for reasons unknown, has locally been called "scoria." This is definitely a misnomer, as this material is in no way related to volcanic scoria. One such fire in the area began when in 1951 a lightning strike started a prairie fire that ignited a lignite bed that burned until 1977.

We have merely glimpsed some of the features in the "fantastically beautiful" badlands that Roosevelt admired so much. Although it would involve some zigzagging, you might consider visiting the Tertiary parks of the Great Plains in geologic sequence, in order to better appreciate the physical, biological, and climatic changes that took place there during the past 70 million years. You would start at Theodore Roosevelt (Paleocene), then visit Fossil Butte (Eocene), the Badlands of South Dakota (Oligocene), and finish with Agate Fossil Beds (Miocene). It would then be only a short run down to Scottsbluff National Monument, in case you would like to see the "ruts," now deep gullies, made by the prairie schooners on their westward trek over the famous Oregon Trail.

FIGURE 13-18
The old Oregon Trail, Scotts Bluff National Monument, western Nebraska. Erosion has deepened the ruts made by countless wagons on their westward treks. (Photo by S. A. Schumm.)

FIGURE 14-1 In the Central Lowlands particularly, state parks largely take the place of national parks. Relict plants left over from the Ice Age still thrive in the cool recesses and overhangs of Rocky Hollow. Turkey Run State Park, western Indiana. Sugar Creek and its tributaries cut deeply into the Mansfield Sandstone, thus forming this unique environment.

CHAPTER 14

CENTRAL LOWLANDS PROVINCE

The Central Lowlands lie east of the Great Plains, from Canada southward into northern Texas. Most of the province has been glaciated, and the bedrock is largely covered by glacial deposits and outwash materials. The rocks beneath the glacial materials are chiefly Paleozoic sedimentaries, but along the western border extensive areas of younger rocks remind us of the vastness of the Cretaceous Geosyncline. In eastern Kansas and farther north, the Dakota Sandstone is locally exposed. When we last saw the Dakota, it was forming the main hogback ridge just east of the Southern Rockies (Figure 10-1). It lies deep beneath most of the Great Plains and reappears in places in the Central Lowlands.

The Osage Plains section, which extends southward through Kansas and Oklahoma into Texas, is the only large area that remained unglaciated. A relatively small area, mainly in southwestern Wisconsin, was by-passed by the glaciers. It is called the Driftless Area.

Like the Great Plains, the Central Lowlands have only locally been subjected to severe tectonic upheavals, and the structures consist largely of broad, gentle upwarps and downwarps such as the Cincinnati Arch and the Michigan Basin. The Arbuckle Mountains, where Chickasaw National Recreation Area is located, is one exception; here the rocks are distinctly folded and locally faulted.

Igneous rocks are not exposed except locally; in the Arbuckle and Wichita mountains in Oklahoma, Precambrian granites cover significant areas. More recent igneous activity that has occurred in many of the provinces has

FIGURE 14-2

View along the top of an esker, near St. Paul, Minnesota, before it was converted into a sand and gravel pit.

not affected the Central Lowlands.

During the Pleistocene, great changes were brought about by huge ice sheets which at least four times pushed down out of Canada. They completely remodeled the drainage systems, the topography, and the landforms of almost all of the Central Lowlands. Most of the deposits, related directly or indirectly to the

FIGURE 14-3

Loess deposits in road cut along U.S. Highway 41 near Princeton, Indiana. Terracing of the loess cuts and establishment of vegetation on benches result in stable embankments.

glaciers, formed excellent parent materials for the soils that developed on them; the Corn Belt of Iowa and Illinois would not have developed otherwise. Distantly related are the loess deposits, wind-laid silt derived in part from glacial outwash. Soils formed from loess are highly productive in Iowa, Illinois, and parts of Indiana.

Within the province, there is a wide variety of natural features of special interest, both geologically and ecologically. State parks are abundant in this section of the country, and they serve essentially the same purpose as the national parks in preserving things natural; examples are Turkey Run State Park in Indiana, Devils Lake in Wisconsin, and Taylors Falls in Minnesota.

Pipestone National Monument, in the southwestern corner of Minnesota, preserves

FIGURE 14-4

Water from Lake Erie plunges 167 feet (51 m) over Niagara Falls on its way to Lake Ontario. (Photo by B. Kiver.)

Indian lore. Here, the Indians quarried the pipestone and shaped the widely known catlinite pipes.

Recently, when seashores and lakeshores were included in the Park Service's preservation program, Indiana Dunes National Lakeshore was established on the shores of Lake Michigan.

CHICKASAW
NATIONAL RECREATION AREA
(Includes Platt National Park)

This recreation area, in the hills near Sulphur in southern Oklahoma, is a pleasant interruption to the plains. It is administered by the National Park Service and was established in 1976 by combining Platt National Park with the Arbuckle Recreation Area.

In the early days, this was Indian country. The Chickasaws loved this particular area because of its many springs, some of "mineral waters" believed to have great curative powers. With the encroachment of the white man, the Indians feared that they would be driven out, so they conveyed the land containing about 30 springs to the U.S. government. By so doing, they were assured that they would have access to the springs and to the little oasis surrounding them. The area was designated Sulphur Springs Reservation in 1902, but the name was changed to Platt National Park three years later. The park, now designated the Travertine District, occupies the northeastern part of the recreation area. The following discussion applies specifically to the Travertine District.

The Travertine District is small, but its size in no way measures its significance in the National Park System. True, its waterfalls are neither deafening nor overpowering, nor are its cliffs breath taking; nevertheless it is a delightful place, a restful spot in the shade of the sycamores, cottonwoods, and oaks.

Ecologically, this is a fascinating area because it is a transition zone between the East and the West—between the hardwoods of the East and the short-grass prairies of the West. Here the road runner from the drylands associates with the eastern bluebird, the cardinal, and the bluejay, and more than a hundred other species. You may see a bison herd, particularly from Bison Viewpoint on the Perimeter Drive east of Bromide Hill. Watch carefully for armadillos, foxes, bobcats, raccoons and opossums, and watch out for skunks.

Geologically, the Travertine District is unique. Here we have waterfalls formed by stream deposition rather than by stream erosion. And waterfall building is continuing today; slowly the little waterfalls are getting bigger and higher. Not long ago, geologically speaking, there were no waterfalls; instead there were merely rapids along Travertine Creek. The rapids cause turbulence, releasing carbon dioxide from the calcium bicarbonate in solution; as a result, travertine, a porous form of limestone, has gradually been deposited, forming low dams across the stream. Now these dams are several feet high, as seen in Figure 14-5. Since such travertine dams are not common, Travertine Creek must be carrying an extraordinarily heavy load of calcium bicarbonate in solution. Take the foot path upstream from Travertine Nature Center, and at Antelope Springs and Buffalo Springs you will see the source of the bicarbonate waters.

Except during times of storm runoff, these two springs are the source of the water in Travertine Creek. Normally the flow is several million gallons a day, but during prolonged drought periods the springs and the creek are dry. Before emerging at the springs, the underground water flows through the 650-foot-thick (200 m) Vanoss Conglomerate which is composed largely of limestone pebbles and cobbles, the probable source of the calcium bicarbonate.

These two springs are fed from a relatively shallow plumbing system and are "freshwater springs." Most of the other springs in the park are located in the western part and are "mineral springs"—springs producing water containing significant amounts of dissolved mineral materials. These waters contain small amounts of rare elements such as bromine, lithium, iodine, and sulfur, and large quantities of sodium and chlorine, evidently a watered-down solution of sea water. Is it conceivable that water from the Paleozoic sea was trapped in some of these porous limestone formations and that it is now being forced out at the surface? It has been so determined; surface water penetrates down into rocks at elevations higher than the springs and flushes out the old sea water from Ordovician rocks. The diluted sea water moves down-dip and is then forced up to the surface by the hydrostatic head provided by the difference in elevation of the intake area and the springs. These mineral springs are therefore *artesian* springs. Due to the structure of the rocks in the artesian system, however, the water does not penetrate to great depths. Consequently, Bromide Spring, Medicine Spring, Black Sulphur Spring, and the others are cold springs.

The National Park Service does not claim medicinal value for the mineral water; furthermore, they warn against taking it in quantity except on the advice of a physician.

The Vanoss Conglomerate that supplies the calcium bicarbonate for the travertine dams is unusual because of the size of the cobbles it contains. Under what conditions were these coarse materials deposited? Obviously the streams that transported these rocks were fast flowing, with high gradients found only in mountainous areas. Therefore mountains must have been here during the Pennsylvanian Period when these deposits were laid down. The limestone cobbles came from the tilted Ordovician and Cambrian limestone formations that were exposed by stream erosion. Pebbles of Precambrian basement rocks are also found in the conglomerate; even the basement rocks had been exposed at high elevations during mountain building. It is apparent that from mountains to the south, coarse sediments were carried into the park area to form the Vanoss Conglomerate. Much later, prolonged erosion all but destroyed the mountains, leaving the gently rolling topography of today. Thus, in this small area, we can reenact a significant number of geologic events that led to the formation of the Travertine District.

Rock Creek, which flows westward through the district, was dammed a few miles downstream — to impound water in the Lake of the Arbuckles. The lake, about 2350 acres in extent, became the Arbuckle National Recreation Area — for camping, boating, and fishing. In 1976, the area was enlarged and joined with Platt National Park to form a single unit called the Chickasaw National Recreation Area. With Turner Falls State Park nearby, tourists and local people have a wide variety of recreation facilities available.

INDIANA DUNES NATIONAL LAKESHORE

The southeastern shore of Lake Michigan is lined with sand dunes. Mt. Tom, one of the few mountains built of sand, rises 192 feet (59 m) above the lake.

In 1919, Stephen T. Mather, the Park Service's first director, recommended that a 40-mile (65 km) stretch of the dunes be established as a national park. Although his suggestion was not approved, the state of Indiana set aside a small area as Indiana Dunes State Park in 1923. Finally, in 1966, after most of the dunes in the surrounding area had been flattened for housing and commercial developments, Indiana Dunes National Lakeshore was created to preserve the remaining dunes, less than one-third of the original area. The Lakeshore, which surrounds the state park, consists of about 8200 acres of dunes and marshlands.

The geologic story begins with the Pleistocene glaciers that advanced over and retreated from the area. Late in the Pleistocene, a lobe of the Wisconsin ice sheet dug deep in the basin

now occupied by Lake Michigan and deposited morainal ridges around the southern end. When the glacier melted, large blocks of ice were left partially buried along the lakeshore; later, when the ice blocks melted, kettle lakes formed in the depressions. In time, the lakes and ponds were filled with sediment, forming marshes and bogs; Cowles and Pinhook bogs are the two that lie within lakeshore boundaries.

The dunes were piled up later, after the waves moved large quantities of beach sand southward along the western shore of Lake Michigan. As the sand accumulated on the beach at the south end of the lake, westerly winds moved it eastward to the dune area. There, the wind, retarded by trees and shrubs, piled up the sand in the form of dunes. Once in place the dunes themselves broke the wind currents and caused additional sand deposition. The important role of vegetation is obvious as you observe the shifting sands on a windy day; it is also clear when you see the dead forests now exposed in the "blowouts" formed by extraordinarily strong winds.

Ecologists are intrigued with the unusual assemblage of plants, some of which are relicts of the glacial period. Udall (1971) suggests that some of the plants that were forced to migrate southward as the glaciers slowly advanced have survived the warmer climate of the present. The Arctic barberry is here, along with the jack pine, tamarack, and birch, all of which are usually found at least 100 miles (160 km) farther north.

Animals that humans either killed or drove from the area include the bear, wolf, lynx, bison, elk, deer, and beaver, but the raccoon, opossum, fox, mink, and squirrel have been able to survive. Migratory birds stop by in droves to feed in the marshes, and shorebirds are on the beaches much of the time.

Thus, the Lakeshore and Dunes State Park provide a haven for wild animals and birds, as well as for weary city-dwellers, almost in the shadow of the steel mills in Gary and East Chicago.

CHAPTER 15

OUACHITA PROVINCE

The Ouachita Province, in eastern Oklahoma and western Arkansas, lies south of the Ozark Plateaus and north of the Gulf Coastal Plain. On the west is the Arbuckle Mountains section of the Central Lowlands. The province is readily divisible into two sections, the Arkansas Valley and the Ouachita Mountains to the south. It is in the latter section that Hot Springs National Park is located.

Here in the Midwest we catch our first glimpse of Appalachian geology. Appalachian structures are continuous westward from Alabama, surfacing only in the Ouachita Mountains and in Texas. Valley and ridge topography, resulting from erosion of intensely folded and faulted Paleozoic sedimentaries, characterizes both areas.

The geologic history began with a geosyncline early in the Paleozoic, with the deposition of shales, sandstones, conglomerates, and cherts. Notable is a variety of chert called novaculite, a dense, even-textured siliceous rock which, when pure, is as white as new snow. It was highly prized by the Indians who used it for arrowheads, spearpoints, and other tools; later it was cut into perhaps the best whetstones in the world. Geologists have long pondered the problem of its origin—how thick layers of pure or nearly pure silica could have developed. Several hundred feet accumulated in the geosyncline during the Devonian and early Mississippian. Shale, however, is the most abundant rock, with as much as 8500 feet (2591 m) accumulating in a single formation.

During and at the end of the Paleozoic, the geosynclinal rocks were intensely folded and thrust faulted, forming the Ouachita Mountains. Where orogenic movements were most intense, shales were metamorphosed to form slates, and sandstones were changed to quartzites.

The mountains were eroded as they were being uplifted, and this erosion continued after mountain-building ceased. Remnants of two erosion surfaces indicate that the area was at least twice worn down and twice uplifted prior to the development of the valley and ridge topography that we see today. Igneous activity was not widespread, judging from the general lack of exposed igneous rocks. Intrusive igneous rocks are found in a few areas; all were emplaced during the Cretaceous.

The topography of the Ouachitas is solely the result of weathering, mass-wasting, and stream erosion; continental glaciers did not push southward into any part of the province. The ridges were formed because of the superior resistances of certain formations, particularly the quartzites and novaculite beds, as is readily observed in the Hot Springs National Park area.

HOT SPRINGS NATIONAL PARK

Within the National Park System, parks and monuments have been set aside for several purposes. Some were established primarily to preserve classic geologic features, others for their archeological and historic values, still others for ecological reasons. Hot Springs National Park is unique in that it was established to preserve hot springs that were believed by some to have extraordinary healing qualities. According to legend, the area was used as a healing ground by the Indians; here, regardless of tribe, their weary and wounded could come in peace to recuperate in this quiet area — the Valley of the Vapors.

As the fame of the springs spread among the westward-moving immigrants, so did the opportunity for the exploitation of those who came. Therefore, in 1832 Congress set aside a 4-square-mile area surrounding the springs. When the Park Service was created in 1916,

Hot Springs Reservation was transferred to its jurisdiction. The name was changed to Hot Springs National Park in 1921.

Originally the city of Hot Springs nearly encircled the park, but, by extensive enlargment, the park now surrounds the northern part of the city.

The Waters

While surface waters in the form of streams have played an important role in shaping the valley and ridge topography of the area, the subsurface waters are of greater significance in the park. They surface as springs in many places where faults and fractures are exposed. Numerous cold springs and warm springs occur in the area, but the hot springs are the main attraction. At the present there are 47 hot springs in the park, of which 45 are covered to prevent pollution. Water from these is cooled to a bearable temperature, then piped to bathhouses where visitors can avail themselves of their benefits. The remaining two hot springs, called Display Springs, are left in essentially their natural state so that visitors can observe them and the calcareous tufa deposits surrounding them. Although the water is too hot for most forms of life, certain types of blue-green algae thrive in it. [Figure 15-1 is in color section.]

Flow from the springs fluctuates from a maximum during the winter and spring to a minimum in the fall. The maximum flow from the 47 springs is about a million gallons per day, and the average temperature is approximately 143°F (62°C). Gradual changes have been recorded in both temperature and flow; both are declining, but at a rate too slow to threaten the bathing industry.

The waters are unusual in that unlike those of many thermal springs, they are essentially free of objectionable odor or taste, and many people fill their jugs from the fountain near the visitor center. The content of dissolved minerals is relatively low, lower actually than that of most groundwaters of Arkansas. Visitors are impressed with the bubbles that grow and then break from the surface of the pools, resulting from a concentration of dissolved gases, mainly carbon dioxide, nitrogen and oxygen. Impressed also by the presence of small amounts of radium and radon gas in the waters, many health seekers are convinced of their curative powers.

The Mountains

The park is located in the ZigZag Mountains section of the Ouachitas. The name reflects the zigzag character of the main ridges. In turn, the ridges reflect the structure and lithology, intensely folded and faulted beds of extremely resistant rocks, mainly novaculite, which are interbedded with poorly resistant valley-forming shales. The mountains rise abruptly above the lowland areas to heights as great as 1410 feet (427 m) on Music Mountain. Although from a distance they appear to have a rounded, rolling topography, there are numerous precipitous cliffs.

There are several drives and hiking trails in the nearby mountains into country usually considered more typical of national parks. One road leads through dense forests of oak, hickory, and short-leaf pines to the top of Hot Springs Mountain and North Mountain, which are east of the city. You may enjoy seeing the abandoned novaculite quarries where Indians used the stone for the shaping of knives and projectile points. A conducted hike takes you up on Indian Mountain during the summer months. The West Mountain section is equally attractive, with scenic overlooks of the various ridges of the Ouachitas.

Rock outcrops are abundant along the roads and trails, even in this humid climate. Many are novaculite, which resists weathering, regardless of humidity. Be sure to see the excellent exposure of white novaculite at the turnaround at the top of West Mountain.

Don't pass up the opportunity to hike some

of the 18 miles of trails in the park. You will not see the bison or elk or bear that once inhabited the area, but where there are hickory nuts and acorns there are squirrels who will keep track of you, chattering and scolding you for invading their domain. You are not likely to see the resident raccoons or opossums unless you also are nocturnal, and you will be lucky to catch a glimpse of a fox or wild turkey.

The vegetation in the park is one of its finest assets, with wildflowers blooming the year around. The most flamboyant of the many trees is the southern magnolia, which has been introduced in the lower areas of the park.

There is much of interest in the country surrounding the park. When you have exhausted the immediate area, there are other scenic drives, notably Highway 7 north from Hot Springs. You will wind through the Ouachitas in Ouachita National Forest; perhaps you will stop at the crystal mine where fine rock crystal quartz and smoky quartz are on display. Also, for a small sum you can dig for real diamonds at the Crater of Diamonds near Murfreesboro, about 60 miles (100 km) southwest of Hot Springs.

FIGURE 15-2

View north from Hot Springs Mountain out over dissected erosion surface.

CHAPTER 16

SUPERIOR UPLAND PROVINCE

The Superior Upland is a small part of the 2-million-square-mile Canadian Shield, probably the largest area of exposed Precambrian rocks in the world. Although Thornbury (1965) views the Superior Upland as a section of the Laurentian Upland Province, it is here regarded as a separate geomorphic province. In the United States, the Superior Upland occupies the Precambrian areas of Minnesota, Michigan, and Wisconsin, although its boundary with the Central Lowland Province is locally indefinite because of the cover of glacial drift.

Geologic Sequence

The Lake Superior District has been studied in detail by many geologists, in part because of the valuable ore deposits, especially iron and copper. In addition, uniquely recorded in this area are important chapters of Precambrian history. Only the broad framework is presented here, into which Isle Royale and Voyageurs national parks fit.

The oldest known rock in northern Minnesota is the Ely Greenstone, more than 2700 million years old.[1] This greenstone was formed by the metamorphism of basaltic flows. (On Isle Royale, the greenstone is much younger —an essentially unaltered basalt; there, the name originated because of a faint greenish

---
[1] For the ages of these Precambrian rocks, see Goldich, 1968.

cast. In some cases, the term is used for certain greenish gemstones). Other Early Precambrian rocks, intrusives, and metamorphosed sedimentary rocks are also exposed in the area. Middle Precambrian rocks occur farther east; the Gunflint Iron Formation, the source of iron ore, is one of the more important members of this sequence.

Later events, such as the eruption of huge quantities of lava, are well recorded in the Lake Superior District. These Keweenawan (Late Precambrian) flows, almost entirely basaltic, have a total thickness estimated to be as much as 30,000 feet (9146 m). Under what conditions could such an enormous volume of lava be extruded? In the Columbia Intermontane Province, large amounts of lava poured out of rifts formed by the pulling apart of the crust. Today, lavas are periodically extruded along the Mid-Atlantic Ridge where seafloor spreading is taking place. Therefore, there is reason to consider the possibility that the crust was being pulled apart in the Lake Superior District about a billion years ago. If true, the case for plate tectonics is strengthened, by extending the time of operation back into the Precambrian.

The sequence of events as outlined by Huber (1975) requires the spreading of the crust away from a long fracture or series of parallel fractures. The first lava flows poured out of long fissures that were parallel to the axis of the now-existing syncline. The lavas contain ellipsoidal masses called pillow structures and thus must have extruded into a shallow body of water. Evidently the water was eventually replaced by lava because the later lava flowed over land. During quiet periods, gravels were carried out and deposited on the lava flows.

At or near the end of the eruption period, a huge mass of gabbro was emplaced within the basalt flows. The Duluth gabbro is exposed in a number of places north and east of Duluth. Associated with the intrusion are valuable copper-nickel deposits.

During the long period of igneous activity,

as more and more magma was removed from below the surface and piled up as lava flows, the Lake Superior area gradually subsided, particularly along the axis where the flows were thickest. Thus, the Lake Superior Syncline was formed. In addition to the downfolding, faulting occurred parallel to the fold axis; one fault of large displacement lies between Isle Royale and the north shore of the lake. Cross-fractures and small faults also developed as the basin subsided; many are parallel and cut across the linear ridges of basalt.

After the eruptions ceased, streams carried large quantities of coarse debris southward across the basin and deposited them on the youngest lava. These conglomerates are now exposed extensively along the south side of the west end of Isle Royale.

Paleozoic events are poorly recorded, but small remnants of Paleozoic rocks indicate that the area was beneath the sea at certain times during the era. During the Late Paleozoic, the Mesozoic, and the Cenozoic, the province stood above sea level, at about its present position.

During the Pleistocene, glaciers pushed down out of Canada and relandscaped the area. Lake Superior's basin was scooped out by glaciers which left large deposits of morainal debris around the southwestern end. Areas where the rocks were unusually resistant to glacial abrasion were left as prominent hills, some of which, including Isle Royale, rise above the present lake level.

The surface of Lake Superior is 602 feet (184 m) above sea level, but at an earlier time the depth of water was greater. When the lake level remained constant for a significant period, wave erosion formed a terrace or beach. In order to distinguish the stillstands of the past, separate names are used. Lake Duluth is an example: The results of Lake Duluth wave erosion are clearly marked on the north side of Lake Superior and on Isle Royale.

When the ice sheet covered the Great Lakes area, the tremendous weight of the ice caused significant depression of the crust; when the ice melted, uplift to its original position began. The uplift is continuing today.

ISLE ROYALE NATIONAL PARK

Isle Royale National Park in Michigan is composed mainly of water; with its boundaries about 4.5 miles (7.2 km) offshore, the park consists of about 600 square miles of Lake Superior and about 210 square miles of land. Isle Royale can be reached by boat from Grand Portage, Minnesota, 22 miles (35 km) away, or from Copper Harbor and Houghton, Michigan, 56 and 73 miles (90 and 118 km) distant, respectively. Sea planes also bring visitors into the park. Isle Royale is a hikers' park, for there are no roads and no automobiles. The moose, timber wolves, foxes, and smaller animals live in peace, largely free from noise and pollution. Two trails extend from Rock Harbor, near the eastern end of the main island, to Windigo, 40 miles (65 km) to the west. These trails are for those who are physically fit and well equipped; even for the hardy, they would probably seem to be somewhat more than 40 miles (65 km) long in the fog and rain. But there are also many shorter trails leading into the back country.

Isle Royale was explored by Indians at least 4500 years ago. No habitation sites have been found, according to Rakestraw (1965), and except for one thing we probably would have no record of their being on the island. Native copper occurs in the rocks of Isle Royale, and the Indians mined it from more than a thousand pits. The copper occurs as *amygdule* fillings of the vesicles (gas-bubble holes) in some of the lavas and as fillings around pebbles in the conglomerates. By using hard beach cobbles as hammerstones, the Indians freed the malleable copper for ornaments and knives. Charred wood from their fires in the mining pits has been dated by radiocarbon. The mines had been idle for a long time when copper was "discovered" once more about 1840. From

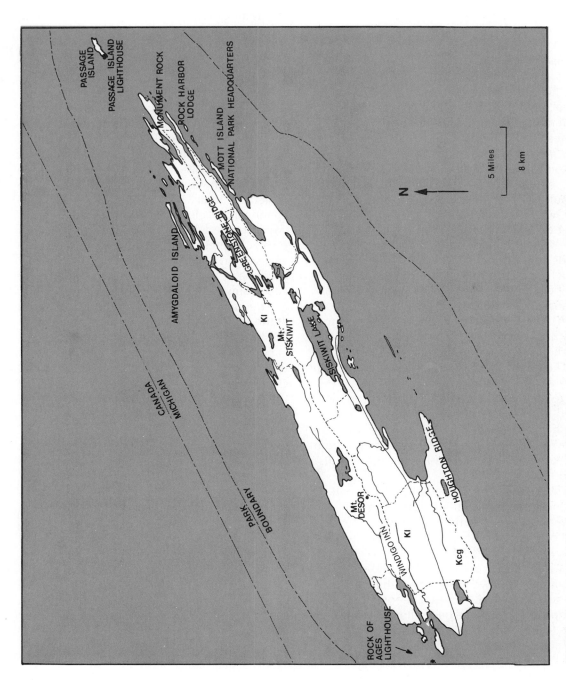

FIGURE 16-1   Map of Isle Royale National Park. Precambrian Keweenawan lava flows (Kl), with younger Keweenawan conglomerate (Kcg).

1843 to 1899 the mines were operated sporadically, but all were abandoned long before the park was established in 1940.

Geologic Setting

Isle Royale is located on the north limb of the Lake Superior Syncline, and the lava flows and sedimentaries — sandstones and conglomerates — dip beneath the water along the south shore. Between Isle Royale and the Minnesota shore the water is in places more than 900 feet (275 m) deep, thus extending down to about 300 feet (91 m) below sea level. With the water at its present level of 602 feet (184 m), there are more than 200 islands of the Isle Royale Archipelago. Almost all of the islands are distinctly elongate, reflecting the structure of the rocks. Amygdaloid Island on the north side and Mott Island, where the park headquarters are located, are typical. It is more than 50 miles (80 km) from the easternmost lighthouse on Passage Island to the one on the Rock of Ages west of the main island.

The large island is about 45 miles (72 km) long and as much as 9 miles (15 km) across. Large peninsulas extend out at each end, with fiord-like harbors between them. Mt. Desor, the highest point, reaches 1394 feet (425 m) above sea level. Elongate lakes, some of them large, are prominent features on the main island.

The corrugated topography of the park is distinctive. In a distant view from the air, the ridges appear to continue unbroken from one end of the main island to the other. A closer view reveals many breaks and offsets related to fractures and faults. The valley and ridge topography is the work of glaciers that overrode the area at least four times. Greenstone Ridge, composed of massive basalt flows, was better able to withstand the onslaught of the glaciers and forms the high backbone of the island.

Bedrock Geology

Keweenawan (Late Precambrian) basalts are the dominant rocks, but conglomerates are extensively exposed along the south shore of the big island. The conglomerates, also Precambrian, are younger than the basalts. All formations dip south-southeastward, in most places at about 15 to 20 degrees.

As outlined earlier, basaltic lavas flooded a subsiding basin periodically during mid-Keweenawan time. The flows are not identical — some are equigranular and massive, others are porphyritic, while many are amygdaloidal. In young lava flows, the upper layer is likely to be vesicular (that is, it contains *vesicles* — spherical or elongate cavities where gases were trapped beneath an already solidified crust). In these ancient flows, however, the vesicles have long since been filled with some secondary mineral. The vesicle fillings are called amygdules and the rock an amygdaloid.

The amygdaloid layers are best developed at the top of a flow, but they are also found at the base. These layers are of particular significance because they are more susceptible to weathering and erosion than massive rocks. Consequently, valleys have been carved out of the amygdaloidal zones; thus, what happened a billion years ago is reflected in the topography of the present.

On Isle Royale, many of the amygdules are composed of an unusual mineral, prehnite, a semiprecious stone. Originally it was believed to be thomsonite, a more highly prized gemstone. Even on Thomsonite Beach, on the north side of the island, most of the amygdules are prehnite. Chlorastrolite, sometimes called "Isle Royale Greenstone," also occurs as amygdules. Polished specimens are usually chatoyant, like the eye of a cat. Much more common, however, are the agates that are found on most of the pebble beaches. They are composed of concentric bands of chalcedony, a translucent variety of quartz. Mainly they form as amygdules in the lavas; later, some were incorporated into the conglomerates, others in the glacial drift. Still later they were freed by wave action and deposited on the beaches.

Native copper also occurs as amygdules but mainly it either fills fractures or acts as the matrix in conglomerate. One copper "nugget" weighing 5720 pounds was unearthed by prehistoric Indians; frustrated by their inability to cut it into workable pieces, they abandoned it.

The conglomerates that overlie the lavas record an ancient environment far different from that which exists today. At the western tip of the island, boulders as large as two feet in diameter are found at the base of the Copper Harbor Conglomerate. Because the size of the materials decreases toward the east, Huber (1975) concludes that the source area lay to the west, the north shore of Lake Superior. In order to transport boulders, the streams must have had considerable velocity. Perhaps flash floods such as occur in the West today were at work in the Lake Superior area more than half a billion years ago. Regardless, a huge alluvial fan deposit was spread over the Keweenawan lava flows.

The Gap in the Record

No rocks of Paleozoic, Mesozoic, or pre-Pleistocene Cenozoic age are found in the area. The evidence in adjacent areas, however, suggests that Early Paleozoic sediments were deposited here but were later eroded away. From the late Paleozoic to the present, the Lake Superior area was apparently stable and remained at essentially its present position above sea level.

The Work of the Ice Sheets

During the Pleistocene Epoch, huge ice sheets (continental glaciers) developed in Canada, expanded southward, and covered nearly all of the United States north of the Missouri and Ohio rivers. There were at least four major advances separated by long periods of warm climate, called interglacials. As the glaciers advanced, they essentially obliterated the evidence of former advances within the area covered; therefore, the records of one or more advances may have been completely destroyed.

Although earlier glaciations undoubtedly affected the Lake Superior area, only the latest glaciation, called the Wisconsin, left records for us to decipher. During the Wisconsin advance, the Lake Superior Basin was scoured out to its present depth which, in places, reaches about 300 feet (91 m) below sea level. The glaciers dug deeper into the rocks in some places and formed many rock-basin lakes. Lake Desor and Siskiwit Lake are two of the larger of this type. Some of the shallow lakes have since been converted to swamps.

The main island is a huge roche moutonnée; if Lake Superior were emptied we would become aware of its true size. Superimposed on the big one are smaller roches moutonnées which are hosts to still smaller ones. In places, the ice overran mounds of till and formed elongate drumlins. Most of them are small, but one is almost 2 miles (3 km) long (Huber, 1975).

Glacier Recession

About 11,000 years ago the climate became warmer and the ice sheet began to retreat. However, at the beginning of the recession, the ice in the Lake Superior Basin was thousands of feet thick and the ice front was many miles to the southwest. Therefore, it was not until much later that any appreciable part of the area was freed from ice.

The retreat was not continuous. The climate, although warming generally, was fluctuating, and at times melting merely equaled the forward movement of the ice; at such times

FIGURE 16-3

Pothole Falls on the Brule River near Grand Marais, north shore of Lake Superior.

the ice front remained almost stationary for extended periods. When the western end of Lake Superior was ice free, retreat ceased and meltwater formed Lake Duluth at the glacier front. Lake Duluth was much smaller than Lake Superior, but the water surface was higher, more than 1100 feet (335 m) above sea level. At this time — about 11,000 years ago — Isle Royale was still buried beneath the glacier.

Lake Duluth remained at this level for a period sufficient for wave action to develop a wave-cut terrace which is more than 500 feet (152 m) above the present lake level. In like manner, stage by stage as the glacier withdrew, Lake Superior was finally formed. As the ice melted off of Isle Royale, morainal deposits — relatively thin ground moraines and irregular recessional moraines — were laid down. In several places huge boulders — erratics — were left high on the ridges, as Huber (1975) relates.

The Work of the Waves

A wave-cut platform with wave-cut cliffs rising above it encircles each of the islands at present lake level. The waves also cut through a headland on Amygdaloid Island when the lake level was higher, thus forming a "sea arch." At a still higher stage, when the big island was about half its present size, the waves almost destroyed a rock hill; however, the lake level was lowered, probably by the downcutting of the outlet stream, in time to save Monument Rock, a stack that rises about 70 feet (21 m) above Greenstone Ridge north of Rock Harbor Lodge.

Your Visit to Isle Royale

Isle Royale provides a good view of Late Precambrian and Pleistocene history. The glaciers relandscaped the islands and formed the lakes and swamps, ideal habitats for the moose. Timber wolves, seldom seen by visitors, prey on the moose and prevent their numbers from

FIGURE 16-5

Monument Rock, a lake stack formed when lake level was much higher than now. (Photo by National Park Service.)

increasing above the levels the islands can support. This isolated unit of the Park System is an excellent place to study plants and animals in an almost undisturbed environment.

Your visit must be planned for, including advance reservations. A two-day tour on a boat such as the *Ranger* or perhaps the *Voyageur,* with an overnight at Rock Harbor or Windigo, will take you around the main island. With the stops at various landings, you get a

FIGURE 16-6

Moose grazing in one of many small lakes on Isle Royale. (Photo by National Park Service.)

reasonably good view of the shore area. But the park experience comes as a result of some foot-work on the trails, for a day or two, at least. To walk alone, deep into the woods, is to know Isle Royale.

FIGURE 16-7

Arm of Rainy Lake from Bear Pass, Voyageurs National Park. (Photo by National Park Service.)

VOYAGEURS NATIONAL PARK

About 219,000 acres of northern Minnesota's wilderness was established as Voyageurs National Park in 1971. The park is in the lake country along the international boundary, about 15 miles (24 km) east of International Falls. Kabetogama Peninsula, south of Rainy Lake and north of Kabetogama Lake, is the main land area. Parts of Rainy Lake and Kabetogama Lake, together with many smaller lakes, make up more than one-third of the park area.

Roads lead up to the boundary but there are no roads in the park itself. For more than a century, however, the Voyageurs traversed the area in their canoes, carrying their furs eastward to Grand Portage at the eastern tip of Minnesota. Here, they traded their furs for supplies brought there from Montreal. This colorful, boisterous breed of men faded from the scene sometime during the 1830s, leaving behind their story written in song.

Park visitors will relive the life of the Voyageurs, canoeing and portaging from one lake to the next, and sleeping under the stars. The master plan calls for a wilderness park. Thus, the Park Service salutes the deeds of those men who contributed much in the development of this country. Another reason for the park is to preserve a small patch of the vast area of virgin timber — fir, spruce, pine, aspen and birch — and its moose, deer, wolves and beaver. Perhaps in time the caribou will return.

The geologic story that long preceded the Voyageurs was written by the glaciers that carved out the lake basins, thousands of them, linking the west with the east. Carved out of the hard Precambrian rocks of this part of the Canadian Shield, the lakes have remained essentially unchanged. Once the forests became established, with a thick blanket of litter on the forest floor, the rain that has fallen has been largely absorbed; runoff and erosion which have destroyed many lakes are at a minimum here. By the invasion of vegetation, certain of the shallow lakes have been converted to swamps, thus creating another facet of the North Woods ecosystem.

In time, when all of the surrounding area has been conquered by man and converted to his alleged needs, Voyageurs National Park with all of its natural beauty and historical significance will hopefully be one of our truly outstanding national treasures. Voyageurs was set aside as a wilderness park, but here as elsewhere, park boundaries do not exclude all outside influences. Paper-pulp mills near International Falls, a short distance west of the park, are sources of a nauseating odor that is carried for long distances by the wind. Control devices have been installed to filter out the particulate matter, but apparently it is virtually impossible to "quench the stench." In addition, a coal-fired power plant may come on line at Atikokan, Canada, close enough to damage the air quality in Voyageurs National Park. For how long will Voyageurs be a wilderness park?

CHAPTER 17

INTERIOR LOW PLATEAUS PROVINCE

The Interior Low Plateaus lie between the Central Lowlands and the higher Appalachian Plateaus on the southeast. In elevation, this province ranges from about 500 feet (150 m) to 1100 feet (335 m) above sea level. Lying mostly in Kentucky and Tennessee, it extends into the southern sections of Illinois, Indiana, and Ohio, and southward into Alabama.

The geologic structures, generally similar to those in the Central Lowlands, are broad uplifts with low dips on their flanks. The Nashville Dome in Tennessee is of special interest because, although it is structurally a dome, it is a topographic basin. The lowest part of the basin coincides with the highest part of the dome, as it was originally formed. The top of the dome was first eroded away, and further erosion deepened the basin. Thus, here we have a classic example of inversion of topography.

Paleozoic rocks are exposed throughout the province. In large areas Mississippian limestones either outcrop or occur at shallow depth beneath beds of sandstone. Green River and its tributaries have cut deep valleys into the limestone in western Kentucky, providing ideal conditions for the development of caverns and other *karst* features.

MAMMOTH CAVE NATIONAL PARK

About 51,000 acres in the karst area of western Kentucky were designated Mammoth Cave National Park in 1941. A distant view of the area suggests that the plateau surface is essentially unbroken, except for Green River Valley; a closer view reveals that it is broken by countless *sinkholes* that lead down into the world's largest known cavern system.

Mammoth Cave was known to the Indians, and many artifacts including sandals, woven cord, and wooden bowls have been found. According to legend, a bear hunter named Houchins rediscovered the cave in 1799. What is known is that a land certificate dated August 18, 1799 mentions "petre caves." The "petre dirt" of the caves was processed to form saltpetre which was used in the manufacture of gunpowder during the War of 1812 — and the Civil War. After the war, a flourishing tourist trade developed and cave explorers, *speleologists,* flocked in. As more and more passageways were explored, it became evident that here was one of the larger caves in the world. Then, one day late in 1972, it was suddenly the largest, when a party of Cave Research Foundation explorers discovered that another large cavern system, the Flint Ridge, is actually a part of Mammoth Cave. The combined system is easily the world's longest, about twice as long as its nearest rival.

Caves are common geologic features; in fact, caves of one type or another occur in every one of the 50 states. Some are lava tubes, others are merely overhangs, but all of the large cavern systems were formed by solution of some soluble rock — generally limestone but sometimes marble or even dolomite. Of the hundreds of national parklands, seven were established primarily to preserve caverns and their fascinating features.

The prerequisites for the formation of caves and the processes involved in cavern development are discussed in some detail in the Introduction and in Chapter 4, where the first cav-

FIGURE 17-1

Frozen Niagara, the largest travertine flowstone deposit in Mammoth Cave, Kentucky. (Photo by National Park Service.)

FIGURE 17-2

A prominent fracture in the roof of Mammoth Cave was responsible for the alignment of these large stalactites. The one to the left has joined with a stalagmite to form a column. (Photo by Robert B. Johnson.)

ern park, Oregon Caves, is presented. Here in Mammoth Cave, we are exploring our last of the Park Service's underworld wonders. The following discussion is confined in large part to the features unique to Mammoth Cave — the only national park within a large and highly developed karst area.

Karst Features — Caverns and Sinkholes

Several Paleozoic formations underlie the Interior Low Plateaus, but the two in which the caverns of the Flint-Mammoth Cave System are developed are the Girkin and Ste. Genevieve limestones of Mississippian age. The Ste. Genevieve that lies beneath the Girkin contains the greatest number of passageways. The Big Clifty Sandstone rests on the Girkin Limestone, except for the areas where the sandstone has been eroded away.

Surface waters penetrate down through the sandstone beds or, where absent, directly into vertical fractures in the limestones. Charged with carbon dioxide, the water is actually a weak acid — carbonic acid — which reacts with the calcite in the limestone and takes it into solution. Thus, fractures are widened and small openings are enlarged, forming small caverns.

With the enlargement of the caverns, the roof rocks eventually collapse, forming surface depressions known as sinkholes or *collapse sinks*. Some of the collapse sinks are very large, and it might be inferred that a huge mass of rock crashed down at one time. However, this seldom occurs; instead, as Fenneman (1938) explains, collapse begins early when the cavern is small. Later as more of the limestone support is removed, the roof rocks fall in, block by block, eventually enlarging the sinkholes to their present size. When a large number of sinkholes have formed, the topography is sometimes referred to as a *pitted plain*.

Solution of the limestone occurs mainly at or near the water table — the upper surface of the saturated zone. In the Mammoth Cave

FIGURE 17-3
Stream flowing down into one of many ``swallow holes.''

area, Green River controls the position of the water table; within the cave, Echo River eventually emerges and flows into the Green.

The much-simplified cross-sections shown below, taken from the Park Service brochure, indicate the general sequence of events. In the upper cross-section the water table is near the surface. The layer of limestone that has been taken into solution is more soluble than those

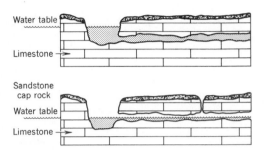

above and below it. In the lower cross-section the water table has been lowered by the down-cutting of the master stream. Note that, by collapse of the roof rocks, a collapse sink has been formed. In Mammoth Cave, as in many large cavern systems, this sequence has been repeated; passageways were formed at different levels as the master stream continued to downcut.

Cavern Formations: Speleothems

The caverns are adorned with many formations that have developed by the precipitation of calcium carbonate. Stalactites and stalagmites are abundant, and huge columns with ornate flutings are common features. In places, large flowstone deposits on the walls resemble waterfalls. Frozen Niagara, not far from the Frozen Niagara Entrance, is one example.

Many slender soda-straw stalactites ornament the Onyx Chamber, which has been dubbed The Macaroni Factory. Nearby is the Onyx Colonnade, a series of stalactites, stalagmites and columns, many highly colored.

Points of Interest

A number of trips, both underground and on the surface, are available to you. If you wish to see the Giant's Coffin, a 50-foot-(15 m)-long collapse block of limestone, take the "Historic Trip." Possibly at some point your guide will turn off all of the lights so you will know what total darkness is really like. During the early years, the highlight of visiting Mammoth Cave was the boat trip down in the cave; visitors had the opportunity to observe the blind fish whose eyes had atrophied in the darkness of the subterranean waters. This unique experience is once again available as part of the Echo River Tour. Current information on this and other trips can be obtained at the visitor center.

A field trip guide prepared by J. F. Quinlan and others (1983) will help those who visit the Mammoth Cave area in their attempts to visualize the intricate underground drainage system. It also contains information on the research that is in progress, particularly on the problem of groundwater pollution—both in the park and in the surrounding area.

Pollution

Problems of water pollution in karst areas are much more difficult to analyze and to deal with than in other terrains. Waste materials that get into one part of the system are certain to move to other parts, perhaps eventually into the entire subterranean system.

Within the park the most serious environmental problem is the sewage from nearby populated areas. Not only has it created an unforgivable stench, it has also threatened the extinction of some of the unique forms of aquatic life, including the blind fish and blind shrimp.

Hidden River Cave, not far from Mammoth Cave, was shown commercially until 1943 when, because of the concentration of sewage and also nauseous waste from a nearby cream-

FIGURE 17-4

The old ferry on Green River, the stream that controls the local water table.

ery, it was forced to close down. A major effort has been launched to prevent Mammoth Cave from meeting a similar fate — a project that deserves vigorous support. It has been the largest cave in the world for a very long time and a World Heritage Site since 1981.

FIGURE 17-5

Typical karst topography with sinkholes and sinkhole ponds near Mammoth Cave. (Photo by National Park Service.)

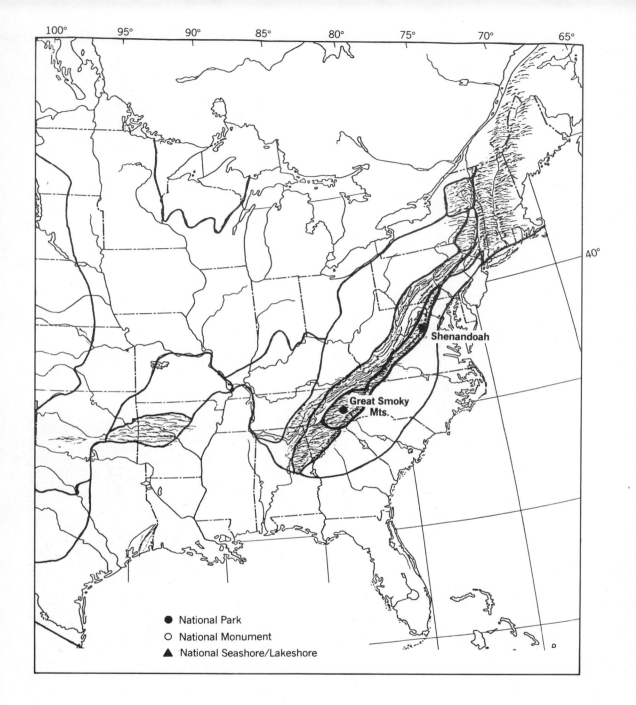

PLATE 18
Appalachian Provinces.

CHAPTER 18

THE
APPALACHIAN
PROVINCES

The term *Appalachians,* as used here, includes the four geomorphic provinces that extend southwestward from the New England Province. From west to east they are: (1) Appala-chian Plateaus, (2) Valley and Ridge (Folded Appalachians), (3) Blue Ridge, and (4) Piedmont provinces.

Other designations such as the Appalachian Highlands, Appalachian Mountains, and Appalachian Mountain System are also in use, in most cases including the New England area. Because the sequence of geologic events is generally similar throughout, the following discussion includes New England.

The Fall Line—an important boundary in eastern United States—as Thornbury (1965) points out, is the boundary between the hard rocks of the Piedmont and the weaker, easily eroded rocks of the Coastal Plain. On a num-

FIGURE 18-1

Falls of the Potomac, at the boundary of the Piedmont and Coastal Plain provinces, near Washington, D.C. (Photo by B. Kiver.)

ber of streams, notably the Delaware and the Potomac, the Fall Line is marked by falls and rapids. In some places, the Fall Line is less distinct, where rapids are found over reaches of several miles. In essentially all cases, whether falls or rapids, boat transportation was interrupted at the Fall Line; consequently, towns — which grew into cities like Baltimore, Richmond, and Raleigh — were located at the "end of the run." This is an excellent example of effect of bedrock geology on the topography, which in turn affected human use and development of the region.

The low, rolling Piedmont, like the topographically rugged New England Province to the north, is composed of similar Precambrian and Paleozoic metamorphic and plutonic rocks. In addition to the schists, gneisses, slates, and granite plutons, some of the elongate grabens contain Triassic sandstones, shales, and basalt. Although the Piedmont is generally a lowland area, it rises as much as 1800 feet (552 m) above the sea in the Dahlonega Plateau of Georgia. When the Piedmont is seen from the Blue Ridge, it appears as a monotonous plain. Actually, the topography holds interest both esthetically and geologically. Much of the area is a gently rolling plain with numerous hills rising prominently above. One well-known hill is the site of Monticello, near Charlottesville, Virginia; another is the granite knob, Stone Mountain, in Georgia. These and similar hills are interpreted by some as monadnocks rising above an erosion surface or peneplain. John Hack (1982) points out that all of the higher hills that he examined owe their elevation to the more resistant rocks that underlie them and are therefore the result of a continuously eroding topography. According to this interpretation, weathering and erosion are continuous, and differences in rock characteristics account for local differences in elevation. Whatever the exact processes of erosion may be, only by being spared rapid erosion for a long period could the Piedmont

have developed more than 100 feet (30 m) of deeply weathered soil (Thornbury, 1965; Hack, 1982).

The Blue Ridge rises abruptly above the Piedmont and forms the backbone of the Appalachians. Although somewhat subdued as compared to the mountains of the West and Alaska, the Blue Ridge contains some exciting topography — precipitous cliffs, spires, and waterfalls, particularly in the southern part. Because the only national parks in the Appalachians are in the Blue Ridge Province, a separate section is devoted to it.

West of the Blue Ridge lies the Valley and Ridge Province and beyond that the Appalachian Plateaus Province. Both areas contain folded, unmetamorphosed rocks and are underlain by large thrust faults produced during a collision of continents during Late Paleozoic time. The intensity of deformation decreased westward away from the Blue Ridge and Piedmont. Rocks thrust westward onto the more stable interior of the continent are tightly folded in the Valley and Ridge Province and have broad, open folds in the Appalachian Plateaus. Most of the rocks in the Valley and Ridge are Early Paleozoic sandstones, limestones, and shales, although Pennsylvanian sandstones and coal beds are common at the north end. The topography is related directly to rock type and structure. The anticlines and synclines are truncated by erosion — valleys are carved into the shales and limestones; the resistant sandstones, quartzites, and dolomites stand high as ridges.

Rock structures predetermined early transportation routes, making travel in the Valley and Ridge Province both tedious and tortuous. The roads follow the valleys between the linear ridges; roads connecting those valleys are widely spaced and confined mainly to the water gaps where streams have cut through the ridges. Now, however, a few major highways such as the Pennsylvania Turnpike cut across the ridges. Some of the rock structures,

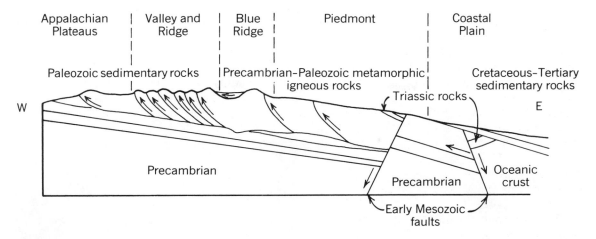

FIGURE 18-2

Generalized west-to-east cross section, showing major structures and rock types. Taken from various sources by E. Kiver.

as shown in the much-simplified cross-section of Figure 18-2, can be seen in the huge road cuts.

Broad, flat valleys are common, such as the Cumberland Valley and the famed Shenandoah farther south. The broad loops of both forks of the Shenandoah are classic examples of meandering streams.

Caverns and other karst features are abundant in the limestone valleys in the province. As a reminder, although limestones generally are ridge-formers in the arid West, here in this humid climate they weather and erode rapidly, and thus form valleys. Here also, subsurface waters are actively dissolving out the calcium carbonate, forming large cavern systems in the limestones. Famous caverns such as Luray Caverns are numerous. Also widely known are Natural Bridge, which supports highway traffic near Lexington, Virginia, and Natural Tunnel, which is the route of the Southern Railway near Clinchport, Virginia. Both were developed by subsurface water. Natural Bridge represents a more advanced stage in the collapse of cavern roofs than Natural Tunnel.

The Appalachian Plateaus on the western side consist mainly of flat-lying sandstones, conglomerates, shales, and coal beds of Middle and Late Paleozoic age. Several broad folds and faults are present, but this part of the geosyncline was least affected by the mountain building and metamorphism that so profoundly changed the rocks in most of the geosynclinal area.

Small but interesting areas containing unusual plutonic rocks called *kimberlites* occur in the plateau as well as in other parts of the Appalachians. These dark-colored, dense rocks moved upward from the earth's mantle through deep fractures produced by tension following the Appalachian Orogeny. Dennison (1983) believes the tensional fractures probably formed because of tectonic uplift or isostatic uplift after some of the overlying sedimentary rocks were eroded away, whereas Parrish and Lavin (1982, 1983) suggest that separation of North America, Europe, and Africa by plate movements in Late Paleozoic and Mesozoic times created the necessary tension. Kimberlites — such as those in South Africa, in the Murfreesboro, Arkansas area, and

FIGURE 18-4

View over symmetrical meanders of the Shenandoah River near Bentonville, Virginia. (Photo by Norfolk and Western Railway in Thornbury, 1965. Reprinted by permission of John Wiley and Sons, Inc.)

in northern Colorado — often have diamonds associated with them. Many gem-quality stones have been found in stream deposits in widely separated areas of the Appalachians; however, their source localities remain a mystery (Sinkankas, 1959).

With but local exceptions, outfacing escarpments bound the Appalachian Plateaus, as described by Thornbury (1965). A well-developed dendritic drainage pattern characterizes the area and easily distinguishes it from the linear, *trellised* pattern of the Valley and Ridge Province to the east. Along the eastern border, dissection is sufficiently complete to form "mountains of erosion" such as the Alleghenies and the Catskills.

The Appalachian Geosyncline played an important role in the formation of the Appalachian Mountains. Although the concept of geosynclines was first recognized by James Hall[1], who worked in the Appalachian region in the 1850s, a more thorough understanding of the mountains and the forces creating them had to await many more detailed investigations and the acceptance of plate tectonics

---

[1] Hall's work on geosynclines is discussed by Press and Siever (1974).

concepts. Hall realized that the accumulation of sedimentary rocks thickened markedly as one approached the Appalachian Mountains from the west, thereby recognizing that areas with thick sedimentary sequences later become mountains. He did not specify adequately the mechanics of how these extremely thick accumulations of sediments become high mountains. This led some geologists to remark that his was a theory of mountain-building with the mountains left out! Others proposed that the earth was cooling and shrinking, thus accounting for the depression, or geosyncline, and for the horizontal forces so evident in most structural features in mountainous areas. With the realization that the earth is not losing appreciable heat and that plates of the earth's crust move laterally along its surface, a new interpretation of the geologic development of the Appalachian area became necessary.

The broad picture of plate movements recently recognized in the Appalachians is both exciting and startling. We know that certain continents and plates are today moving toward each other, while others are splitting and moving apart in other parts of the world. It seems likely that many collisions and much breaking apart of plates must have occurred in the past. Indeed, it is a complex sequence that was involved in the location, timing, and deformation recorded in the rocks and structures in the Appalachians. The following summary is greatly simplified, but more detail can be found in numerous journal articles (see Cook et al., 1979; Cook, 1983; Taylor and Toksöz, 1982, and Memoir 158 of the Geological Society of America).

Following a Late Precambrian Orogeny that metamorphosed and deformed the rocks some 1 billion years ago, a rift developed in what is now eastern North America, one that led to the formation of the Appalachian Geosyncline. Marine sediments were deposited in the resulting ocean, and volcanic rocks were emplaced near the rift and subduction zones.

Although this ocean lay off the east coast of North America, it was not the present Atlantic Ocean — much was yet to happen before the present Atlantic would come into being. After hundreds of millions of years, the widening between the plates ceased during Early Paleozoic time, and the plates again moved toward each other. With unimaginable slowness by human standards, huge plates containing North America, Africa, Europe, and South America crept ever closer to each other. The closing of the giant vise occurred episodically,

FIGURE 18-5

Natural Bridge in the Valley and Ridge Province near Lexington, Virginia. Because of its historic and geologic interest, Natural Bridge compares favorably with many of our national monuments. (Photo by B. Kiver.)

first in the northern Appalachians in mid-Ordovician time and later in the southern Appalachians. By Late Paleozoic time the continents had joined into a supercontinent, called Pangaea. The collisions left their marks on the plate edges, and the original rocks that were subjected to intense pressure and heat changed to the gneisses, schists, and other metamorphic rocks of the New England, Blue Ridge, and Piedmont provinces. Extensive melting took place at depth, and the resulting magma intruded, complicating even further the geology of these areas. The sedimentary rocks away from the plate edges in the Valley and Ridge and Appalachian Plateaus were less affected, but they too were changed — deformed by folds and faults.

Although structures at the surface were mapped many years ago, controversy arose concerning what happened to these structures at depth. All agreed that a relatively thin slice of crust (thin-skinned tectonics) was thrust westward over more rigid basement rock in the Appalachian Plateaus and Valley and Ridge provinces. Most workers a few years ago believed that these thrusts were analogous to sheets of toothpaste and that the tube or root of the mountains underlies the Blue Ridge and Piedmont. Information from recent drilling and geophysical studies strongly indicates that thin-skinned deformation also extends under the entire Blue Ridge and at least part of the Piedmont. Whether shallow thrust faults end under the eastern part of the Piedmont or continue under the Coastal Plain cannot be determined without better information than is now available (Harris and Bayer, 1979; Cook, 1983). How far the thrusting continues could have important economic consequences: If some of the oil-rich rocks in the Valley and Ridge extend eastward beneath the thrust slices, then the potential for oil in what was assumed to be barren areas could be high.

At the end of Paleozoic and during Early Mesozoic time, what is now the east edge of the North American plate was an impressive range of mountains in the interior of Pangaea. Again the thermal forces that drive the plate-tectonics machine began to pull the plates apart. Large fault-block valleys, or grabens, extend from Canada to Florida in the Piedmont, in New England, and beneath the Coastal Plain Province. Late Triassic rivers washed debris into the troughs, and large numbers of dinosaurs roamed the Connecticut Valley, New Jersey, and other areas; they left their tracks in the muddy sediments, particularly on the edges of freshwater lakes. Deep-reaching fractures channeled basaltic magma toward the surface where it formed dikes, sills, and lava flows in the grabens. The Palisade Sill along the Hudson River west of New York City is one of the most striking and famous of these intrusions. The splitting apart of the continents continued, and by mid-Mesozoic time marine sediments were being deposited on the coastal plains along the newly opened ocean that would later be named the Atlantic.

Eastern North America became the trailing edge of the westward-moving continent, thus ushering in the tectonic quiet that persists there today. However, the relentless forces of erosion will continue to tear away at the mountains, and the stresses that move the plates will again produce another geosyncline and ultimately another generation of mountains.

BLUE RIDGE PROVINCE

The Blue Ridge forms the backbone that rises distinctly above the adjacent provinces, extending southwestward from southern Pennsylvania into northern Georgia. The term *Blue Ridge* was first used in the northern section where it is a single prominent ridge, in places flanked by lower ridges. South of the Roanoke River in Virginia, the Blue Ridge widens and consists of several irregular ridges, of which the Great Smoky Mountains are outstanding.

For the early settlers who were intent on

pushing westward, the Blue Ridge was the "front range," the first real physical barrier encountered on their westward trek. It rises abruptly above the gently rolling hills of the Piedmont, in places as much as half a mile. With an essentially impenetrable forest, water gaps such as the Shenandoah and the Potomac were all-important gates to exploration. Therefore, the geologic development of the water gaps had a profound effect on the exploration and settlement of this area which at that time was "The West." But there are no water gaps through the Blue Ridge south of Roanoke, Virginia. Consequently, penetration into the southern Blue Ridge was slow; and once there, having given so much of themselves in establishing their homes, the settlers and their offspring remained, forming an island of tranquillity surrounded by the swirling turmoil known as progress. It was an exercise in survival not just for a weekend but for a lifetime, and these isolated mountain people were extremely ingenious in making their own tools and in developing their own culture. The life of the past continues today in certain areas, and park visitors should not fail to take advantage of the opportunity to see it in Cades Cove in Great Smoky Mountains National Park.

The Geologic Story of the Blue Ridge

Early interpretations of Blue Ridge geology were controversial, chiefly because of the lack of data on the ages of the older rocks. Today we have radiometric dates on many of the rocks, but even so, the precise sequence of development of the older rocks — those more than a billion years old — has yet to be established. Therefore, the metamorphics and granitics are together referred to as the Basement Complex (King et al, 1968).

About a billion years ago, granitic magmas intruded into the Basement Complex in several places in the Appalachian area; one example is the Old Rag granite, now known to be about 1.1 billion years old. This determination

was made using uranium-lead analyses of zircon, one of the accessory minerals in the granite (Lukert, 1982). Later, after prolonged erosion, sedimentary rocks were laid down in many places; elsewhere, basaltic lavas and volcanic ash covered wide areas. These sedimentaries and volcanics formed the floor of the Appalachian Geosyncline, which developed at or near the end of the Precambrian, when the landmass rifted and a narrow ocean formed along what is now the eastern edge of North America.

As the Cambrian seas advanced, coarse *clastics* were laid down near shore while the silts and clays were carried farther out and deposited in deeper water. Later in the Cambrian and during the Ordovician, several thousand feet of limestone and dolomite were laid down.

Early geologists believed that deposition continued without interruption until near the end of the Paleozoic, when the Appalachian Orogeny formed the Appalachian Mountains. Later, it became evident that significant tectonic disturbances occurred periodically, beginning in Late Ordovician time. Thus folding, faulting, igneous intrusion, and metamorphism were interspersed with deposition prior to the culmination of the orogeny near the end of the Paleozoic.

So intense were the pressures that most of the Blue Ridge rocks were highly metamorphosed, thus forming schists, phyllites, quartzites, greenstones, and gneisses. In places, however, the original shales, sandstones, lava flows, and granites were less altered, even essentially unaltered.

The late Paleozoic Appalachian Orogeny was particularly severe because it involved the collision of two major crustal plates: the North American and Gondwanan (Africa, South America, Australia) plates. The ocean between the plates was eliminated, and large-scale thrusting of geosynclinal rocks occurred toward the interior of the continent. Both the Blue Ridge and at least part of the Piedmont are part of a giant thrust sheet that is believed to have been shoved at least 163 miles (260 km) to the west (Cook et al., 1979; Harris and Bayer, 1979). The greater elevation of the Blue Ridge as compared to other Appalachian areas may be due in part to subsidiary faults that branch from the main thrust and bring more resistant crystalline rocks to the surface.

As the mountains were being thrust up, erosion was at work tearing them down. The topography of the Blue Ridge is the result of prolonged stream erosion. Flattish summit and subsummit areas have been interpreted by most workers to represent the final products of stream erosion cycles. However, the number of cycles has long been a matter of controversy. Those interested in the full account are referred to Fenneman (1938) and to Thornbury (1965). The classical interpretation involves three erosion cycles, the last of which is still in progress. Remnants of two peneplains are recognized by many workers, but there is diversity as to the degree of perfection and the age of each of the two. More recently, even the cyclic development concept was challenged. Hack (1960) proposed instead a "dynamic equilibrium" theory that emphasizes the significance of rock resistance to weathering and erosion; in this way, he attempts to explain the flattish areas at different elevations.

In a recent summary of his observations and ideas concerning the Appalachian areas, Hack (1982) not only reaffirms the correlation of rock resistance and elevation but also stresses that recent, probably continuous uplift is necessary to maintain significant elevation of a landmass. Although many geologists do not question the importance of rock resistance, there remains the question as to whether this concept replaces or merely supplements the cyclic development concept.

There is almost a complete absence of sharp peaks in the Blue Ridge, even though in many places the side slopes are very steep. Instead, the tops are characteristically rounded, and the term *dome* is widely used. In this warm, humid climate, chemical weathering is effective in rounding off the sharp corners and edges.

Glaciation, which profoundly changed the mountains of New England, did not affect the Blue Ridge. The colder climate that prevailed at times during the Pleistocene, however, left its mark on the Blue Ridge in the form of frost-heaved boulders and other features.

Within the Blue Ridge, there are two national parks—Shenandoah in the northern section and Great Smoky Mountains in the southern part. Another beautiful unit of the National Park System is the Blue Ridge Parkway, a 469-mile-long (756 km) scenic highway that connects the two parks.

SHENANDOAH NATIONAL PARK

In north-central Virginia a long and generally narrow section of the Blue Ridge Mountains was set aside as a national park in 1926. At that time, many of Shenandoah's 300 square miles were in poor condition. Most of the mountain people who had eked out a bare living on the steep Blue Ridge slopes had finally moved out, abandoning their worn-out, badly eroded farmlands. Lumber companies had stripped off much of the timber, leaving the steep slopes unprotected from erosion; and most of the wild game was gone. Parks are generally created to preserve the pristine environment; at Shenandoah the immediate objective was the restoration of a badly misused area. The hardiest of the mountaineers were relocated in more favorable areas, including those people who, in secluded Nicholson Holler (Hollow), had been operating a brisk spirituous business!

In the protective hands of the Park Service, the area is now in amazingly good condition. Essentially all of the old wounds have healed to the extent that many visitors find it difficult to accept the descriptions of conditions that obtained in the early 1900s. High precipitation and a long growing season are favorable for rapid revegetation, and in a short time, locusts, pines, and other trees began to reclaim the cutover forests and abandoned fields. The trees, together with vines and shrubs, held the soil and rocks in place, greatly reducing the erosion that had been going on unchecked. And the animals — black bear, white-tailed deer, raccoons, opossums, skunks, squirrels, and others — have returned along with a host of songbirds, to restore in large part the original ecosystem.

The north entrance is near Front Royal, and the south entrance is about 75 air miles (120 km) to the southwest, near Waynesboro. Along Skyline Drive, there are about 75 overlooks from which you can look eastward out over the Piedmont and westward down into the broad Shenandoah Valley, where the South Fork of Shenandoah River loops back and forth on its way northward to the Potomac. On a clear day, you can look eastward across the Piedmont to the Coastal Plain beyond, and westward across the Valley and Ridge Province to the Appalachian Plateaus. The Appalachian Trail from Maine to Georgia also follows the ridge, crisscrossing the highway in many places. This trail affords an opportunity to turn the clock back and to see this part of the Blue Ridge as people saw it long before Skyline Drive was there. Even an hour or two on this well-beaten path through the forest will be long remembered. For long hikes involving the use of shelter cabins, advance reservations are necessary.

The Geologic Story of Shenandoah

The general sequence for the Blue Ridge applies in the Shenandoah National Park area. However, certain events were more significant in the northern section than elsewhere. Fortunately, a good source of information is available. In 1976, Gathright and his co-workers prepared an account that makes Shenandoah geology understandable.

Beginning slightly more than a billion years

FIGURE 18-8

Trail marker on the famous Appalachian Trail in Shenandoah National Park.

FIGURE 18-9

Residual boulders of Old Rag Granite on top of Old Rag Mountain ("Ole Raggedy"), with the Piedmont far below. (Photo by M. T. Lukert.)

ago, granitic magmas intruded into older rocks — gneisses of the Pedlar Formation. The granite, which Lukert (1982) determined to be about 1.1 billion years old, was named the Old Rag Granite, after Old Rag Mountain ("Old Raggedy"), where the granite is well exposed.

Prolonged erosion finally stripped off the older rocks into which the magma was intruded, exposing the Old Rag and other granitics. On this Late Precambrian erosion surface, basaltic lavas poured out and, with interbeds of volcanic ash and clastic sediments, reached a thickness of as much as 2000 feet (610 m) and covered all but the higher hills. This Late Precambrian Catoctin Formation, now mainly metamorphosed basalts, is of special interest because, as Gathright (1976) reports, it is resistant and caps many of the higher mountains in the park. Until recently, lacking evidence to the contrary, it was assumed that the lavas flowed out over land areas. However, pillow lavas that were recently found indicate that at least at times the eruptions occurred beneath the sea (Michael Lukert, personal communication, 1982).

The Precambrian landmass then rifted, and these volcanics formed the floor of the developing Appalachian Geosyncline. During the Cambrian, sediments from nearby highlands washed into the resulting ocean, forming the conglomerates, sandstones, and shales of the Chilhowee group. These rocks were later metamorphosed to metaconglomerates, quartzites, and phyllites but are usually not as resistant to erosion as the Precambrian granite and greenstones that underlie some of the higher ridges and peaks in the park. Deposition of later Paleozoic rocks in the area was periodically interrupted by tectonic disturbances, as already noted. During and after the folding, thrust faulting, and metamorphism, near the end of the Paleozoic, erosion stripped off most of the Paleozoic rocks from the Blue Ridge. There are, however, a number of Cambrian quartzite ridges along the western border; associated shales are generally covered by alluvium. Also, Cambrian beds outcrop in a few places along Skyline Drive, notably near Jeremys Run Overlook in the northern part and for some distance north of the south entrance.

Triassic events are sometimes overlooked because much of the evidence is found beyond the Blue Ridge. Streams carried boulders and gravel down the steep east slopes of the Blue Ridge and deposited them in various basins, mainly in the Piedmont. Igneous activity, which was extensive during the Triassic, continued into the Jurassic (Gathright, 1976). Diabase dikes formed in the fractures in Blue Ridge rocks, although few have been found within the park.

Since Triassic time, erosion and isostatic — and perhaps tectonic — uplift have been the dominant processes. As indicated earlier, there are alternative interpretations of the erosional sequence. Several high-level flattish areas, notably at Big Meadows, have through the years been regarded as remnants of a widespread erosion surface (peneplain), one that was uplifted and largely destroyed during later cycles of erosion. The evidence, chiefly in the Valley and Ridge and in the Piedmont, appears to support this time-honored interpretation — that the Appalachians have been eroded down and then uplifted and eroded, at least twice. Blue Ridge rocks are considerably more resistant to erosion than the adjacent rocks; therefore, the Blue Ridge stands high above the Piedmont and the Valley and Ridge.

Pleistocene glaciers did not reach the Blue Ridge Province; however, the cold climate is clearly recorded in Shenandoah. Large boulder fields on the slopes and huge taluses at the base of cliffs are the products of frost wedging.

Visitors may elect to see the park from Skyline Drive, stopping at the many overlooks. Those interested in the geology should obtain Gathright's *Geology of Shenandoah National Park,* which contains a detailed road log beginning at the north entrance and ending where

FIGURE 18-10

Gneiss outcrop at Hazel Mountain Overlook on Skyline Drive.

Skyline Drive changes its name to Blue Ridge Parkway, 105.2 road miles (169.3 km) to the south. At Signal Knob Overlook (mileage 5.7; 9.2 km), red sandstone overlies vesicular basalt in which the pyroxenes have been partially altered to pistachio-green epidote. Also, good columnar jointing in a younger basalt flow is well exposed. At Hogback Overlook, you will see typical Precambrian granodiorite, in which ancient fractures are outlined with iron-oxide stains. From Thorofare Mountain Overlook, you will see Old Rag Mountain (3291 feet; 1003 m) which stands alone east of the crest of the Blue Ridge. These few stops provide merely an introduction to the many geologic points of interest along Skyline Drive.

Although some visitors are content with the views from the overlooks, others wish to really become a part of Shenandoah by hiking the trails that lead off from Skyline Drive. In addition to the Appalachian Trail, there are many others that test, to varying degrees, such devotion. The trails up to the top of Old Rag Mountain are perhaps the most strenuous, and advice from a park ranger will be helpful in selecting the best route. If you are on the Ridge Trail, you will climb a natural stairway. The steps are columnar-joint blocks in a dike. Here, the sheer walls that rise above you are of Old Rag Granite, much more resistant than the diabase dike.

Of the others, the Stony Man Self-Guiding Nature Trail is especially interesting, as the many geological and botanical features are identified by numbered markers. At one point, a prospect pit marks the spot where an early optimist had vain hopes of mining copper. From the cliffs at the end of the trail you have sweeping views of the Shenandoah Valley far below.

FIGURE 18-11

Recumbent profile of Stony Man, at the end of Stony Man Trail. (Photo by National Park Service.)

Shenandoah is now a beautifully forested area, consisting mainly of hardwoods of many species. When you are there in the forest, perhaps on the Appalachian Trail, observing the animals and the birds, recall that without the Park Service this would be a disaster area.

BLUE RIDGE PARKWAY

Our discussion of park areas thus far has been confined almost entirely to the national parks and monuments. Because there are many geologically interesting features along or near the Blue Ridge Parkway and because it links Shenandoah with Great Smoky Mountains National Park, brief mention of the parkway seems appropriate here.

For most of its 469 miles (756 km) southwestward from Shenandoah National Park, the parkway follows the main Blue Ridge; in the southern section, where there are many ridges, it skirts around the south side of Mt. Mitchell (6684 feet; 2038 m), the highest point in eastern United States, and ends at the Oconaluftee entrance of Great Smoky Mountains National Park, in western North Carolina.

The parkway, with its visitor centers and many overlooks, provides an excellent opportunity to become acquainted with both the natural and human history of the region. There are restored homesteads with bear-proof hog pens, water-powered flour mills, and centers where the old handcrafts are taught and practiced. Near Lexington, Virginia, a short distance from the parkway, is the famous Natural Bridge, already mentioned; it holds historic and geologic interest. George Washington made the first survey of the land where it is located, and it was later owned by Thomas Jefferson. Like the Luray Caverns farther north, Natural Bridge is in the Valley and Ridge Province west of the Blue Ridge.

Farther south, at milepost 331, the Park Service maintains the Museum of North Carolina Minerals. Mineral specimens are attractively displayed here, many of them from nearby Spruce Pine, where reputedly there are more different minerals (including rare gemstones) than anywhere else in the United States.

The mountains, and consequently the parkway, are higher here — more than 6000 feet (1829 m) in places — and the several ridges are in full view as you drive along. You may become aware of the bluish haze that gives the Blue Ridge its name. The source of much of the haze is apparently the forests of hardwoods; the moisture given off by the trees and other plants is carried upslope by the winds, where it condenses.

A large dome will loom up on the south side of the parkway in the Mt. Pisgah area. This is Looking Glass Rock, which was rounded off by exfoliation; when wet the reflections suggest a looking glass. Highway cuts in this area are generally larger than in the northern section, and a display of distinctly banded gneiss can

FIGURE 18-12

Spectacular exposure of banded gneiss, near Milepost 419 on Blue Ridge Parkway.

be seen near milepost 419. It is worth a stop and a picture.

Allow sufficient time for the parkway. Although expertly engineered, it is a winding road and is speed controlled. With the aid of the Park Service's map and road guide, your parkway experience could be one of the best.

If you are there in the spring or early summer, the rhododendrons, mountain laurels, and azaleas will be in bloom. If the trees are in full color in the fall, your extra rolls of film may not be enough.

GREAT SMOKY MOUNTAINS NATIONAL PARK

The park sits astride the Tennessee – North Carolina border in the Great Smoky Mountains, the most rugged and spectacular range of the Blue Ridge. The range forms the drainage divide between the Atlantic and the Gulf of Mexico and contains 16 peaks over 6000 feet in altitude, making this area the highest in the eastern United States. Air masses moving inland from the sea cool as they rise over the

FIGURE 18-13

Typical Blue Ridge log house, restored at Oconaluftee Visitor Center, Great Smoky Mountains National Park.

range and release large amounts of precipitation; thus, this area has the highest rainfall in the eastern United States. Precipitation ranges from 50 inches in sheltered valleys to over 82 inches on Clingmans Dome.

The park, consisting of almost 159,000 acres, was established in 1926. It is open throughout the year and is not far removed from heavily populated areas; consequently, more people visit Great Smokies than any other national park. If you drive the Blue Ridge Parkway you enter the park on the south side, at the Oconaluftee entrance and visitor center, in North Carolina. On the north side of the park, near Gatlinburg, Tennessee, is the Sugarland visitor center and park headquarters, on the West Prong of Little Pigeon River. U.S. Highway 441, which climbs up and over Newfound Gap, is the connecting link between the two centers.

Forney Ridge Road winds southwestward along the divide from Newfound Gap to Clingmans Dome (6643 feet; 2026 m), the high point in the park. This side trip is a "must"

FIGURE 18-14

View from Maloney Point on Tennessee Highway 73, east of Fighting Creek Gap. The cross section shows the Greenbriar Fault, an important structure in the Great Smokies. (Sketch by Philip B. King, U.S. Geological Survey Professional Paper 587, 1968).

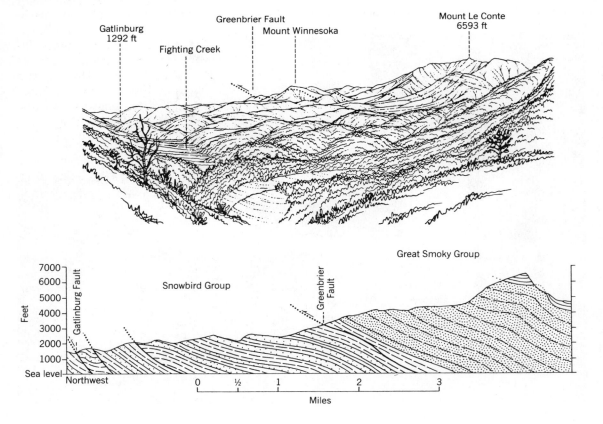

because of the superb view from the tower, where you look out over the treetops. Most of the park's forests are hardwoods, but on this mountaintop the climate is similar to that of central Canada, with similar balsam forests. On this Clingmans Dome side trip, allow an extra half-hour for a "bear-jam," where the little black moochers hold up traffic. Feeding the bears, however, is prohibited, as it is ultimately harmful to them.

Cades Cove, in the Tennessee section of the park, is of special interest, from both the human and geologic standpoints. An unusual set of geologic circumstances produced a broad, level valley surrounded by mountains and underlain by fertile soils ideal for farming. Early settlers discovered this and other remote areas in the Smoky Mountains and developed a nearly self-sufficient life-style and their own culture. In Cades Cove the life-style of the early mountain people is preserved, with original log houses, an overshot wheel still grinding corn at Cable Mill, and a general store. During the summer, you will have the opportunity, rare in these times, to see a walking plow in operation, pulled by a mule.

The Motor Nature Trail south of Gatlinburg follows one of the old sled roads. It is mainly a low-gear, 10-miles-an-hour road that crosses many steep, narrow ridges. From parking areas, trails lead off to Grotto Falls, Rainbow Falls, and other points of geologic interest.

The Geologic Story of the Great Smokies

Deep soils and thick forests conceal a complex assemblage of rocks and geologic structures in the Great Smokies. Many years of intensive study were required before the geologic story could be pieced together. Although pioneer investigations by Keith (1895, 1896) and others laid the groundwork, it was not until the 1960s that King and others of the U.S. Geological Survey published the first comprehensive geologic history of the region.

The broad picture of geologic events parallels that of Shenandoah Park about 300 miles (480 km) to the north. The timing and type of deformation to affect both areas are similar except that much broader and more intense deformation occurred in the Great Smoky Mountains. The oldest rocks, designated by King and others (1968) as the Earlier Precambrian Basement Complex, consist of gneisses, schists, and granitic rocks exposed along the south-

FIGURE 18-15

View across the valley of Cades Cove (see Figure 18-16). (Photo by R. J. Larson.)

FIGURE 18-16

Generalized cross section across Cades Cove, showing thrust fault. (Adapted from King, 1964.)

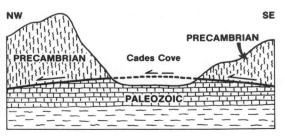

FIGURE 18-17

The Sinks on Little River near Gatlinburg, Tennessee. (Photo by National Park Service.)

eastern boundary of the park. These rocks have experienced multiple episodes of metamorphism, making the original relationships difficult to determine. The "radiometric clocks" in minerals have also been reset and reflect the younger episodes of deformation; some minerals that were not recrystallized yield dates of earlier metamorphic events, about a billion years ago.

By far the most abundant rocks in the park are the younger Precambrian clastic sedimentary rocks and their metamorphic equivalents. Virtually all of the summits and waterfalls in the park are on this extensive group of rocks — the Ocoee Series. Although all have been metamorphosed in varying degrees, the original sedimentary formations are recognizable, and the mapping was done on this basis. Metamorphism was most intense to the southeast

where the deforming forces originated. For example, shales that were converted to slates in the northwestern part of the park grade to phyllites and even mica schists elsewhere. The prominent Thunderhead Sandstone formation is locally a full-fledged quartzite, as in the parking area below Clingmans Dome.

Cambrian-age marine clastic rocks unconformably overlying the Ocoee rocks make up the Chilhowee Group that is exposed on Chilhowee Mountain. These same rocks can be followed northward to Shenandoah National Park, indicating that similar conditions existed in many areas of the geosyncline. The oldest known fossils from the Great Smoky Mountains — trilobites and other invertebrates — have been collected from the Chilhowee Group. Fossils, however, need not be the actual remains or impressions of the once-living organisms but can include any physical record left by an organism. Thus, the remains of vertical tubes infilled by sediment in the sandstones and quartzites are evidence that a primitive Cambrian-age sea worm (*Scolithus*) burrowed through these sediments. Cambrian and Ordovician limestones record an increasingly deep and extensive sea in the geosyn-

FIGURE 18-18

Thunderhead quartzite, Clingman Dome Parking Area on Forney Ridge.

cline and on the North American continent. These rocks are abundantly exposed in the Valley and Ridge Province just west of the park but occur only in Cades Cove within the park. Younger Paleozoic rocks have been removed by erosion or perhaps were never deposited in the area.

Faults are the dominant structures in the Great Smoky Mountains. A fault system, a series of large thrust faults, underlies the mountains and surfaces along the northwestern flank, generally paralleling the crest of the range. The mapping of the faults was a formidable task because almost everywhere they are buried beneath alluvium or slope wash. However, it is clear that the Precambrian rocks were thrust north-northwestward up and over the Paleozoics which, in the adjacent Valley and Ridge Province, have been reexposed by erosion.

Faulting occurred here at different times. Some of the faults, like the Greenbrier Thrust, affect only rocks of Ocoee age and are believed to be Early Paleozoic. Part of the movement along the Great Smoky Fault may also date from this same Ordovician Orogeny that affected the Blue Ridge and Piedmont provinces, but most of the movement occurred in Late Paleozoic time.

Convincing support for thrust faulting is found within the Smokies, in Cades Cove in the park, and in Tuckaleechee and Wear coves north of the boundary. Here, the thrust sheet has been locally stripped off, exposing the Paleozoic limestones and dolomites beneath, as shown in the cross-section of Figure 18-16. In Cades Cove, we can look down through a *fenster* or window and see the rocks that are elsewhere in the park covered by the Precambrian rocks of the thrust sheet.

The Great Smoky fault system is unquestionably the big structure in the park and in the entire Blue Ridge Province. The fault can be traced for hundreds of miles, from Virginia to Alabama (Rodgers, 1953). Thornbury (1965) reported as much as 35 miles of displacement;

FIGURE 18-19

Smoky Mountain Fault (arrow) beneath overhang; Precambrian resting on Ordovician rocks. (Photo by R. J. Larson.)

more recently Cook et al. (1979) indicate that at least 74 miles (120 km) of Late Paleozoic displacement must have occurred. The incredible amount of mechanical work accomplished during this orogeny would seem to be an appropriate consequence of the collision of major plates like those containing the continents of North America and Africa.

Folding both preceded and followed the thrust faulting, in places warping the thrust planes. Further complicating the pattern, there was additional faulting, this time by steep-angled faults that offset the older structures. Then erosion began to carve the rugged terrain of today. Many of the streams developed their valleys in the fractured rocks of the fault zones; others eroded their valleys in the less resistant formations. A number of high ridges are held up by quartzites, the most resistant of rocks.

Clingmans Dome, like most of the other ridges, is rounded rather than sharp. Peregrine Peak and the Chimneys, with their sharp, craggy tops, stand out in contrast.

Erosion cannot be the only process operating for the past 225 million years that accounts for today's rugged mountain topography. Erosion without additional uplift should reduce the landscape to 10 percent of its former height in 18 to 20 million years (Ahnert, 1970,

FIGURE 18-20

Close folding in late Precambrian rocks. (Photo by R. J. Larson.)

and Hack, 1982). Thus, the Appalachians would be a rolling plain at or close to sea level if vertical adjustments, either tectonic or isostatic, had not occurred. Hack (1982) believes that these adjustments must be going on now or must have occurred very recently in order to account for the high elevations in the Blue Ridge.

Geomorphic features of special interest include the Sinks on the Little River west of Gatlinburg. Ordinarily the term *sink* refers to solution sinkholes, generally in limestone; here, the Sinks are plunge-pits formed when the river shortened its course by cutting through a meander neck.

Alum Cave Bluffs south of Mt. Le Conte is one of the many points of interest available to the hiker. It is not a true cave formed by solution but is merely a large overhang of highly resistant metamorphic rocks. The trail to Alum Cave Bluffs also leads to one of the *balds,* conspicuous but incompletely understood features in the Great Smokies. Some of the balds are grassy and seem to have a degree of permanency about them, perhaps maintained by exposure to prevailing winds. Other "heath balds" or "laurel slicks" are covered by an impenetrable tangle of shrubs, mainly mountain laurel and rhododendron. The heath balds become smaller as the surrounding forest encroaches. But why are they treeless in this land of trees? Although some are on the tops of mountains, others are lower down. The highest mountain in the park, Clingmans Dome, is heavily forested. Therefore, a climatic cause of the balds, involving a colder, more rigorous climate during the Ice Age, appears untenable. Perhaps they are related to landslides, blowdowns, or fires, either lightning fires or fires set by Indians for lookouts. Whatever their origin, *bald* is a complimentary term here in the Smokies where the wild azaleas bloom.

Although glaciers did not affect the Smokies, the climate of that time was significantly colder than now. Snowline never lowered far enough to intercept the high peaks in the

FIGURE 18-21

Phyllite in road cut near Look Rock, Foothills Parkway, west of Cades Cove.

Smoky Mountains, but colder temperatures lowered timberline (the elevation above which the climate is too rigorous for trees to survive). The entire area, and especially the regions of exposed rock above timberline, were subjected to intense frost wedging that loosened blocks of bedrock and concentrated them in the boulder fields and talus slopes of today. Just as at Shenandoah, the colder climates allowed more northern varieties of trees like spruce and fir to inhabit the area, many of which survive today in the higher areas as relics of the Ice Age.

FIGURE 18-22

Chimney Tops Mountain, one of the true peaks in the Great Smokies.

Human History

The earliest known occupants of the Great Smoky Mountains were the Cherokee Indians, who roamed the area for centuries before the white man came. The settlers moved in and occupied the choice areas, mainly the valleys and coves. Later, the U.S. government relo-cated most of the Cherokees and confined the remainder to the Cherokee Reservation in the vicinity of Cherokee, North Carolina, south of the national park. Here, park visitors enjoy the displays of Cherokee dancing, crafts, and arti-facts. Thus, they see the Smokies of today, of yesterday in Cades Cove, and still earlier, the Smokies of the Cherokees.

View across Frenchman Bay from the top of Cadillac Mountain, in Acadia National Park, Maine.

CHAPTER 19

THE
NEW
ENGLAND
PROVINCE

The New England Province is related to and is sometimes regarded as an extension of the Appalachian Mountain provinces. Therefore, the sequence of plate movements outlined in Chapter 18 is not repeated here. However, because of more frequent and more extensive igneous activity, the landforms are different, and Fenneman (1938) designated the area east of the Hudson River and Lake Champlain valleys as a separate province. Also included in the New England Province is the Reading

FIGURE 19-1

Beginning 2 to 3 million years ago, huge glaciers moved down out of Canada and relandscaped the New England area. With each advance, more and more Canadian rocks and boulders were left strewn over the land. In clearing their fields, the early settlers piled the stones along the borders; thus the stone fences are of geologic and historic as well as scenic interest.

As Robert Frost's neighbor said, "Good fences make good neighbors".

FIGURE 19-2

Mt. Monadnock, southwestern New Hampshire. (Photo by J. S. Shelton; reprinted from Thornbury, 1965, by permission of John Wiley & Sons, Inc.)

Prong which extends across northern New Jersey into eastern Pennsylvania, ending at Reading.

The province is mainly an upland area, a low plateau that has been dissected to varying degrees. This uplifted peneplain has several monadnocks rising above it, including Mt. Monadnock in southwestern New Hampshire.

Many of the world's famous mountains — such as Everest, McKinley and Whitney — are modestly or vastly higher than 14,000 feet (4242 m). What then makes Mt. Monadnock, considerably less than 4000 feet high, a mountain of distinction? Apparently, it is "because it was there" when the early settlers arrived in the area. According to newspaper accounts, more than 400 people were tramping up the mountain on a single day in August of 1860. Earlier, as reported by Carter (1831) in his *Geography of New Hampshire,* people hike to the top "for the sake of the wide prospect which they can there behold."

Ralph Waldo Emerson paid tribute to the mountain in his 1847 poem "Monadnoc".[1] Henry David Thoreau hiked to the top on four occasions and kept a fascinating "natural history journal." It was Thoreau who, pointing out that the Indians regarded the mountain as a sacred place, suggested that it be set aside for the benefit and enjoyment of everyone.

---

[1] One of several spellings of the Algonquin Indian word that means "a mountain that stands alone."

In his *Annals of the Grand Monadnock*, Chamberlain (1936) also mentions that William Morris Davis selected Mt. Monadnock to represent all mountains that stand prominently above old erosion surfaces. Davis was an eminent geographer (now called geomorphologist), and his designation was followed.

Many monadnocks are composed of rock that is harder and more resistant to erosion than the surrounding rocks. Interestingly, the prototype is different; the top of Mt. Monadnock is composed of schist that is *less* durable than the granite on the lower slopes: This was pointed out by Professor Joseph H. Perry in the 1904 *Journal of Geology*. Perry maintained that this and other mountains of the area owe their existence to their position (location) and not to the superior resistance of the rocks topping the mountains. The area that lies at the headwaters of several streams is the area least eroded and therefore remains high. Consequently, although the tops of some monadnocks are made up of highly resistant rocks, this is by no means always the case.

The Taconic Mountains, the Green Mountains, and the White Mountains are prominent ranges. The Connecticut River flows southward in a broad valley through the western part of the province. Extending from Rhode Island northeastward through Maine, the Seaboard Lowland section includes low hills, coastal lowlands, estuaries, peninsulas, and islands. On Mt. Desert Island off the Maine coast, resistant rocks rise to 1530 feet above the Atlantic in Cadillac Mountain, the highest point on the eastern coast.

The bedrock geology of the province is complex, indicating a long and involved sequence of events during the Precambrian and the Paleozoic. Although Precambrian granites and gneisses are extensively exposed in places, they are not as widespread as was once thought. Billings (1933) and others have concluded that many of the "Precambrian" igneous rocks are Paleozoic in age and, furthermore, that intrusions occurred at different

FIGURE 19-3

Pillow lava, exposed near Amherst, Massachusetts, was formed when lava flowed into a lake about 190 million years ago. (Pillow lavas recently formed in much the same way when lavas poured into the sea along the southern coast of the island of Hawaii.) (Photo by E. Erslev.)

times, beginning perhaps as early as Late Ordovician. These intrusions, coupled with the compressive forces of mountain building, metamorphosed many of the sedimentary rocks; slates, schists, quartzites, and marbles are widespread, in addition to the granites and gneisses related to igneous activity.

The erosional history of the province is exceedingly complicated. It is clear that in order to expose deep-seated rocks of both Precambrian and Paleozoic age, a vast amount of erosion took place and that with each orogeny, a new cycle of erosion was initiated. Precisely how many peneplains were formed and later destroyed is not known. Furthermore, the number of erosion surfaces still in evidence is not a matter of complete agreement. In addition, Barrell (1913) suggested that it was marine erosion, rather than stream erosion, that formed the "peneplain" surfaces. Although his interpretation received some support, it is no longer acceptable to most geologists. There appears to be general agreement that in the New England area one erosion surface or peneplain is well documented. Uplift of the region caused the rejuvenation of the streams and the development of the topography which was to be remodeled later by the glaciers.

New England is not far removed from the Labrador center of accumulation of ice during the Pleistocene, and each time there was a major advance of the ice, New England was

FIGURE 19-4

Mt. Washington, New Hampshire's highest mountain, is only 6288 feet high, but its weather may be the worst on Earth — it has winds up to 231 miles per hour and temperatures as low as — 60°F. Not clearly shown here are the effects of glaciation; it was overridden by continental ice sheets and later modified locally by cirque glaciers. (Photo by Fairchild Aerial Survey, Inc.; reprinted from Thornbury, 1965, by permission of John Wiley & Sons, Inc.)

one of the areas overrun by the continental glaciers. Locally, the mountains were high enough to support alpine glaciers, as attested to by well-developed cirques. In the White Mountains, one cirque, called the Great Gulf, has a head wall that rises abruptly about 1500 feet (455 m) above its floor.

Glacial erosion was dominant in many parts, particularly in Maine and New Hampshire where rock basin lakes and roches moutonnées abound. Glacial and fluvioglacial de-posits are extensive, including some that were submerged when sea level rose during the last major retreat of the ice. Moraines, *kames, kame terraces,* eskers, and outwash deposits are widespread. Although drumlins are not numerous in most areas, they are abundant in Massachusetts and southeastern New Hampshire.

Fjord-like embayments are characteristic of the Maine coastline. In almost all cases, however, they lack the depth and the near-vertical

FIGURE 19-5

Glacial erratic, so delicately balanced on smooth bedrock that it can easily be rocked back and forth. One of New Hampshire's famous "rocking stones."

walls of true fjords and are interpreted to be drowned river valleys. One true fjord is Somes Sound, on Mt. Desert Island in Acadia National Park.

Stone fences, the hallmark of New England, are indirect results of glaciation. They represent the long labors of the early settlers in their struggle to convert the boulder-strewn landscape into farms. Once cleared of the boulders, in many fields the stony soils were found to be too shallow for good production. Fortunately, there are many exceptions, particularly in the river valleys where the alluvial soils are deep and fertile. Glaciation probably affected the lives of the New Englanders more than any other geologic process.

New England and eastern Canada are facing a major environmental crisis. Many streams and lakes, until recently highly prized for fishing, no longer support fish life. In fact, certain lakes are now devoid of all life. Acid rain is one cause — in some cases the only one — of the problem. The main source of the acid is the sulfur in the coal burned in industrial plants, mainly in the Middle West. Actually, the acid rain problem is widespread; it now appears that all of the United States has been affected to some extent. And serious though it is, the snuffing out of life is only one part of the problem. Man-made structures, of metal and masonry, are being adversely affected.

Although considerable time and money have been spent on studies of the acid rain problem and on possible remedial measures, essentially nothing has actually been done to improve the situation. As this book goes to press, the impression (admittedly unconfirmed) is that those responsible are assuming that if enough studies are made, the acid condition will somehow be neutralized. Consequently, numerous additional studies have been authorized.

The National Academy of Sciences predicts that the number of lakes affected by acid rain will double by 1990, unless remedial measures are taken. For more information on this issue

and on the activities of the National Clean Air Coalition, read Kaufman's article, "Acid Rain Costs U.S. $5 Billion Annually," in *National Parks Magazine* (July 1983).

ACADIA NATIONAL PARK

Most of Acadia National Park is on Mt. Desert Island just off the Maine coast, southeast of Bangor. State Highway 3 south of Trenton crosses the Mount Desert Narrows to the north shore of the island, thence around the north side to Bar Harbor where the park headquarters are located. Large areas outside the park are privately owned and there are private holdings here and there within the park boundaries. Numerous roads lead to points of interest around the craggy coast and inland, including a spur road to the high point, the top of 1530-foot (466-m) Cadillac Mountain. From here, you can see the southern tip of Schoodic Peninsula, about 5 miles to the east, the only part of the park on the mainland. About 25 miles (40 km) southwest of Mt. Desert Island is Isle au Haut, which can be reached by boat from Stonington. For those seeking solitude, this section of the park is a wilderness area. Four other islands near Mt. Desert Island are also parts of the Acadia National Park complex.

Mt. Desert Island, Isle au Haut, the Cranberry Islands, and a host of others form the Acadia Archipelago. Prior to the submergence resulting from the melting of the glaciers, parts of the area formed a coastal plain which is now beneath the Gulf of Maine. The ocean waves have since shaped and reshaped the shorelines to form the seascapes for which Acadia is well known.

Mt. Desert Island is in no sense a desert, as spruce-fir and hardwood forests cover all but the mountaintops. The name originated in 1604 when the explorer Samuel de Champlain first saw the bare-topped mountains which he named I'Isle des Monts-deserts, Isle of Solitary Mountains. Sieur de Monts National Monu-

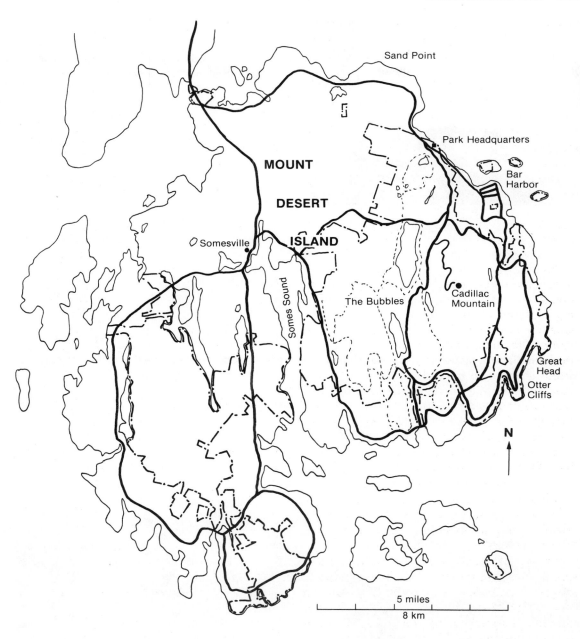

FIGURE 19-6

Map of Acadia National Park.

FIGURE 19-7

Bald Porcupine, like other islands in Acadia National Park, is being eroded by never-ending wave action. (Photo by National Park Service.)

FIGURE 19-8

View from Cadillac Mountain across Frenchman Bay to the Schoodic Peninsula section of Acadia National Park.

ment, established in 1916, became Lafayette National Park in 1919, and then Acadia National Park in 1929.

Unlike Shenandoah and Great Smokies, where most of the rocks are buried beneath residual weathered material and soil, Acadia's rocks are well exposed, not only in the high sea cliffs but inland. This is due mainly to the glaciers, which not only scraped off all the residual material but dug deep into the unweathered bedrock.

For an in-depth look into the park and its geology, read Chapman's comprehensive booklet (1970).

The Geologic Story

Most of Acadia's rocks are granites and diorites that intruded at various times during the Devonian. Perhaps the highlight was a great invasion of magma, late in the sequence, one that so weakened the roof rocks that they began to sag and then to sink, en masse, into the magma. As the sinking progressed, there were upsurges of magma to fill in the spaces thus created (Chapman, 1970). Around the borders of the intrusion the rocks were badly shattered and brecciated. Granite magma intruded into the fractures and formed large dikes. Where the fractures were curved, so-called *ring dikes* formed; several are now exposed on both the west and east sides of Mt. Desert Island. Apparently the latest igneous phase was the intrusion of basaltic magmas that formed the many black dikes that cut across the granites.

The heat from the magmas metamorphosed the rocks into which they intruded. In places, large masses of early Paleozoic siltstones were converted to the dense, flinty rock *hornfels*. Numerous other examples of contact metamorphism are seen in the park.

The geologic record for more than 300 million years prior to the Pleistocene is incomplete. Apparently by Cretaceous time the New England area had been reduced to a peneplain, and the deep-seated rocks discussed

above were exposed. According to Chapman (1970), these rocks were sufficiently more resistant than those in the surrounding area that they remained higher — as the Mt. Desert Range, actually a large monadnock.

During the Tertiary, uplift rejuvenated the streams and they deepened their valleys. Most of the valleys were aligned in a north-south direction, generally parallel to the main fractures. Thus the stage was set for the coming of the glaciers that formed today's landscapes.

At times during the past 3 million years, continental glaciers developed around three main centers in Canada and pushed southward into northern United States. The Labrador center was the source of the ice that covered the New England Province. Pleistocene glaciers rode over the Mt. Desert Island area and rounded off the high ridges. Late in Wisconsin time, according to Chapman (1970), the ice piled up on the north side of the Mt. Desert Range and only tongues or lobes pushed farther south. It is his interpretation that these ice lobes that followed the valleys are responsible for much of the erosional topography of today — the deep, steep-sided valleys and the elongate rock-basin lakes. One lobe pushed all the way across what is now Mt. Desert Island and scoured deep into the rocks, thus forming a fjord known as Somes Sound. At one point, the wall of the fjord rises more than 1000 feet (303 m) above the floor.

Other ice lobes excavated many elongate rock-basin lakes such as Eagle Lake and Jordan Pond east of Somes Sound, and Echo Lake, Long Pond, and Seal Cove Pond to the west. If the excavations had been deeper and longer by a few miles, they too would have been converted into fjords.

All of Mt. Desert Island was glaciated during one advance or another. As the glacier ground its way across the mountains, it not only rounded off the tops but also sheared off the sides, forming elongate hills called roches moutonnées. Typically they are asymmetrical in long dimension, with a gradual slope formed

FIGURE 19-9

Somes Sound, believed to be the only true fjord along the New England coast. It was excavated by a lobe of the Wisconsin glacier that extended far beyond the main ice front.

FIGURE 19-10

South Bubble, a roche moutonnée reshaped by ice that rode over its top, from right to left. Note large erratic on skyline which was carried almost to the top of South Bubble.

by abrasion as the ice moved up and over the top; the down-glacier end is steep and jagged where, by quarrying, the glacier lifted out and removed blocks of rock. North and South Bubbles north of Jordan Pond are outstanding examples.

Glacial grooves, or striations, and glacial polish are seen in a number of places, particularly on the bare roches moutonnées. Glacial erratics, boulders carried far from where they were quarried out of the bedrock, are abundant. Look for the one perched precariously on the slope of South Bubble, plainly visible from the Loop Road.

Morainal deposits are found here and there, but, as most of the glaciers pushed beyond what is now Mt. Desert Island, many of the moraines are now covered by sea water. Some of those on the island are not readily seen because they are obscured by dense vegetation.

The Battle with the Sea

The glaciers came, did their work, and vanished, but the sea is ever present. It batters the Acadian shoreline century after century, relentlessly. By hydraulic action, blocks of granite are pulled from the cliffs and used as abrasive tools to pulverize other rocks. Weak fracture zones are converted into sea caves. The more resistant rocks are left extending out into the water as "headlands," or "heads," or "points." Great Head and Schooner Head on the east coast are typical. A large sea cave that extends back into the east face of Great Head can be readily seen from a boat. A short distance south of Great Head is Thunder Hole, which can be reached by a short hike from the parking area. When the seas are heavy and a giant wave suddenly compresses the air and then displaces it from the cave, the sound is much like a clap of thunder.

FIGURE 19-12

Thunder Hole, a sea cave formed by the battering of waves, Acadia National Park.

FIGURE 19-11

A "northeaster" hits the shores of Acadia National Park. (Photo by A. Hathaway.)

In places, sea caves form along the sides of headlands and when enlarged sufficiently penetrate entirely through the rock wall, forming a sea arch. At Sand Point on the north shore, the sea cave was extended back along a prominent vertical fracture, forming the high, narrow arch (Figure 19-13 in color section).

As the waves continue to undercut the cliffs, the wave-cut platform at the base is enlarged; as the cliffs retreat with each rockfall, they become more prominent. At Otter Cliffs south of Thunder Hole the 107-foot (33 m)-high cliffs rise vertically above a well-developed wave-cut bench or platform.

For most park visitors, the life in the ocean remains largely a mystery. At low tide, however, the tide pools and part of the wave-cut platform afford glimpses into this fascinating world. The sea snails, periwinkles, crabs and barnacles are but a few of the myriad forms that live in the tide zone.

The glaciers of the past and ocean of today have fashioned Acadia National Park. It was very attractively done, especially so when decorated in fall colors.

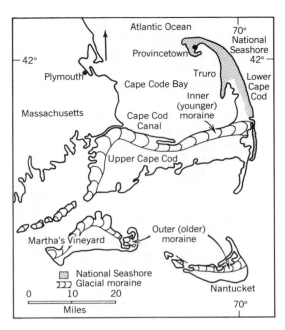

FIGURE 19-14

Map of Cape Cod area.

CAPE COD NATIONAL SEASHORE

Cape Cod National Seashore is located along the eastern end of Massachusetts' prominent, oddly shaped peninsula, about 65 air miles (110 km) southeast of Boston. It is within easy access of a large population, as the 5-million yearly visitation suggests. Quaint and picturesque villages — some very old — shipwrecks, lighthouses, and sailboats add both flavor and variety to the beaches, dunes, and nearby cranberry bogs.

The attraction of Cape Cod was so strong that private and commercial development was proceeding at an alarming rate, and even many of the fiercely independent residents whose ancestors had lived there for hundreds of years realized that some form of protection was needed. In 1959 a bill was proposed by Massa-

chusetts Senators John F. Kennedy and Leverett Saltonstall which would protect a 40-mile (64-km) stretch of coast as a national seashore, but would also allow recreational activities to continue and homeowners to remain within its boundaries. On August 7, 1961, President Kennedy signed into law the park he had proposed three years earlier!

Cape Cod is a uniquely shaped appendage of land that, as we shall see, also has a unique geologic history. On maps, this land area resembles an upturned right arm with a flexed muscle. The "bicep" connected to the mainland is upper Cape Cod, and the "forearm" and "fist" are lower Cape Cod, which is the location of the national seashore. The "fist" is a sandy extension of land called a *spit* that curves around to create the ideal harbor near the town of Provincetown. Here the *May-flower* stopped in 1620 before proceeding to nearby Plymouth on the mainland, and here also was anchorage for some of the Yankee whaling and fishing fleets. Other pieces of the geologic puzzle of this area are nearby Nantucket and Martha's Vineyard, islands lying south of Cape Cod (Figure 19-14).[2]

Before the Glaciers

The basement rocks, a few hundred feet below sea level, are granites, schists, and gneisses. Like the rocks of parts of the northern Appalachians, they are related to, and were once connected with, the rocks of Morocco in northern Africa (Rodgers, 1972). According to Schenk (1971), these rock masses — along with parts of Nova Scotia — are "microcontinental fragments broken off Africa." When the huge landmass split apart, perhaps 200 million years ago, sea water filled the gaps.

By Late Cretaceous time, deltaic, swamp, and marine clay deposits, such as those exposed on Martha's Vineyard, were being deposited on the newly formed coastal plain. Similar Cretaceous deposits probably underlie Nantucket and Cape Cod. Tertiary coal deposits exposed beneath the sands near Provincetown, in combination with other deposits from nearby areas, indicate that alternate marine and nonmarine conditions persisted through the Tertiary Period. At the beginning of the Pleistocene there probably were low hills parallel to the present coastline. These hills would significantly influence the glaciers that were soon to cover the Cape Cod area.

Arrival of the Glaciers

Although vast continental glaciers originating in Labrador moved southward across New England and retreated a number of times, the

---

[2] Strahler's *A Geologist's View of Cape Cod* is a widely used reference. See also Melham, 1975; Kaye, 1980; and Oldale, 1980.

Wisconsin glaciers left the best record and created most of the major geologic features of this area. As the glaciers accumulated additional ice, sea level was lowered, gradually exposing much of the continental shelf. Between 25,000 and 16,000 years ago, the ice sheet moved across Cape Cod and temporarily stabilized on what is now Nantucket and Martha's Vineyard (Oldale, 1980). Topographic irregularities produced different rates of ice movement and resulted in the extension of some sections of the glaciers — called lobes — farther than others, creating a scalloped ice front. Thus, some of the arcuate features here were glacier-caused but others were reshaped by the sea.

Moraines, some of them large, were formed at the front of the glacier as the frontal section melted and the rock material was freed from the ice. The prominent morainal ridges across Martha's Vineyard and Nantucket indicate that the ice front was essentially stationary for a long period. Meltwater carried materials out away from the ice front and deposited them on the *outwash plain*. Here, isolated blocks of ice that were partially buried in the outwash deposits eventually melted, leaving depressions called *kettles*. Those that were later filled with water are now *kettle lakes*.

The glacier began to retreat, and about 15,500 years ago tundra plants grew on Martha's Vineyard (Oldale, 1980). Later, the ice front was located on Upper Cape Cod where large morainal ridges were built up. Much of lower Cape Cod (the "forearm" part) was located between the two ice lobes and received mostly sandy outwash debris. Additional retreat left a depression between Cape Cod and the ice front (now Cape Cod Bay) that filled with meltwater and formed a glacial lake that spilled through what is now Cape Cod Canal.

When the glacial ice retreated farther, it left a vast, rolling landscape on the continental shelf. Tundra vegetation grew initially, later followed by pine and spruce, and finally by the maples and black oak hardwoods of today.

Return of the Sea

Shrinkage of glaciers freed vast quantities of water, and once again the level of the sea began to rise; the outer shelf was inundated first. Still the water levels rose, rapidly at first and then at gradually reducing rates. Lower hills became islands and then *shoals* as rising water levels and erosion by waves overwhelmed these areas. Whole forests were submerged, as drowned stumps off Provincetown and other areas indicate. Mammoth and mastodon teeth dredged from the continental shelf indicate that these giant animals roamed these woodlands several thousand years ago.

About 6000 years ago, a morainal area southeast of Cape Cod was inundated and became the important fishing area known as Georges Bank. The wave action then concentrated on the next land area to the west, Cape Cod. At this time Cape Cod must have extended at least two miles farther into the Atlantic (Strahler, 1966). Presently, Great Beach, the continuous beach facing the sea on the east side of lower Cape Cod, is retreating at a rate of about three feet each year.

Storm waves that crash onto Cape Cod's beach wash away the loose glacial sediment that makes up the sea cliffs. Some of the sediment is carried offshore out of reach of the

FIGURE 19-15

Wave-cut cliff in glacial deposit, illustrating two of the geological processes involved in Cape Cod National Seashore. (Photo by M. W. Williams, National Park Service.)

waves, but much is moved laterally by the longshore drift generated when waves impinge at an angle rather than parallel to the shore. Sand washed up onto the beach is carried laterally a short distance along the beach before backwashing into the ocean again. Each wave repeats the process, and the net effect is that large volumes of sand are moved each year and, where the water deepens, longshore drift builds extensions of the beach as shoals and spits. Longshore transport near Truro on the north end of Cape Cod carries sand northward, where it is added onto the impressive spit that helps form the harbor at Provincetown. South of Truro, sediment moves southward, building Nauset and other spits.

Unfortunately for the longevity of Cape Cod, for each acre of land added by the sea, two acres are destroyed each year. At the present rate of erosion, the northern part of Cape Code will become an island in 4000 to 6000 years when the narrow section south of Truro is severed by the sea (Kaye, 1980). More islands will form, and each will in time be reduced to shoals unless glaciers reverse the process by causing sea level to lower. Fortunately, for many generations we will be able to enjoy the rich cultural and natural features of this changing landscape.

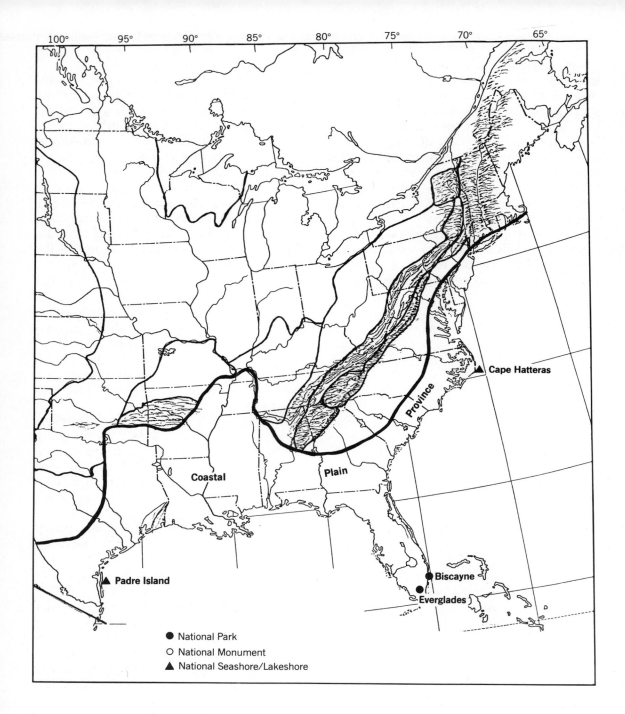

PLATE 20
Coastal Plain Province.

CHAPTER 20

COASTAL PLAIN PROVINCE

The Coastal Plain Province consists of the seaward-sloping lowlands along the Atlantic Ocean and Gulf of Mexico, and the submerged section, the continental shelf. Fluctuations in sea level and uplifts of the land have both affected the area of the two sections.

In this province, the Park Service extends its domain beyond the continent to include Cape Hatteras and Padre Island seashores, Biscayne National Park and the Florida Bay area of the Everglades — all parts of the continental shelf.

At the beginning of the Cenozoic Era, the shoreline of the continent was inland from its present position, particularly in the Mississippi Embayment, which extended northward to the southern tip of Illinois. Part of the coastline is extremely irregular, with deep indentations where submergence has caused the drowning of river mouths; Chesapeake Bay is a good example.

Essentially all of the rocks of the Coastal Plain fall into one of three groups. Discontinuously around the inner border of the province are marine sedimentary rocks deposited when the Cretaceous sea invaded this part of the continent. They rest on Paleozoics in most places and on Precambrian metamorphics elsewhere. In the middle section, marine Tertiary rocks rest on the Cretaceous and dip gently toward the sea. Along the coast, sediments of Quaternary age form a more or less continuous band of varying width from southern Texas to Long Island.

Salt domes, structures unknown in most areas, are abundant in the coastal section of Texas, Louisiana, and Mississippi, and offshore. (The principles involved in their formation were discussed in connection with the salt anticlines in the Paradox Basin in the Colorado Plateaus Province.) Salt domes are fascinating to geologists, but they are even more intriguing to oil companies that have obtained millions of barrels of oil from these structures.

The development of the vast Mississippi delta is an interesting story summarized by Thornbury (1965). By delta building, this great river has been extending the shoreline out into the Gulf, and it is continuing its work today. Here is a unique environment, one of the very few not now included in the National Park System.

Glaciation affected only the northern area; many of the glacial deposits are beneath the sea, but large moraines cover extensive sections of Cape Cod and Long Island. Farther south, terraces generally believed to be of marine origin are prominent features of the subdued landscape.

Karst features, essentially absent elsewhere in the province, are abundant in central and northern Florida. Sinkholes and the caverns beneath them are widespread features north of Lake Okeechobee. Only in the higher areas are the caverns open; most of the vast cavern system is below sea level.

Coral reefs are attractive features around the southern coast of Florida, and offshore. Extending for about 150 miles (242 km) south and then west, the Florida Keys, an arcuate chain of islands, ends at Key West.

Long, narrow barrier islands, formerly called offshore bars, extend discontinuously around the Gulf Coastal Plain and from New Jersey to South Carolina, along the Atlantic Coast. Farther south, from near Charleston, South Carolina, down to the southern coast of Georgia, a chain of irregularly shaped "sea islands" occupy similar positions offshore, with narrow lagoons separating them from the mainland.

Regarding the origin of these offshore islands, elongate or irregular, the only consensus at present is that there is no consensus, except that the classical interpretation of Douglas Johnson of emergent shorelines is untenable. Papers by Fisher (1968), Hoyt (1970), Otvos (1970), Swift (1975), and others indicate that the problem is not being neglected. Possibilities that have been considered include the formation of coastal dune ridges that are later submerged; the development of complex *spits* on a shoreline of submergence; the emergence of offshore bars; and the shoreward migration of offshore bars. Whatever the theory on the mechanism or combination of processes, it must take into account the rise in sea level that resulted from the melting of Wisconsin glaciers, beginning about 11,000 years ago. Significant also are the changes that the barrier islands are undergoing at the present time.

Clearly they are the results of the work of the waves and the wind. All coasts are lashed periodically by storms, sometimes with hurricane force. The dune ridge superimposed on all barrier islands may be locally shifted landward; both the seaward and the landward beaches are likely to be eroded; in some cases the waves break through, forming new inlets or *passes.*

The Park Service had two objectives in mind when they established National Seashores such as Cape Cod, Cape Hatteras, and Padre Island. One purpose was to provide the public with additional recreation; the other was to preserve the beach areas in their natural state. It soon became apparent that their newly adopted areas were changing — in some cases drastically — with each violent storm. Their response was to initiate a program of stabilizing the dunes by building sand fences, and the shorelines by installing elaborate erosion-control structures. After spending large amounts of money, in certain cases without overwhelming success, the policy was re-examined. The Park Service had not attempted to thwart nature by trying to stop the erosion in Grand Canyon or in the Badlands. The decision was made in 1973 to let nature take its course, even with

this highly dynamic natural system (Dolan and Hayden, 1974). One bonus result from the new policy — it provides the opportunity for scientists to study natural processes at work, without any modifying influences by man.

PADRE ISLAND NATIONAL SEASHORE

The National Park Service broadened its coverage of natural environments by creating national seashores, beginning with Cape Hatteras in North Carolina in 1937. They now dot our shores, from Point Reyes near San Francisco to Cape Cod on the East Coast.

Padre Island National Seashore, established in 1962, occupies the 80-mile-long midsection of the 113-mile-long Padre Island near Corpus Christi, Texas. First called the White Islands because of its almost-white sands, this seashore was renamed around 1800 when Padre Nicolas Balli received a large Spanish land grant that included the island. Actually, Padre Island is a chain of elongate islands separated by narrow "passes" that were broken through during violent storms, only to be sealed shut by sand deposited by longshore currents. But new passes were broken through to once again connect the Gulf of Mexico with Laguna Madre, the long lagoon that lies between Padre Island and the Texas mainland.

The Intracoastal Waterway extends through the length of Laguna Madre; when the channel was dredged in 1949, the dredged material was piled up alongside. These spoil piles became island nesting sites for such waterbirds as the laughing gull, foresters tern, great blue heron, roseate spoonbill, and the white pelican. Because Laguna Madre is shallow and essentially cut off from the gulf, its waters have become increasingly saltier; according to Chief Naturalist Robert G. Whistler (personal communication, 1981) the salinity is at times almost twice that of gulf waters. Therefore, only salt-tolerant plants such as shoalgrass can survive in the marshlands near the lagoon.

FIGURE 20-1a

Padre Island National Seashore: air view southwest across Padre Island. (Photo by R. Whistler, National Park Service.)

FIGURE 20-1b

Padre Island National Seashore: explanatory sketch showing zones and features.

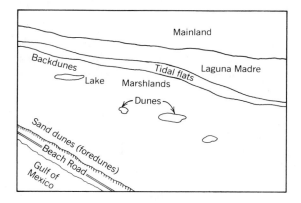

Padre Island ranges in width from a few hundred yards to about three miles. The northern end of the seashore is sufficiently wide for the various geomorphic features to be clearly represented. High sand dunes, the foredunes, form a ridge that rises abruptly above the beach, as shown in Figure 20-1a. In the area beyond the foredunes there are many clusters of dunes, some temporarily stabilized by low-growing vegetation. Many of the interdunal areas are marshlands, but small lakes occupy some of the depressions. Farther west are the back-island dunes, with the tidal flats and Laguna Madre beyond. The best way to see a representative section of the western part of the seashore is to take the Bird Island Basin Road, which leads west a short distance north of the ranger station.

Your first stop should be the visitor center where you can obtain valuable information, including warnings of hazards such as the Portuguese man-of-war jellyfish, stingrays, rattlesnakes, and unexploded ammunition (relics of the days when the U. S. Navy used parts of the island as target sites). Next, get an overview of the island from the observation tower on Malaquite Beach. With binoculars, you can see all of the features mentioned above.

Origin and Development of Padre Island

Padre Island is young, less than 5000 years old (Fisk, 1959). Several mechanisms have been proposed for the origin of barrier islands but no one explanation appears to be satisfactory for all of the world's offshore islands.

For Padre Island, Weise and White (1980) favor the offshore bar origin. They picture a sandbar, a ridge of sand piled up by the waves, in shallow water near shore. With continued wave action the bar is built up above sea level, thus forming the embryonic island that will be widened and heightened by subsequent wave and wind action. But problems arise when we examine this sequence carefully. Anything exposed to the lashing of the waves will be destroyed in time; a low ridge of loose sand would not withstand the onslaught of even one of the hurricanes that occur here every few years. Also, it is significant to note that, although there are submerged bars offshore now, at no point have they been built up to the surface.

Others workers visualize an entirely different sequence. Frank Ethridge (personal communication, 1981) begins the sequence with a long, high ridge of sand bordering the mainland coast — coastal dunes like those found in many places today. Sea level has been rising since the great glaciers began to melt back some 11,000 years ago; eventually the waters of the gulf invaded the area back of the dune ridge, thus forming Laguna Madre and isolating Padre Island from the mainland. Hoyt (1967) and others agree that this explanation is the logical one for the origin of Padre Island, at the same time recognizing that other barrier islands may have been formed in other ways.

As is true of all barrier islands, Padre Island is being reshaped. During heavy storms, the waves erode the sand away from the beach side of the foredunes, thus shifting the beach-dune boundary westward and enlarging the beach area. From time to time, hurricane-driven water overwashes the dunes in places and forms new passes through the ridge. The floodwaters rush across the island and deposit sand as washover fans where the water spreads out over the tidal flats, even out in the lagoon. Thus, the island is being extended westward at the expense of Laguna Madre.

On the gulf side of the island, the wind picks up the sand from the beach and carries it up onto the ridge, where much of it is deposited. Some, however, is carried far beyond the ridge and is deposited on the flats, where it is shaped into dunes. Until the dunes are stabilized, they are moved about by the wind, as "wandering" dunes. The back-island dunes form a low, irregular ridge along the east side of Laguna Madre. In many places the dunes are at least partially stabilized by grasses, low-growing shrubs, and by the long railroad vine. In the interdunal areas, grasses thrive in most places, providing good grazing for cattle prior to the time that the national seashore was established. Live oak trees, once fairly widespread, are almost entirely gone, having succumbed to droughts and to burning by the ranchers — to improve the grass.

When fierce winds blow, the least stabilized sand is carried away, forming "blowouts," bowl-shaped "wounds" on the windward side of the dunes. Once the deflation process

starts, it is likely to continue for some considerable period, sometimes enlarging the blowout to enormous size. A blowout that will probably be there for years is along the Grassland Nature Trail in the northern section of the seashore. An hour or two on this nature trail is time well spent, assuming that you keep a sharp eye out for rattlesnakes.

The *Padre Island Field Guide* by Hunter et al. (1972) contains a road log and other information that will help you make the best possible use of your time. The Malaquite Beach section is off-limits for all vehicles. Most of the roads are travelable by road car, but a 4-wheel-drive vehicle is a must if you want to go far to the south, with stops on Little Shell and Big Shell beaches along the way. You will be well rewarded if you spend two or three hours on the beach near the northern end of the Seashore, at the end of the North Beach Access Road. To the south of the parking area, the beach is for walking only; as a result, burrowing animals such as the ghost crabs, mole crabs, and ghost shrimps have the opportunity to live their lives without being disturbed by cars, trucks, and motorcycles. If you are there at the right time, you will see the beautiful blue Portuguese man-of-war jellyfish. *Do not touch;* if you get stung, *do not rub!* Go immediately to the nearest first-aid station for treatment.

Padre Island attracts birds of many kinds, especially during the winter. One big bird that probably will not be there is the whooping crane. To see it, you will have to go to Aransas National Wildlife Refuge about 50 miles (83 km) north of Corpus Christi. To see these big fellows "gal-loping" low over the water is an experience to long remember.

The animals on and around Padre Island include the coyote, jackrabbit, the Mexican freetail bat, several kinds of rodents, a few Ridley sea turtles, lizards, and snakes. There are also fish in the Gulf and in Laguna Madre. There are many fascinating things to see at Padre Island; surely some of them will be of sufficient interest to distract you from swimming and fishing, at least for part of the time. Padre Island is open throughout the year; winters are delightful except when a "norther" moves in to chill the air.

Most barrier islands are "graveyards for ships." Since hurricanes are a common occurrence, Padre Island has had its share of shipwrecks. Long ago, when the island was the home of cannibalistic Indians, the hazard was much greater. Of the many who were "fortunate" in surviving the storm in 1533, when a Spanish fleet ran aground, only two escaped from the Indians.

EVERGLADES NATIONAL PARK

At the southern tip of Florida, where the waters of the Gulf of Mexico meet those of the Atlantic, is a vast, flat lowland known as the Everglades. Large areas are covered by shallow water — in lakes, freshwater sloughs, estuaries and bays. This aquatic, subtropical environment of sawgrass and mangrove trees is alive with crocodiles, alligators, manatees, roseate spoonbills, anhingas, wood storks, and a host of more common animals and birds. It was for the purpose of preserving this unique environment that about 1,400,000 acres of land and water were established as Everglades National Park in 1947.

The park extends westward from Homestead, where the park headquarters are located, to Everglades on the Gulf of Mexico. On the west side the park extends out into the Gulf and includes the Ten Thousand Islands; on the south it extends across Florida Bay almost to the Florida Keys. Highway 27 leads west from Homestead and then south to Flamingo at the southern tip, with stub roads to Royal Palm, Pa-hay-okee Overlook, Mahogany Hammock and other points of interest.

The land that is now the Everglades was beneath the sea until recent geologic time. Now, after being above the sea for a short period, the Everglades are gradually losing ground (Glea-

son, 1972). Although the boundaries of the park remain constant, the land area is gradually shrinking.

The geology of Everglades National Park is relatively simple. Along the western side, limestones of the Miocene Tamiami Formation lie beneath the swamps and bogs (Gleason, 1972). Apparently no Pliocene sediments were laid down in this area. Most of the park is underlain by the Pleistocene Miami *Oolite,* or oolitic limestone, composed largely of tiny, spherical concretions that resemble fish roe.

The oolite is porous and permeable, and the surface layers are unusually susceptible to chemical weathering. Consequently, the surface is in many places pitted and irregular. On this surface, peat accumulation has been rapid because of the luxuriant vegetation — mangrove trees, water-lilies, and sawgrass. Perhaps this is a present-day coal-forming swamp.

If we probe deeply into the past, we find that beneath the surface rocks are Paleozoic metamorphic rocks similar to those exposed

FIGURE 20-2

Everglades National Park headquarters, from along Anhinga Trail.

north of the Florida Peninsula. This information, obtained from wells, leads to the realization that long before the Miami oolite was laid down, a projection of the continent occupied the same general area as the present Florida Peninsula. That landmass was submerged and deposited upon, most recently by the oolite, and then became emergent again in relatively recent times.

Three words — low, limy, and flat — describe the Everglades of today. The only features that break that level line as you look out across the sawgrass flats are the *hammocks* which rise several feet above the swamps. They are elongate, tree-covered mounds

FIGURE 20-4

Sawgrass flats with tree-covered hammocks rising above the vast plains. (Photo by National Park Service.)

FIGURE 20-3

The Anhinga, or snake-bird or water turkey; also called the "drip-dry" bird. Because of lack of oil in its feathers, it must sit with wings spread in order to dry out. (Photo by National Park Service.)

aligned in a north-south direction. Although their origin is puzzling, it is probable that the trees play an important role in hammock development. Once established, possibly along fracture zones in the limestone, the trees protected these "islands" from devastating floods which, until recently, periodically rushed southward from Lake Okeechobee. Then, by solution and reprecipitation, "rock reefs" developed into elevated mounds. Additional investigations will probably reveal the solution to this intriguing problem. Regardless, visitors will not want to miss Mahogany Hammock, which is along the road to Flamingo.

Along the coastal areas, beach deposits and low, wave-cut cliffs are the main features. In places, the results of recent hurricanes that sweep across the Everglades will remind the park visitor of this unique aspect of the environment.

Many of the interesting features can be seen along the highway to Flamingo. Others may be observed along the Tamiami Trail from Miami west to the town of Everglades, and on Western Water Gateway to the Ten Thousand Islands. Parts of the park must be seen from a boat, preferably with an experienced guide. If you are manning your own boat, navigation charts are essential. And be sure to file a "float plan" so that the park ranger will know which swamp to search in case you do not report in as scheduled. Needless to say, the Everglades may be visited at any time of the year.

FIGURE 20-5

Sawgrass, a sedge, has teeth of opaline silica capable of inflicting severe flesh wounds.

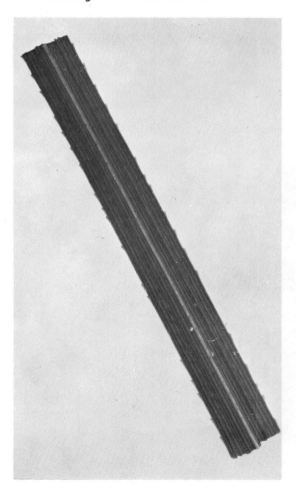

The environment as nature designed it has been changed by humans. Originally, Lake Okeechobee, a large lake about 75 miles (125 km) north of the Everglades, overflowed periodically and its floodwaters swept southward across Big Cypress Swamp and then the Everglades. Thus the water in the lakes and swamps was replenished, and the habitats for fish, alligators, and water birds were secure. Reclamation projects included the construction of canals that carry Lake Okeechobee's excess water southeastward to the Atlantic. Consequently, the Everglades is now deprived of the floodwaters from the north, and during prolonged droughts many of the lakes and swamps dry up and large numbers of animals and birds are lost. It is ironic that this "conservation measure" so impairs the park established specifically to preserve a remarkable environment.

Finally, a new chapter is being written on the Everglades. The present board of directors of the South Florida Water Management District recognized that the century-long program of canalization of the Kissimmee and other rivers had seriously damaged the environment of the Everglades. Consequently, in July of 1984 they began the lengthy process of filling in the canals in order to restore the original meandering character of the streams. According to W. J. Schneider and J. H. Hartwell, in their article in the November 1984 issue of *Natural History,* the original environment of the Everglades will eventually be restored.

BISCAYNE NATIONAL PARK

Biscayne National Park—175,000 acres of blue water, mangrove shoreline, barrier islands and coral reefs—is located about 20 miles (33 km) south of Key Biscayne and immediately north of Key Largo, Florida, within view of downtown Miami. Established as a national monument in 1968, it gained park status in 1980. Park headquarters are at Convoy Point, about 9 miles (15 km) east of the city of

Homestead. Nearby areas of interest are Everglades National Park several miles west and John Pennekamp Coral Reef State Park immediately to the south.

The park consists of the mainland mangrove shoreline, Biscayne Bay, with several elongate islands — called "keys" — along its east side, then Hawk Channel beyond which living coral reefs rise above a shallow platform. Beyond the reefs lies the eastern boundary of the park, just west of the Florida Gulf Stream. Elliott Key, the largest of the islands, stretches for eight miles along the east side of Biscayne Bay; much smaller, Sands Key extends two miles to the northern boundary. Just beyond, Boca Chita Key, the Ragged Keys, and Soldier Key are marked for addition to the park, in order to make the set of keys complete. South of Elliott

Key, across Caesar Creek (a channel or passageway between the islands), Old Rhodes and Totten Keys are the main islands; farther west, beyond the Intracoastal Waterway, are the Arsenicker Keys, which rise from the shallows of the bay.

The days of the pirates are long gone, but Caesar Creek is a reminder that Black Caesar and his rogues once struck fear into the hearts of the unfortunates who were shipwrecked in these dangerous waters. On Turkey Point, just west of the park, the stacks of a nuclear power plant rise 415 feet (126 m) into the sky, lending an ominous air to the otherwise tranquil natural scene.

The life forms in the park are highly varied and seemingly endless, both beneath its waters and on land. The northernmost coral

FIGURE 20-6

Headquarters, Biscayne National Park.

reefs in the continental United States reveal colorful flower, star, brain, staghorn, elkhorn and finger corals that are like plants in that they are "fixed." Free to move are fish of many colors, including the four-eyed butterfly and the French angelfish, the loggerhead turtle, and the bottle-nosed dolphin; and on land, the raccoon, the marsh hare and the exotic Mexican red-bellied squirrel.

Those who think that life in the city is complicated and perilous should don snorkel and fins and learn what life is like on one of Florida's coral reefs — one of the most complex ecosystems known. The inexperienced are well advised to obtain professional guidance in order to make it a thrilling rather than a painful experience. Cuts from jagged coral can be serious, and the punctures from the spiny sea urchin are slow to heal. *Look only, touch nothing* is good advice. For an intriguing story on reef life, read Dr. Starck's article in *National Geographic* (November 1966), which is based on his experiences as a marine biologist.

Unlike at Padre Island, where trees are essentially absent, the Biscayne keys support a dense and luxuriant forest. Tropical hardwoods of many types, including the mahoganies, dominate the interior of the islands, which are ringed by the spraddle-legged mangroves that defend the shoreline. A quiet few minutes in the Hurricane Creek passageway, with the mangroves hovering over you, is a delightful, relaxing interlude — when the winds are gentle breezes. In many places stately palms sway gently when the wind is light and bend well when gales are blowing.

Park visitors find it amazing that the dense forest on Elliott Key is not virgin and that most of the island was under cultivation for a time — until about 50 years ago. Oddly, as reported in the book *Everglades* (Time/Life, 1973) the existing forest is believed to be similar in type and diversity to the one that was destroyed. What mars the forest of today, when viewed from the air, is the gash that extends from one end almost to the other. It was the work of the developers' bulldozers, immediately before Biscayne National Monument was established. But even this ghastly wound will heal; and even the Sargent's palm, known also as the buccaneer palm, rare and almost wiped out, will recover under the protection of the National Park Service.

The bird life of the park is abundant, changing with the seasons. Herons and cormorants are year-round residents. There are many nesting areas in Biscayne; particularly popular are the Arsenicker Keys in the southwestern part of the park.

Biscayne's Past

The geologic story involves only the last part of the last chapter in the geologic history book — the latter part of the Pleistocene Epoch. The Key Largo Limestone, with type-section on Key Largo, outcrops on the chain of keys and lies beneath the waters of the eastern part of the park. It was a living coral reef over 100,000 years ago (Hoffmeister, 1974). Although corals

FIGURE 20-7

Flower coral *(Eusmilia fastigiata),* one of many fascinating corals on the reef at Biscayne National Park. (Photo by National Park Service.)

were probably dominant, many other marine invertebrate animals and plants (mainly algae) contributed their energies and their skeletons in the reef-building process.

Beneath Biscayne Bay and exposed along its western shores is the Miami Oolite, a limestone only slightly younger than the Key Largo. Although other materials, fossils and fragments of fossils ("hash"), are present in varying amounts, the Miami is largely made up of tiny spherical concretions (oolites). The oolites, resembling fish roe, are formed by accretion of one thin layer upon another, generally around a tiny nucleus, as the limy ooze on the sea floor is agitated. The snow-white sand on the bottom of Biscayne Bay is derived in large part from the Miami Formation.

Reef building was interrupted when sea level was lowered significantly during the Wisconsin glacial advance; left high and dry, the corals and other fixed life forms died, leaving the reef exposed to the elements. The top part of the reef was destroyed by weathering and wave action, leaving an eroded reef platform.

Later, about 11,000 years ago, when the glaciers began to melt, the seas rose again. Soon, a new generation of corals began to build new reefs, particularly in the eastern section of the park. As before, essentially the same assemblage of animals and plants assisted in the reef-building process. *Symbiosis* was again a way of life on the reef: One animal or plant paired off with some other species, for their mutual benefit. For example, corals provided protection for the algae (plants) and the algae provide life-sustaining oxygen to the corals. It is a beautiful system, one that is difficult to improve upon. It will continue to work well until some major change takes place, either as a result of natural causes or by human ignorance, perhaps plain carelessness.

At Biscayne we have the dead reefs — the keys — and living reefs abuilding on top of the eroded reef platform. In time they will be as large as those of the past, unless sea level is lowered again — or instead if all of the water

tied up in existing glaciers melts and returns to the sea.

Access to the wonders of Biscayne is easy if you have your own boat, detailed navigational charts, and a good radio for storm warnings. However, on weekends when the weather and the tides are right, you can enjoy the four-hour boat tour out onto Elliott Key, hike the Nature Trail and go through the visitor center, courtesy of the National Park Service.

CAPE HATTERAS AND CAPE LOOKOUT NATIONAL SEASHORES

Cape Hatteras, established in 1937, is the oldest of our national seashores. Cape Lookout, which extends southwestward from Cape Hatteras, was established in 1966. Together these elongate barrier islands bracket about 125 miles (210 km) of North Carolina's coastal waters — part of a much longer chain of islands that extends discontinuously from New Jersey southward to Florida. The two seashores form the Outer Banks, where the struggle between sand, sea, and wind produces "one of the most dynamic areas in the National Park System" (Dolan et al., 1973). A short distance north of Cape Hatteras, the Wright Brothers National Memorial and nearby Fort Raleigh, first but unsuccessful English settlement in the New World, are well worth visiting.

Origin of the Barrier Islands

The geologic story begins only 10,000 to 11,000 years ago when the huge continental ice sheets melted, returning vast quantities of water to the sea. As sea level rose, the waters encroached over the land, inundating extensive areas of the continental shelf. As at Padre Island in Texas, the sequence begins with a coastal dune ridge that became an elongate island when the shoreward area — now Pamlico Sound — was covered with sea water. Pamlico Sound now separates the barrier islands by as much as 25 miles (40 km). In places, the islands are as much as 2.5 miles (4.0 km) wide, but there are long stretches less than a quarter of a mile in width. Some sections of the dune ridges are low, but others rise as high as 40 feet (12 m) above the sea.

Storms have broken through the ridges in places, and these inlets permit tidal currents to carry sediment out into the ocean, where it is deposited on the *shoals* that extend as much as 10 miles (16 km) out from the barrier islands. Many ships have foundered on the shoals, and some of their masts are visible, reminding today's mariner to beware.

Barrier Islands in Dynamic Balance

On a quiet day, when the breeze is light and the sea is calm, here is a truly tranquil scene. It is the time to see what a barrier island is made of. As shown in Figure 20-8, back of the beach area is the dune ridge, partially stabilized by a scattering of hardy grasses. Considerably more vegetation grows on the overwash terrace that slopes westward toward the salt marshes along Pamlico Sound. On such a day it seems that nothing is changing except on the beach where the surf quietly saps the sand.

But even one storm, particularly one of hurricane force, can profoundly alter the landscape. Gigantic waves override the ridge at low points and cut new inlets across the island; terrific winds move the sand about, depositing it on the lee side or in Pamlico Sound. The static situation cannot weather the storm.

On unmodified barrier islands such as Cape Lookout, the wave energy that is not utilized by washing up the wide beaches is spent washing over the dune ridge and down into the salt marsh behind. Fine sands brought into the marsh by overwash slowly build the area upward and toward the mainland. The net effect is that the barrier island will slowly migrate and thousands of years hence will become a mainland beach, unless more sand is brought to the beaches by longshore currents than is removed by overwash and other processes. This is not presently the case (Lisle and Dolan, 1983). Also contributing to this mainland migration is the slow rise of sea level as glaciers continue to adjust to present-day climates.

People as Geologic Agents

Early settlers used the Outer Banks mostly for grazing and built few permanent structures. Their occupations included whaling and the salvaging of the numerous shipwrecks. Development was heavier in the northern section, although permanent roads were impossible because of the occasional oceanic overwash during storms.

In the 1930s, the decision was made to establish a boundary along the northern islands across which the sea was not to pass (Dolan et al., 1973). Sand fences were constructed on the dunes along the northern islands from 1936 to 1940, creating what was then thought to be an inexpensive, artificial, protective dune system. Special grasses were planted and

FIGURE 20-8

Idealized cross section of natural barrier island, Cape Hatteras National Seashore. (Adapted from Dolan et al., 1973.)

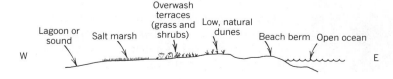

by the late 1950s a much higher dune system and a continuous mass of vegetation covered the northern Outer Banks. A new but unjustifiable feeling of security led to construction of permanent roads, inter-island bridges, and buildings on privately owned land behind the barrier dunes.

The artificial barrier dune no longer permitted the big waves to expend their excess energy by overwashing into the sound; instead, all the energy was concentrated between the

FIGURE 20-9

Historic Cape Hatteras Lighthouse, more than 200 feet high, tallest on the continent. (Photo by E. H. Wrenn.)

dune and the ocean. The wide beaches that were one of the main attractions at Cape Hatteras began to disappear. In contrast to Cape Lookout, where no artificial dune system exists and where the beaches are over 410 feet (125 m) wide, many of Hatteras's beaches are now only 100 feet (30 m) or less in width. The big waves now crash into the barrier dune, which is slowly being destroyed. Dune breaches and overwashes are becoming more common each year (Dolan et al., 1980). The destruction of the structures behind the dune is inevitable, even if many millions of dollars are spent on temporary stabilization measures.

One area where wave erosion is particularly well documented, and an area whose historic values have been deemed important enough to preserve, is the Cape Hatteras Lighthouse, the symbol of Cape Hatteras. The first lighthouse was built in 1802 but was razed after it was severely damaged by cannon fire during the Civil War. Although the original lighthouse stood about a mile from the water's edge, because of the westward migration of the barrier island, the foundation was washed into the sea by the big storm of 1980. This same storm washed around the steel groin protecting the present lighthouse and allowed the ocean within 50 feet (15 m) of its foundation. This lighthouse, built in 1870, was originally about 1500 feet (457 m) from the ocean! Although conditions are temporarily stabilized, plans are under way to build a concrete retaining wall as a longer-term solution. It is anticipated by the Park Service that in 50 to 70 years, boats instead of cars will take visitors to "Lighthouse Island."

The lessons are clear: what humans regard as a disruptive change in nature is often essential to the maintenance of a particular ecosystem, and interference with these natural processes can be very costly. Meanwhile, we can enjoy the delightful seascapes and moods of the ocean, knowing that each process and plant community has a place in the continually changing picture at Cape Hatteras.

FIGURE 20-10

Meandering river, the result of low stream gradient on the Coastal Plain, Congaree Swamp National Monument, near Columbia, South Carolina. (Photo by National Park Service.)

FIGURE 20-11

Shells of marine animals on the beach in Cumberland Island National Seashore, off the southern coast of Georgia. (Photo by National Park Service.)

FIGURE 20-12

"Sweetwater lake" among sand dunes in Cumberland Island National Seashore; local rainfall is the source of the fresh water. (Photo by National Park Service.)

FIGURE 21-1
Coral sand beaches in coves on the island of St. John. Wave erosion accentuates the irregular shorelines of the volcanic islands of the West Indies. (Photo by George E. Fischer.)

CHAPTER 21

WEST
INDIES
ARCHIPELAGO

The West Indies, also called the Antilles, form an arc more than 1000 miles long that encloses the Gulf of Mexico and the Caribbean Sea. The Greater Antilles lie to the northwest and are made up of large islands — Cuba, Jamaica, Hispaniola (Haiti and the Dominican Republic), and Puerto Rico. The Lesser Antilles are composed of smaller islands of the windward and leeward groups, Barbados, Trinidad and Tobago.

Here, as in the Hawaiian Archipelago, the islands are mountaintops; however, the geologic history, although imperfectly known, is more complex and is probably more nearly comparable to that of the Aleutian Archipelago. Detailed studies have been made on a number of the islands; several papers are presented in *Caribbean Geological Investigations,* edited by Hess (1966).

Aside from the Jurassic rocks of Trinidad, near South America, the oldest rocks on which age determinations are considered reliable are of Cretaceous age. Schists and related metamorphic rocks on the islands of Cuba and Hispaniola are believed to be older, but to our knowledge, their precise age is not yet known.

The archipelago began to form in mid-Tertiary time, probably during the Oligocene. Folding and faulting, accompanied and followed by igneous activity, formed the basic structures that led to the development of today's island chain. It was above the sea at times and perhaps connected with the Florida peninsula; by erosion and crustal movements,

it was slightly below the sea at other times, as the marine limestones indicate.

As in the other archipelagos glimpsed in our travels — the Hawaiian and the Aleutian — volcanic activity played an important role. It began during the Oligocene and continued intermittently to the present. The violent and destructive eruption of Mount Pelée in 1902, which destroyed St. Pierre on the island of Martinique, was a forceful reminder that the fires are not dead. Early in 1979, Soufriere blew its top and covered much of the island of St. Vincent with ash, as pictured in *National Geographic* (September 1979).

As is the fate of all islands, the waves are at work attempting to destroy them; in the process, wave-cut cliffs and platforms are carved out of the rocks. In these tropical waters, marine life is abundant. In places, coral reefs and coral sand beaches are tourist attractions; one such place is Virgin Islands National Park.

VIRGIN ISLANDS NATIONAL PARK

This tropical park occupies much of the island of St. John, one of the three main islands of the U.S. Virgin Islands Territory. The colorful human history of the Virgin Islands begins with their discovery by Columbus on his second voyage in 1493. At that time, they were inhabited only by Indians and were truly "virgin islands." In 1717, the Danes gained control of the Virgin Islands; they established large plantations and the sugar cane industry flourished for a long period. However, beset by slave uprisings, Denmark abolished slavery in 1848, and the industry declined rapidly on St. John. Here in the tropics, natural revegetation took place rapidly, and a reasonable facsimile of the original virgin conditions was restored.

In 1917, St. Thomas, St. John, and St. Croix, together with about 50 smaller islands, were purchased by the United States. In 1956, Virgin Islands National Park was established on St. John, which lies about 50 miles (80 km) east of

Puerto Rico. A large part of the island, about 15,150 acres, lies within park boundaries.

The terrain of most of the island is moderately rough, particularly in the Bordeaux Mountains near the eastern end. Bordeaux Mountain, the highest of the range, rises 1277 feet (390 m) above the sea—an excellent lookout point for those who climb up the trail to the summit. Shallow waters cover beautiful coral reefs around the island and provide excellent swimming and snorkeling.

The climate is ideal throughout the year. Easterly trade winds that blow in from the ocean temper the sun's heat, making the days pleasant and the nights delightful. The Park Service brochure mentions that the lowest temperature on record is 61°F and the highest 98°F. Tropical fruits are abundant in places, some of which are poisonous.

The geologic story is still only imperfectly pieced together, due in part to the lack of good outcrops of bedrock. Here in this tropical climate, both chemical and physical weathering are taking place at a rapid rate; also, the tropical vegetation is effective in holding the loose material in place. Good exposures of the bedrock are found where wave erosion is active and in artificial cuts along some of the roads.

The oldest rocks are believed to be Cretaceous in age. They are mainly submarine lava flows and volcanic tuffs that were laid down prior to the crustal deformation that gave birth to the archipelago. Later in the Cretaceous, the area was above the sea, as indicated by stream-deposited conglomerates and subaerial volcanic breccias. On top of these rocks are Late Cretaceous marine limestones, indicating that the sea floor was unstable.

Although the record is less than clear, it was apparently in mid-Tertiary time that the main deformation began to shape the archipelago. Upfolding was accompanied by extensive faulting and igneous activity. Large masses of magma intruded and the heat from the magma metamorphosed the adjacent rocks. Extensive recrystallization of the limestones produced the metamorphic rock known as marble. Dikes were formed where the magma was forced out along faults and related fractures. Some of the magma reached the surface, and volcanoes erupted lavas and pyroclastic materials. For those interested in the details of the igneous activity here, see Donnely (1966).

After a series of uplifts during the latter part of the Tertiary, the present configuration of the Virgin Islands was finally attained. The landscapes of St. John were shaped by stream erosion—less rugged and less jagged than those in some of the parks, but very attractive nonetheless. Wave erosion has produced wave-cut cliffs and wave-cut platforms, particularly on the windward side of the island.

The latest adornment at Virgin Islands National Park was added when the environment was right for the development of the coral reefs for which the park is famous. They form a "fringe" around the island and are referred to as *fringing reefs*. Here, many animals—some fixed like the corals, others mobile like the starfish—and plants, such as the coralline algae, are building up the structure. It is another world, unknown to those who insist on keeping their feet on dry land.

With under-the-sea viewing and with "sight-seeing" from the mountaintop, no one is likely to leave the park except with reluctance.

BUCK ISLAND REEF NATIONAL MONUMENT

"A coral city in a turquoise world" is Udall's description of Buck Island Reef, featured in his *America's Natural Treasures* (1971). This tiny dot in the Caribbean Sea, just off the eastern end of St. Croix, became a national monument in 1961.

Buck Island is a tropical wilderness jungle of entangled turpentine trees, acacias, and the poisonous manchineel, generally alive with birds of many colors. But it is chiefly what is beneath the sea that attracts thousands of people each year.

As in Virgin Islands National Park, the clear tropical waters provide an ideal habitat for corals and other reef-building animals and plants. Corals are fixed animals; they put down "roots" in the form of calcium carbonate, which they secrete. Thus, fixed to the sea floor or to older, dead coral, they depend on the gentle currents to bring them their food — tiny particles of organic matter. They are hosts to green algae, tiny plants that convert carbon dioxide from the corals into oxygen needed by the corals. This mutually beneficial arrangement is one of the many kinds of symbiosis that operate in the biological world. Other kinds of algae, collectively referred to as coralline algae, are also there, along with a host of other marine plants and animals. Their shells and skeletons will be added to those of their ancestors in the reef-building process. Staghorn coral and brain coral are the two main species in the colorful underwater world around Buck Island. Here we see in miniature what led to the building of the huge Permian Reef now exposed in Guadalupe Mountains.

Fish of many colors, some striped and some with "eyes" fore and aft, entrance the snorkeler as he or she moves silently from one underwater sign to another along the Nature Trail around the island. The huge green sea turtles are here, in one of the few places in the world where they are safe.

With its wide variety of natural attractions — living and dead — Buck Island Reef, like Virgin Islands National Park, is a good place to be when most of the other park areas are buried in snow.

CHAPTER 22

IN

RETROSPECT

In our travels through our national parks and monuments we have experienced many different environments, from the tropic to the arctic, from the depths of Death Valley to the heights of Mount McKinley. We have examined the various geomorphic provinces and found a different geologic sequence in each region. These differences, together with climatic factors, have resulted in unique assemblages of landforms. We have noted the geologic relationships between the parks within each province and have seen that the relationship varies from one province to another. In the Colorado Plateaus, for example, one park records a section of geologic history, while others provide other parts of the story. In the Cascades the parks illustrate different stages in the development and destruction of stratovolcanoes. This regional approach highlights the geologic relationships between the park areas.

As park visitors, we have had the opportunity to ponder the geologic changes not just of the past but those which are occurring now. In Hawaii Volcanoes National Park, we saw not only the huge accumulation of lavas that built Mauna Loa, but also the still-warm lavas that have been pouring out of Mauna Ulu and other vents during the past several years. In Alaska's Glacier Bay, we saw the magnificent scenery created by glaciers in the past, along with live glaciers grinding from the bedrock the silt that clouds the fjords below. On our descent into Grand Canyon we observed the rock debris being swept along the Colorado River, and we realized that the river that dug the canyon is deepening it now. The landscapes of the future are being fashioned today.

No more dramatic example is needed than Mt. St. Helens, which, beginning in 1980, drastically altered the landscape of southwestern Washington. Its explosive volcanic eruptions serve as a reminder that although many geologic processes are slow and gentle, others operate rapidly and with deadly violence.

We have found that the plate tectonics concept is providing explanations for major crustal adjustments, past and present, throughout the Hawaiian Islands, Alaska and the remainder of the Pacific Border Province, the Cascades, the Basin and Range, the Appalachians, and other areas.

Although the physical features of the parks have been emphasized, biological evolution is also well recorded. Primitive algae in Glacier National Park and in the older rocks of Grand Canyon depict the simple life of those ancient days. Paleozoic life was more complex, as documented in the younger canyon rocks. Invertebrate animals became abundant and diversified, and tracks show that primitive vertebrates inhabited the canyon area during the latter part of the Paleozoic. Big Bend fossils indicate further advancement of the invertebrates and show that the Mesozoic reptiles prospered there, as in the Dinosaur Monument area. Cenozoic landscapes were dominated by mammals; many remains are found in Big Bend, in Fossil Butte, and in the Badlands. Both the biological and the physical worlds unfold in the national park areas.

We have seen just about every rock type, each with its own characteristics and its own response to local geologic processes. The soil derived from a particular rock in a given climate can support some plant species but not others. And the animals seek out the habitats best suited to their own needs. Thus a system both complex and beautiful has been uniquely preserved.

Our parks and monuments are a precious national heritage. To the John Muirs, the Judge Hedges, the Theodore Roosevelts, and the Stephen T. Mathers who devoted themselves

FIGURE 22-1

This subdued view of the sun, almost hidden behind the mountain, is symbolic. Although the sun is the greatest and most obvious source of energy, we have kept it largely hidden. As a result, pollution is serious and widespread, reaching even into our national parks. This is ironic, because solar energy is nonpolluting and produces no radioactive wastes.

to their creation, and to the Park Service men and women who are dedicated to their preservation — our sincere thanks.

The Park Service people are doing their best to save these national treasures for us and for future generations. However, developments outside the parks are beyond their control, and we have witnessed a serious deterioration of park environments. No longer are the skies clear and the waters pure. With the demand for energy apparently insatiable, we may see strip-mining and oil operations marring park landscapes, and we may perhaps even accept this as commonplace. Will there come a time when our children will learn only from pictures, drawings, and recordings what the parks were like when they were set aside for the enjoyment of all for all time? Even short of complete destruction, will we be satisfied to escape from our smog-ridden cities only to find ourselves in a smoggy national park?

Surely the plants and animals in the parks deserve better than this, even the human animals. Let us not give our children reason to ask us why we allowed our beautiful national parks and monuments to be ruined!

APPENDIX A

DIRECTORY

OF THE

NATIONAL

PARKLANDS

INCLUDED

IN THIS

BOOK

For maps and other information, write to the superintendent of the park areas of your choice. There is no charge for these materials.

Acadia National Park, RFD 1, Box 1, Bar Harbor, Maine 04609.

Agate Fossil Beds National Monument, Scotts Bluff National Monument, P. O. Box 427, Gering, Nebraska 69341.

Aniakchak National Monument and Preserve, Box 7, King Salmon, Alaska 99613.

Arches National Park, Canyonlands National Park, 446 South Main Street, Moab, Utah 84532.

Badlands National Park, P. O. Box 6, Interior, South Dakota 57750.

Bandelier National Monument, Los Alamos, New Mexico 87544.

Bering Land Bridge National Preserve, P. O. Box 220, Nome, Alaska 99762.

Big Bend National Park, Texas 79834.

Biscayne National Park, P. O. Box 1369, Homestead, Florida 33030.

Black Canyon of the Gunnison National Monument, P. O. Box 1648, Montrose, Colorado 81401.

Blue Ridge Parkway, 700 Northwestern Bank Building, Asheville, North Carolina 28801.

Bryce Canyon National Park, Bryce Canyon, Utah 84717.

Buck Island Reef National Monument, P. O. Box 160, Christiansted, St. Croix, Virgin Islands 00820.

Canyon de Chelly National Monument, P. O. Box 588, Chinle, Arizona 86503.

Canyonlands National Park, 446 South Main Street, Moab, Utah 84532.

Cape Cod National Seashore, South Wellfleet, Massachusetts 02663.

Cape Hatteras National Seashore, Route 1, Box 675, Manteo, North Carolina 27954.

Cape Krusenstern National Monument, National Park Service, Kotzebue, Alaska 99752.

Capitol Reef National Park, Torrey, Utah 84775.

Capulin Mountain National Monument, Capulin, New Mexico 88414.

Carlsbad Caverns National Park, 3225 National Parks Highway, Carlsbad, New Mexico 88220.

Cedar Breaks National Monument, P. O. Box 749, Cedar City, Utah 84720.

Channel Islands National Park, 1901 Spinnaker Drive, Ventura, California 93003.

Chickasaw National Recreation Area, P. O. Box 201, Sulphur, Oklahoma 73086.

Chiricahua National Monument, Dos Cabezas Route, Box 6500, Willcox, Arizona 85643.

Colorado National Monument, Fruita, Colorado 81521.

Crater Lake National Park, P. O. Box 7, Crater Lake, Oregon 97604.

Craters of the Moon National Monument, P. O. Box 29, Arco, Idaho 83213.

Death Valley National Monument, Death Valley, California 92328.

Denali National Park and Preserve, P. O. Box 9, McKinley Park, Alaska 99755.

Devils Postpile National Monument, Sequoia and Kings Canyon National Park, Three Rivers, California 93271.

Devils Tower National Monument, Devils Tower, Wyoming 82714.

Dinosaur National Monument, P. O. Box 210, Dinosaur, Colorado 81610.

El Morro National Monument, Ramah, New Mexico 87321.

Everglades National Park, P. O. Box 279, Homestead, Florida 33030.

Florissant Fossil Beds National Monument, P. O. Box 185, Florissant, Colorado 80816.

Fossil Butte National Monument, P. O. Box 527, Kemmerer, Wyoming 83101.

Gates of the Arctic National Park and Preserve, P. O. Box 74680, Fairbanks, Alaska 99707.

Glacier Bay National Park, P. O. Box 1089, Juneau, Alaska 99801.

Glacier National Park, West Glacier, Montana 59936.

Grand Canyon National Park, P. O. Box 129, Grand Canyon, Arizona 86023.

Grand Teton National Park, P. O. Drawer 170, Moose, Wyoming 83012.

Great Sand Dunes National Monument, Mosca, Colorado 81146.

Great Smoky Mountains National Park, Gatlinburg, Tennessee 37738.

Guadalupe Mountains National Park, 3225 National Parks Highway, Carlsbad, New Mexico 88220.

Haleakala National Park, P. O. Box 357, Makawao, Maui, Hawaii 96768.

Hawaii Volcanoes National Park, Hawaii 96718.

Hot Springs National Park, P. O. Box 1860, Hot Springs, Arkansas 71901.

Indiana Dunes National Lakeshore, 1100 North Mineral Springs Road, Porter, Indiana 46304.

Isle Royale National Park, 87 North Ripley Street, Houghton, Michigan 49931.

Jewel Cave National Monument, c/o Wind Cave National Park, Hot Springs, South Dakota 57747.

John Day Fossil Beds National Monument, 420 West Main Street, John Day, Oregon 97845.

Katmai National Park and Preserve, P. O. Box 7, King Salmon, Alaska 99613.

Kenai Fjords National Park, Box 1727, Seward, Alaska 99664.

Klondike Gold Rush National Historical Park, P. O. Box 517, Skagway, Alaska 99840.

Kobuk Valley National Park, National Park Service, Kotzebue, Alaska 99752.

Lake Clark National Park and Preserve, Alaska Regional Office, National Park Service, 540 West 5th Avenue, Anchorage, Alaska 99501.

Lassen Volcanic National Park, Mineral, California 96063.

Lava Beds National Monument, P. O. Box 867, Tulelake, California 96134.

Lehman Caves National Monument, Baker, Nevada 89311.

Mammoth Cave National Park, Mammoth Cave, Kentucky 42259.

Mesa Verde National Park, Colorado 81330

Mount Rainier National Park, Tahoma Woods, Star Route, Ashford, Washington 98304.

Natural Bridges National Monument, c/o Canyonlands National Park, 446 South Main Street, Moab, Utah 84532.

Noatak National Preserve, National Park Service, Kotzebue, Alaska 99752.

North Cascades National Park, 800 State Street, Sedro Woolley, Washington 98284.

Olympic National Park, 600 East Park Avenue, Port Angeles, Washington 98362.

Organ Pipe Cactus National Monument, Route 1, Box 100, Ajo, Arizona 85321.

Oregon Caves National Monument, 19000 Caves Highway, Cave Junction, Oregon 97523.

Padre Island National Seashore, Corpus Christi, Texas 78418.

Petrified Forest National Park, Arizona 86028.

Pinnacles National Monument, Paicines, California 95043.

Point Reyes National Seashore, Point Reyes, California 94956.

Rainbow Bridge National Monument, c/o Glen Canyon National Recreation Area, P. O. Box 1507, Page, Arizona 86040.

Redwood National Park, 1111 Second Street, Crescent City, California 95531.

Rocky Mountain National Park, Estes Park, Colorado 80517.

Sequoia and Kings Canyon National Parks, Three Rivers, California 93271.

Shenandoah National Park, Luray, Virginia 22835.

Sitka National Historical Park, Box 738, Sitka, Alaska 99835.

Sunset Crater National Monument, Route 3, Box 149, Flagstaff, Arizona 86001.

Theodore Roosevelt National Park, Medora, North Dakota 58645.

Timpanogos Cave National Monument, Rural Route 3, Box 200, American Fork, Utah 84003.

Virgin Islands National Park, P. O. Box 7789, Charlotte Amalie, St. Thomas, Virgin Islands 00801.

Voyageurs National Park, P. O. Box 50, International Falls, Minnesota 56649.

White Sands National Monument, Box 458, Alamogordo, New Mexico 88310.

Wind Cave National Park, Hot Springs, South Dakota 57747.

Wrangell–Saint Elias National Park and Preserve, P. O. Box 29, Glennallen, Alaska 99588.

Yellowstone National Park, P. O. Box 168, Yellowstone National Park, Wyoming 82190.

Yosemite National Park, P. O. Box 577, California 95389.

Yukon–Charley Rivers National Preserve, P. O. Box 64, Eagle, Alaska 99738.

Zion National Park, Springdale, Utah 84767.

BIBLIOGRAPHY

Selected Textbooks

Dott, Jr., Robert H., and Roger L. Batten. 1981. *Evolution of the Earth*. New York: McGraw-Hill Book Co.

Dunbar, C. O. 1966. *Historical Geology*. New York: John Wiley & Sons.

Flint, R. F. 1973. *The Earth and Its History*. New York: W. W. Norton and Co.

————. 1971. *Glacial and Quaternary Geology*. New York: John Wiley & Sons.

Gilluly, J., A. C. Woodford, and A. O. Waters. 1975. *Principles of Geology*. San Francisco: W. H. Freeman and Co.

Hamblin, W. K. 1975. *The Earth's Dynamic Systems*. Minneapolis, Minnesota: Burgess Publishing Company.

Hamblin, W. K., and J. D. Howard. 1971. *Physical Geology Laboratory Manual*. Minneapolis, Minnesota: Burgess Publishing Company.

Press, F., and R. Siever. 1974. *Earth*. San Francisco, Calif.: W. H. Freeman and Co.

Shelton, J. S. 1966. *Geology Illustrated*. San Francisco, Calif.: W. H. Freeman and Co.

Stearn, Colin W., Robert L. Caroll, and Thomas H. Clark. 1979. *Geological Evolution of North America*. New York: John Wiley & Sons.

Abbey, Edward. 1968. *Desert Solitaire*. New York: Simon and Schuster.

Adams, J. E., and H. N. Frenzel. 1950. "Capitan barrier reef, Texas and New Mexico." *Jour. Geol.* 58:289–312.

Adkins, L. H. 1975. "Mountaineering." *Alaska Geographic* 2, no. 4:42–45.

Ahnert, Frank. 1970. "Functional relationships between denudation, relief, and uplift in mid-latitude drainage basins." *Am. Jour. Science* 268:243–63.

Alberding, H. 1971. *The geologic story of the Zion National Park area*. Nat. Park Service release.

Albert, E. 1963. *Rocky Mountain National Park*. Rocky Mountain Nature Assoc.

Albert, Gene. 1975. "Glacier: beleaguered park of 1975." *National Parks and Conservation Magazine* 49, no. 11:4–10.

Alden, W. C. 1953. "Physiographic and glacial geology of western Montana and adjacent areas." U.S. Geol. Survey Prof. Paper 231.

Alexander, George. 1981. "Going, going, gone: Why did the dinosaurs die out?" *Science* 81:2, no. 4:64–69.

Alt, D. D., and D. W. Hyndman. 1973. *Rocks, ice, and water: The geology of Waterton-Glacier Park*. Missoula, Montana: Mountain Press Publ. Co.

Alvarez, Luis W., Walter Alvarez, Frank Asaro, and Helen V. Michel. 1980. "Extraterrestrial cause for the Cretaceous-Tertiary extinction." *Science* 208:1095–1108.

American Geological Institute. 1966. *Glossary of Geology and Related Sciences*.

Anderson, C. A. 1935. "Alteration of the lavas surrounding the hot springs in Lassen Volcanic National Park." *Am. Mineralogist* 20:240–52.

Anderson, C. H., and W. R. Halliday. 1969. "The Paradise Ice Caves, Washington: An extensive glacier cave system." *National Speleological Soc. Bull.* 31:55–72.

Anderson, D. D. 1970. "AKMAK — An early archeological assemblage from Onion Portage, northwest Alaska." *Acta Arctica* 16.

Andrews, P. 1933. "Geology of the Pinnacles National Monument." *Calif. Univ. Publ. Geol. Sci.* 24:1–38.

Ash, S. R., and D. D. May. 1969. *Petrified Forest: the story behind the scenery*. Holbrook, Arizona: Petrified Forest Museum Assoc.

Atwater, Brian F. 1983. "Jokulhlaups into the Sanpoil arm of glacial Lake." In *Guidebook for field trip of Rocky Mountain Cell*. Columbia, Washington: Friends of the Pleistocene. 1–19.

Atwater, T. 1970. "Implications of plate tectonics for the Cenozoic tectonic evolution of western North America." *Geol. Soc. Am. Bull*. 81:3513–36.

Atwood W. W., and W. W. Atwood, Jr. 1932. "Working hypothesis for the physiographic history of the Rocky Mountain region." *Geol. Soc. America Bull*. 49:957–80.

Atwood, W. W., Jr. 1935. "The glacial history of an extinct volcano, Crater Lake National Park." *Jour. Geol.*, 42:142–68.

Baars, D. L. 1972. *Red Rock Country: The Geologic History of the Colorado Plateau*. New York: Natural History Press/Doubleday.

Babcock, R. S., E. H. Brown, and M. D. Clark. 1974. *Geology of the Grand Canyon*. Mus. of Northern Arizona and Grand Canyon Nat. Hist. Assoc., Flagstaff, Arizona.

Bacon, C. R. 1983. "Eruptive history of Mount Mazama and Crater Lake caldera, Cascade Range, U.S.A." *Jour. Volcanology and Geothermal Research* 18:57–115.

Badgley, P. C. 1960. *Structural and Tectonic Principles*. New York: Harper and Row.

Baker, C. L. 1935. "Major structural features of Trans-Pecos Texas." In *The Geology of Texas, Vol. II. Structural and Economic Geology*. Univ. Texas Bull. 3401:137–214.

Baldwin, E. M. 1976. *Geology of Oregon*. Ann Arbor, Mich.: J. W. Edwards Publ., Inc.

Bandelier, Adolph. 1890, reprinted 1971. *The Delight Makers*. New York: Harcourt Brace Jovanovich Inc.

Barnes, D. F., and R. D. Watts. 1977. "Geophysical surveys in Glacier Bay National Monument." *U.S. Geol. Survey Circ.* 751-B:93–95.

Barr, T. C. 1968. "Cave ecology and the evolution of troglobites." *Evolutionary Biology* 2:35–102.

Barrell, Joseph. 1913. "Piedmont Terraces of the northern Appalachians and their mode of origin." *Geol. Soc. America Bull*. 24:688–90.

Bateman, P. C., and C. Wahrhaftig. 1966. "Geology of the Sierra Nevada." In *Geology of northern California*. California Div. Mines and Geology Bull. 190:107–72.

Bauer, C. M. 1962. *Yellowstone: Its Underworld—Geology and Historical Anecdotes of Our Oldest National Park*. Albuquerque, New Mexico: Univ. of New Mexico Press.

Bedinger, M. S. 1974. *Valley of the vapors, Hot Springs*. Philadelphia, Pennsylvania: Eastern National Park and Monument Assoc.

Berg, H. C., D. L. Jones, and D. H. Richter. 1972. "Gravina-Nutzotin Belt: Tectonic significance of an Upper Mesozoic sedimentary and volcanic sequence in southern and southeastern Alaska." U.S. Geol. Survey Prof. Paper 800-D:1–24

Berkland, J. O., and L. A. Raymond. 1973. *Cirque in Carolina*. News note in July 1973 *Geotimes*.

———. 1973. "Pleistocene glaciation in the Blue Ridge Province, S. App. Mtns., N. Carolina." *Science* 181:651–53.

Bevan, Arthur. 1925. "Rocky Mountain peneplains northeast of Yellowstone Park." *Jour. Geol.* 33:563–87.

———. 1929. "Rocky Mountain Front in Montana." *Geol. Soc. America Bull*. 40:427–56.

Billings, M. P. 1933. "Paleozoic age of the rocks of Central New England." *Science* 79:55–56.

———. 1938. "Physiographic relations of the Lewis overthrust in northern Montana." *Am. Jour. Sci.* 235:260–72.

Blackwelder, Eliot. 1931. "Pleistocene glaciation in the Sierra Nevada and Basin Ranges." *Geol. Soc. America Bull.* 42, no. 4:895–922.

———. 1932. "The age of Meteor Crater." *Science* 76:557–60.

———. 1933. "Lake Manly: An extinct lake of Death Valley." *Geog. Rev.* 23:464–71.

———. 1934. "Origin of the Colorado River." *Geol. Soc. America Bull.* 45:551–56.

Bluemle, John P., and Arthur F. Jacob. 1973. "Geology along the south loop road." Educational Series 4, North Dakota Geological Survey.

Boehm, W. D. 1975. "Glacier Bay: Old Ice, New Land." *Alaska Geographic* 3:1–134.

Bohn, Dave. 1967. *Glacier Bay: The Land and the Silence.* San Francisco: The Sierra Club.

Breed, W. J., and Evelyn Roat, eds. 1976. *Geology of the Grand Canyon,* 2d ed. Museum of Northern Arizona and Grand Canyon Natural History Assoc.

Bretz, J. H. 1913. "Glaciation of the Puget Sound Region." *Wash. Geol. Survey Bull.* 8.

———. 1923. "Channeled scablands of the Columbia Plateau." *Jour. Geol.* 31:617–49.

———. 1925. "The Spokane flood beyond the channeled scabland." *Jour. Geol.* 33:97–115.

———. 1969. "The Lake Missoula floods and the Channeled Scabland." *Jour. of Geology* 77:505–43.

Brew, D. A. 1976. "Glacier Bay National Monument — mineral-resources studies recommenced." *U.S. Geol. Survey Circ.* 733:58–59.

———. 1976. "Apparent pre-middle Tertiary right-lateral offset on Excursion Inlet Fault, Glacier Bay National Monument." *U.S. Geol. Survey Circ.* 733:59.

———. 1977. "Probable Precambrian or lower Paleozoic rocks in the Fairweather Range, Glacier Bay National Monument, Alaska." *U.S. Geol. Survey Circ.* 751-B:91–93.

———. 1979. "Intrusive rocks in the Fairweather Range, Glacier Bay National Monument, Alaska." *U.S. Geol. Survey Circ.* 804-B:88–90.

Brew, D. A., and R. P. Morrell. 1978. "Tarr Inlet suture zone, Glacier Bay National Monument, Alaska." *U.S. Geol. Survey Circ.* 772-B:90–92.

———. 1979. "The Wrangell Terrane ("Wrangellia") in southeastern Alaska; the Tarr Inlet suture zone with its northern and southern extensions." *U.S. Geol. Survey Circ.* 804-B:121–23.

Brew, D. A., A. B. Ford, D. Grybeck, B. R. Johnson, and C. J. Nutt. 1976. "Key foliated quartz diorite sill along southwest side of Coast Range complex, northern southeastern Alaska." *In* Cobb, E. H. (ed.), *The U.S. Geologic Survey in Alaska: Accomplishments during 1976. U.S. Geol. Survey Circ.* 733:60.

Brew, D. A., D. Grybeck, B. R. Johnson, and C. J. Nutt. 1976. "Glacier Bay National Monument mineral-resource studies recommenced." *In* Cobb, E. H. (ed.), *The U.S. Geologic Survey in Alaska: Accomplishments during 1975. U.S. Geol. Survey Circ.* 733:58–59.

Brew, D. A., Christine Carlson, and C. J. Nutt. 1976. "Apparent pre-middle Tertiary right-lateral offset on Excursion Inlet fault, Glacier Bay National Monument." *In* Cobb, E. H. (ed.), *The U.S. Geologic Survey in Alaska: Accomplishments during 1975. U.S. Geol. Survey Circ.* 733:59.

Brew, D. A., B. R. Johnson, A. B. Ford, and R. P. Morrell. 1978. "Intrusive rocks in the Fair weather Range, Glacier Bay National Monument, Alaska." *In* Johnson, K. M. (ed.), *The U.S. Geologic Survey in Alaska: Accomplishments during 1977. U.S. Geol. Survey Circ.* 772-B:88–90.

Broecker, W. C., and P. C. Orr. 1958. "Radiocarbon chronology of Lake Lahontan and Lake Bonneville." *Geol. Soc. America Bull.* 69:1009–32.

Brooks, A. H., 1906. "The geography and geology of Alaska." U.S. Geol. Survey Prof. Paper 45.

Brosge, W. P., and I. L. Tailleur. "Depositional history of northern Alaska." *In Geological seminar on the North Slope of Alaska,* Palo Alto, Calif., 1970, edited by W. L. Addison and M. M. Brosge. American Association of Petroleum Geologists, Pacific Section meeting, *Proceedings,* D1–D18.

Brown, Dale. 1972. *Wild Alaska.* Time-Life Wilderness Series.

Brown, W. H. 1925. "A probable fossil glacier." *Jour. Geol.* 33:464–66.

Brucher, R. W., J. W. Hess, and W. B. White. 1972. "Role of vertical shafts in movement of ground water in carbonate aquifers." *Ground Water* 10:5–13.

Brucher, R. W., and R. A. Watson. 1976. *The Longest Cave.* New York: Knopf.

Brugman, M. M., and Austin Post. 1980. "Effects of volcanism on the glaciers of Mount St. Helens." *U.S. Geol. Survey Circ.* 0850-D.

Bryan, Kirk. 1922. "Erosion and sedimentation in the Papago country, Arizona." *U.S. Geol. Survey Bull.* 730:19–90.

———. 1946. "Cryopedology—the study of frozen ground and intensive frost action with suggestions of nomenclature." *Am. Jour. Sci.* 244:622–42.

———. 1954. "The geology of Chaco Canyon, New Mexico." *Smithsonian Misc. Collections* 122, no. 7.

Bunnell, David, and Carol Vesely. 1983. "The amazing caves of Santa Cruz Island." *National Speleological Soc. News* 41, no. 2:86–91.

Callaghan, Eugene. 1933. "Some features of the volcanic sequence in the Cascade Range in Oregon." *Am. Geophys. Union Trans.* 14:243–49.

Capps, S. R. 1929. "The Lake Clark-Mulchatna region, Alaska." U.S. Geol. Survey Bull. 824.

Carden, J. R., and A. W. Laughlin. 1974. "Petrochemical variations within the McCartys basalt flow, Valencia County, New Mexico." *Geol. Soc. of Am. Bull.* 85:1479–84.

Carrara, P. E., and R. G. McGimsey. 1981. "The late-Neoglacial histories of the Agassiz and Jackson glaciers, Glacier National Park, Montana." *Arctic and Alpine Research* 13, no. 2:183–96.

Chamberlain, Allen. 1975. *The Annals of the Grand Monadnock.* Concord, New Hampshire: Society for the Protection of New Hampshire Forests.

Chapman, C. A. 1962. *The geology of Mt. Desert Island, Maine (Acadia National Park).* Ann Arbor, Michigan: Edwards Bros., Inc.

———. 1970. *The Geology of Acadia National Park.* Old Greenwich, Connecticut: The Chatham Press.

Chapman, R. M., R. L. Detterman, and M. D. Mangus. 1964. "Geology of the Killik-Etivluk Rivers region, Alaska." U.S. Geol. Survey Prof. Paper 303-F:325–407.

Chidester, A., and R. G. Schmidt. 1975. *Atlas of volcanic phenomena.* U.S. Geol. Survey.

Chittenden, H. 1933. *Yellowstone National Park.* Stanford Univ. Press (reprinted, Univ. of Okla. Press, 1964).

Chronic, John, and Halka Chronic. 1975. "Prairie, Peak and Plateau." *Colorado Geol. Survey Bull.* 32.

Chronic, J., M. E. McCallum, C. S. Ferris, and D. H. Eggler. 1969. "Lower Paleozoic rocks in diatremes, southern Wyoming and northern Colorado." *Geol. Soc. America Bull* 80:149–56.

Churkin, M., Jr., W. J. Nokleberg, and C. Huie. 1979. "Collision-deformed Paleozoic continental margin, western Brooks Range, Alaska." *Geology* 7, no. 6:379–83.

Clements, T. 1958. *Geological story of Death Valley*. Palm Desert, California: Desert Magazine Press.

Coats, R. P. 1950. "Volcanic activity in the Aleutian arc." *U.S. Geol. Survey Bull*. 974-B:35–49.

Colton, H. S. 1937. "The basaltic cinder cones and lava flows of the San Francisco Mountain volcanic field, Arizona." *Museum of Northern Arizona Bull*. 10.

Contor, R. J. 1963. *The underworld of Oregon Caves National Monument*. Crater Lake Natural History Assoc.

Cook, Frederick A. 1983. "Some consequences of palinspastic reconstruction in the southern Appalachians." *Geology* 11:86–89.

Cook, F., D. Albaugh, L. Brown, S. Kaufman, J. Oliver, and R. Hatcher, Jr. 1979. "Thin skinned tectonics in the crystalline southern Appalachians: COCORP seismic reflection profiling of the Blue Ridge and Piedmont." *Geology* 7:563–67.

Cooper, W. S. 1937. "Age and chemistry of Mesozoic and Tertiary plutonic rocks in south-central Alaska." *Geol. Soc. America Bull*. 80:23–44.

Crandell, D. R. 1965. "The glacial history of western Washington and Oregon." In *The Quaternary of the United States*, edited by H. E. Wright, Jr. and D. G. Frey, 341–54. Princeton, N.J.: Princeton Univ. Press.

——— . 1969. "Surficial geology of Mount Rainier National Park, Washington." *U.S. Geol. Survey Bull*. 1288.

——— . 1969. "The geologic story of Mount Rainier." *U.S. Geol. Survey Bull*. 1292.

——— . 1971. "Postglacial lahars from Mt. Rainier volcano, Washington." U.S. Geol. Survey Prof. Paper 677.

——— . 1972. "Glaciation near Lassen Peak, northern California." U.S. Geol. Survey Research 800-C:C179–C188.

Crandell, D. R., and Robert K. Fahnestock.

———. 1965. "Rockfalls and avalanches from Little Tahoma Peak on Mount Rainier, Wash." *U.S. Geol. Survey Bull*. 1221-A.

Crandell, D. R., and R. D. Miller. 1974. "Quaternary stratigraphy and extent of glaciation in the Mount Rainier region, Washington." U.S. Geological Survey Prof. Paper 847.

Crandell, D. R., and D. R. Mullineaux. 1967. "Volcanic hazards at Mount Rainier, Washington." *U.S. Geol. Survey Bull*. 1238.

——— . 1978. "Potential hazards from future eruptions of Mount St. Helens volcano, Washington." *U.S. Geol. Survey Bull*. 1383-C.

Crandell, D. R., and D. J. Varnes. 1960. "Slumgullion earthflow and earth slide near Lake City, Colo." (Abs.) *Geol. Soc. America Bull*. 71, no. 12:1846.

Culver, D. C. 1982. *Cave Life: Evolution and Ecology*. Cambridge, England: Cambridge University Press.

Curry, R. R. 1966. "Glaciation approximately 3,000,000 B.P. in the Sierra Nevadas, California." *Science* 154:770–71.

——— . 1971. *Glacial and Pleistocene history of the Mammoth Lakes, Sierra, California (A geologic guidebook)*. Geol. Series Publ. 11, Univ. of Montana.

Curtis, B. F., ed. 1975. "Cenozoic history of Southern Rocky Mountains." *Geol. Soc. America Memoir* 144.

Curtis, G. 1958. "Katmai." In *Landscapes of Alaska*, edited by H. Williams. Berkeley, Calif: Univ. of Calif. Press.

Daly, R. A. 1910. "Pleistocene glaciation and the coral reef problem." *Am. Jour. Sci*. 180:297–308.

——— . 1926. *Our Mobile Earth*. New York: Charles Scribner's Sons.

Danner, W. R. 1955. *Geology of Olympic National Park*. Seattle, Wash.: Univ. Washington Press.

Dansgaard, W. S., J. Johnsen, J. Moller, and

C. C. Langway, Jr. 1969. "One thousand centuries of climatic record from Camp Century on the Greenland Ice Sheet." *Science* 166:377–81.

Darton, N. H., and C. C. O'Harra. 1907. "Devils Tower Folio." *U.S. Geol. Survey Folio* 150.

Davis, W. M. 1901. "An excursion to the Grand Canyon of the Colorado." *Harvard Mus. Comp. Zool. Bull.* 38:106–201.

Day, A. L., and E. T. Allen. 1925. "The volcanic activity and hot springs of Lassen Peak, California." *Carnegie Inst. Washington Pub.* no. 360.

Dellenbaugh, F. S. 1908. *A Canyon Voyage.* New Haven: Yale Univ. Press (reprinted 1926, 1962, Yale Univ. Press).

Dennison, John M. 1983. Comment on "Tectonic model for kimberlite emplacement in the Appalachian Plateau of Pennsylvania." *Geology* 11:252–53.

Denny, C. S. 1965. "Alluvial fans in the Death Valley Region, California and Nevada." *U.S. Geol. Survey Prof. Paper* 466.

Denton, G. H. 1974. "Quaternary glaciations of the White River Valley, Alaska, with a regional synthesis for the northern St. Elias Mountains, Alaska and Yukon Territory." *Geol. Soc. Am. Bull.* 85:871–92.

Detterman, R. L., and J. K. Hartsock. 1966. "Geology of the Iniskin-Tuxedni region, Alaska." *U.S. Geol. Survey Prof. Paper* 512.

Detterman, R. L., T. Hudson, G. Plafker, R. G. Tysdal, and J. M. Hoare. 1976. "Reconnaissance geologic map along Bruin Bay and Lake Clark Faults in Kenai and Tyonek quadrangles, Alaska." *U.S. Geol. Survey Openfile Map* 76–477 (with explanation).

Detterman, R. L., T. P. Miller, M. E. Yount, and F. H. Wilson. 1981. "Quaternary geologic map of the Chignik and Sutwik Island Quadrangles, Alaska." *U.S. Geol. Survey Misc. Invest. Series Map* 1–1292.

Dibblee, T. W., Jr. 1967. "Areal geology, western Mojave Desert, California." *U.S. Geol. Survey Prof. Paper* 522.

Dickinson, W. R. 1979. "Cenozoic plate tectonic setting of the Cordilleral region in the United States." In *Cenozoic paleogeography of the western United States,* 1–13. Society of Economic Paleontologists and Mineralogists.

———. 1979. "Mesozoic forearm basin in central Oregon." *Geology* 7, no. 4:166–70 (Boulder, Colorado).

Diller, J. S. 1895. "Description of the Lassen Peak sheet, California." *U.S. Geol. Survey Geol. Atlas, Folio* 15.

———. 1902. "The geology of Crater Lake National Park." *U.S. Geol. Survey Prof. Paper* 3.

Dineen, M. P., ed. 1973. *101 Wonders of America.* Waukesha, Wisconsin: Country Beautiful Corp.

Dolan, Robert, Paul J. Godfrey, and William E. Odum. 1973. "Man's impact on the barrier islands of North Carolina." *American Scientist* 61, no. 2:152–62.

Dolan, R., and B. Hayden. 1974. "Adjusting to nature in our national seashores." *National Parks and Conservation Magazine* 48, no. 6:9–14.

Dolan, Robert, Bruce Hayden, and Harry Lins. 1980. "Barrier islands." *American Scientist* 68, no. 1:16–25.

Donnely, T. W. 1966. "Geology of St. Thomas and St. John, U.S. Virgin Islands." In *Caribbean geological investigations,* 85–176. Geol. Soc. America Memoir 98.

Dorf, Erling. 1960. "Tertiary fossil forests of Yellowstone National Park, Wyoming." In *Billings Geol. Society 11th Ann. Field Conf. Guidebook.*

Dott, R. H., Jr., and R. L. Batten, 1981. *Evolution of the Earth.* New York: McGraw-Hill Book Co.

Douglas, David. 1982. "UNESCO's World Heritage program." *National Parks and Conservation Magazine* 56, no. 11–12:4–8.

Douglas, W. O. 1961. *Muir of the Mountains.* Boston: Houghton Mifflin Co.

Drake, C. L., and J. C. Maxwell. 1981. "Geodynamics—Where we are and what lies ahead." *Science* 213, no. 4503:15–22.

Drake, E. T. 1982. "Tectonic evolution of the Oregon continental margin." *Oregon Geology* 44, no. 2:15–21.

Drewes, H. 1958. "Structural geology of Southern Snake Range, Nevada." *Geol. Soc. America Bull.* 69:221–40.

DuToit, A. L. 1937. *Our wandering continents: An hypothesis of continental drifting.* New York: Hafner Publ. Co., Inc.

Dutton, C. E. 1880. "Report on the geology of the High Plateau of Utah." U.S. Geog. and Geol. Surv. of Rocky Mtn. Region (Powell) 32.

Dyson, J. L. 1960. *The geologic story of Glacier National Park.* Glacier Natural History Assoc., Inc.

————. 1962. *Glaciers and glaciation in Glacier National Park.* Glacier Natural History Assoc., Inc.

Eardley, A. J., and J. W. Schaack. 1971. *Zion: the story behind the scenery.* Las Vegas, Nevada: KC Publications.

Eaton, Gordon P., R. L. Christiansen, H. M. Iyer, A. M. Pitt, H. R. Mabey, H. R. Blank, Jr., D. Zietz, and M. E. Gettings. 1975. "Magma beneath Yellowstone National Park." *Science* 188:787–96.

Editors of American West. 1973. *The Magnificent Rockies.* Palo Alto, California: American West Publ. Co.

Eichelberger, J. C. 1975. "Origin of andesite and dacite: Evidence of mixing at Glass Mountain in California and at other circum-Pacific volcanoes." *Geol. Soc. America Bull.* 86:1381–91.

Emery, K. O. 1955. "Submarine topography south of Hawaii." *Pacific Sci.* 9:286–91.

Emmons, S. F. 1879. "The volcanoes of the Pacific Coast of the United States." *Am. Geog. Soc. Jour.* 9:45–65.

Epis, R. C. 1968. "Cenozoic volcanism in the southern Rocky Mountains." *Quart. Colo. School of Mines* 63, no. 3.

Evernden, J. F., G. H. Curtis, and R. Kistler. 1957. "Potassium-argon dating of Pleistocene volcanics." *Quarternaria* 4:13–17.

Fenneman, N. M. 1931. *Physiography of Western United States.* New York: McGraw-Hill Book Co.

————. 1938. *Physiography of Eastern United States.* New York: McGraw-Hill Book Co.

Fernald, A. T. 1964. "Surficial geology of the central Kobuk Valley, northwestern Alaska." *U.S. Geol. Survey Bull.* 1181-K.

Feth, J. H., and Meyer Rubin. 1957. "Radiocarbon dating of wave-formed tufas from the Bonneville Basin." *Geol. Soc. America Bull.* 68:1827 (abs.).

Field, W. O. 1964. "Observations of glacier variations in Glacier Bay, Southeastern Alaska, 1958 and 1961, Prelim. Report." *Am. Geog. Soc.*

Finch, R. H. 1928. "Lassen Report No. 14." *The Volcano Letter,* no. 161:1.

————. 1930. "Activity of a California volcano in 1786." *The Volcano Letter,* no. 162:1 (reprinted in MacDonald, G. A., 1966, *Calif. Div. Mines & Geol. Bull.* 190:76).

Findley, Rowe. 1981. "Eruption of Mt. St. Helens." *National Geographic* 159, no. 1:2–65.

Fisher, John J. 1968. "Barrier island formation: discussion." *Geol. Soc. America Bull.* 69, no. 10:1421–25.

Fisher, Wm. A. 1960. *Yellowstone's Living Geology*. Yellowstone Library and Museum Assoc.

Fisk, H. N. 1959. "Padre Island and the Laguna Madre Flats, coastal south Texas." *In* Russell, R. J. (chair), *2nd Coastal Geography Conference* (April):103–151.

Fiske, R. S., C. A. Hopson, and A. C. Waters. 1963. "Geology of Mount Rainier National Park, Washington." U.S. Geol. Survey Prof. Paper 444.

Flawn, P. T. 1959. "The Ouachita structural belt." In *Geology of the Ouachita Mountains*. A Symposium, Dallas Geol. Soc.-Ardmore Geol. Soc., p. 20–29.

Fleisher, P. J. 1975. *Geology of selected national parks and monuments*. Dubuque, Iowa: Kendall/Hunt Publ. Co.

Flint, R. F. 1935. "Glacial features of the southern Okanogan region." *Geol. Soc. America Bull.* 46:169–94.

———. 1957. *Glacial and Pleistocene geology*. New York: John Wiley & Sons, Inc.

———. 1971. *Glacial and Quaternary geology*. New York: John Wiley & Sons, Inc.

———. 1973. *The Earth and its History*. New York: W. W. Norton & Co.

Ford, Corey. 1966. *Where the sea breaks its back*. Boston: Little, Brown and Company.

Ford, D. C., and R. O. Ewers. 1978. The development of limestone cave systems in the dimensions of length and depth. *Canadian Jour. of Earth Sci.* 15:1783–1798.

Ford, T. D., and W. J. Breed. 1969. "Preliminary report of the Chuar group, Grand Canyon, Arizona." In *Geology and natural history of the Grand Canyon region*. 1968 Field Conference, Four Corners Geol. Soc: 114–22.

Ford, T. D., and C. H. D. Cullingford. 1976. The *Science of Speleology*. London: Academic Press.

Four Corners Geological Society. 1969. *Geology and natural history of the Grand Canyon region*. Fifth Field Conf., Powell Centennial River Expedition.

Foxworthy, B. F., and M. Hill. 1982. "Volcanic eruptions of 1980 at Mt. St. Helens: The first 100 days." U.S. Geol. Surv. Prof. Paper 1249.

Freeman, O. W., J. D. Forrester, and R. L. Lupher. 1945. "Physiographic divisions of the Columbia Intermontane province." Assoc. Amer. Geog., *Annals,* 35, 53–75.

Frome, Michael. 1967. *National Park Guide (with map of Parks and Monuments)*. New York: Rand McNally.

Fryxell, F. M., L. Horberg, and R. Edmund. 1941. "Geomorphology of the Teton Range and adjacent basins, Wyoming-Idaho." *Geol. Soc. America Bull.* 52:1903 (abs.).

Garner, H. F. 1974. *The Origin of Landscapes*. New York: Oxford Univ. Press.

Gathright, T. M. 1976. "Geology of the Shenandoah National Park, Virginia." *Virginia Div. of Min. Resources, Bull.* 86, Charlottesville, Virginia.

Gilbert, G. K. 1877. "Report on the geology of the Henry Mountains." U.S. Geog. and Geol. Survey of Rocky Mountain Region.

———. 1890. "Lake Bonneville."U.S. Geol. Survey Mon. 1.

Gilbert, W. G. 1979. *A geologic guide to Mount McKinley National Park*. Anchorage, Alaska: Alaska Natural Hist. Assoc.

Gilluly, James. 1928. "Basin Range faulting along the Oquirrh Range, Utah." *Geol. Soc. America Bull.* 39:1103–30.

Gilluly, J., S. C. Waters, and A. O. Woodford. 1975. *Principles of Geology*. San Francisco: W. H. Freeman and Co.

Gleason, P. J. 1972. *Holocene sedimentation in the Everglades and Saline Tidal Lands*. AMQUA Guidebook to Everglades National Park.

Goldich, S. S. 1968. "Geochronology in the Lake Superior region." *Canadian Jour. of Earth Sci.*, no. 5:715–25.

Goldstein, M. 1977. *The Magnificent West.* Garden City, N.Y: Doubleday and Co.

Goldthwait, J. W. 1938. "The uncovering of New Hampshire by the last ice sheet." *Am. Jour. Sci.* 236:345–72.

Goldthwait, R. P. 1963. "Dating the little ice age in Glacier Bay, Alaska." Report of the International Geological Conference, 1960, Norden, Part XXVII, 37–46.

Goldthwait, R. P. et al. 1966. "Soil development and ecological succession in a deglaciated area of Muir Inlet (Glacier Bay), southeast Alaska." Inst. of Polar Studies, Report No. 20.

Graves, John. 1974. "Redwood National Park: controversy and compromise." *National Parks and Conservation Magazine* 48, no. 10:14–19.

Greeley, Ronald. 1977. "Basaltic 'plains' volcanism." In *Volcanism of the eastern Snake River Plain, Idaho: Guidebook,* R. Greeley and John S. King. National Aeronautics and Space Administration, Washington, D.C.

Greeley, R., and J. H. Hyde. 1972. "Lava tubes of the Cave Basalt, Mt. St. Helens, Washington." *Geol. Soc. Am. Bull.* 83:2397–98.

Gregory, H. E. 1917. "Geology of the Navajo Country." U.S. Geol. Survey Prof. Paper 93.

————. 1952. "Geology and geography of the Zion Park region, Utah and Arizona." U.S. Geol. Survey Prof. Paper 220.

————. 1956. *Geologic and geographic sketches of Zion and Bryce National Parks.* Zion-Bryce Natural History Assoc.

————. 1957. "Bryce Canyon National Park." (Explanatory material on the back of USGS Topographic Map of Bryce Canyon National Park.)

Gregory, H. E., and J. C. Anderson. 1939. "Geographic and geologic sketch of the Capital Reef Region, Utah." *Geol. Soc. America Bull.* 50:1827–50.

Gregory, H. E., and N. C. Williams. 1949. *Zion National Monument (Park), Utah.* Zion-Bryce Natural History Assoc. (Orig. publ. in *Geol. Soc. America Bull.* 58, 1947.)

Griggs, R. F. 1922. *Valley of Ten Thousand Smokes.* National Geographic Society.

Grybeck, D. 1977. "Geologic map of the Brooks Range, Alaska." U.S. Geol. Survey open-file rpt:77B–166B.

Gryc, G. 1958. "Brooks Range." In *Landscapes of Alaska,* edited by H. Williams, 111–18. Berkeley, Calif.: Univ. of Calif. Press.

Gutenberg, B., J. P. Buwalda, and R. P. Sharp. 1956. "Seismic explorations of the floor of Yosemite Valley, California." *Geol. Soc. America Bull.* 67, no. 8:1051–78.

Guthrie, R. D. 1972. "Recreating a vanished world." *National Geog.* 141, no. 3:294–301.

Hack, J. T. 1960. "Interpretation of erosional topography in humid temperate regions." *Am. Jour. Sci.* 258-A:80–97.

————. 1982. "Physiographic divisions and differential uplift in the Piedmont and Blue Ridge." U.S. Geological Survey Prof. Paper 1265.

Hadley, J. B., and R. Goldsmith. 1963. "Geology of the eastern Great Smoky Mountains, North Carolina and Tennessee." U.S. Geol. Survey Prof. Paper 349-B:B1–B118.

Hagood, Allen. 1971. *Dinosaur: The story behind the scenery.* Las Vegas, Nevada: KC Publications.

Hamblin, W. K. 1975. *The Earth's Dynamic Systems.* Minneapolis, Minn.: Burgess Publ. Co.

————. 1976. "Intracanyon flows." In *Geology of the Grand Canyon,* 2d ed. Museum of Northern Arizona and Grand Canyon Natural History Assoc.

Hamilton, W. 1960. *Late Cenozoic tectonics and volcanism of the Yellowstone region, Wyoming-Montana-Idaho.* Billings Geol. Soc. Guidebook of 11th Ann. Field Conf.: 92–105.

———. 1961. "Geology of the Richardson Cove and Jones Cove Quadrangles, Tennessee." U.S. Geol. Survey Prof. Paper 349A.

———. 1969. "Mesozoic California and the underflow of Pacific mantle." *Geol. Soc. Am. Bull.* 80:2409–30.

Hamilton, W., and W. B. Myers. 1966. "Cenozoic tectonics of the western United States." *Rev. Geophysics* 4:509–549.

Hamilton, W. L. 1981. *Guide to the Geology of Zion National Park.* Zion Natural History Assn.

Hammond, A. L. 1980. "The Yellowstone Bulge." 1 *Science* 3:68–73.

Hammond, P. E. 1979. "A tectonic model for evolution of the Cascade Range. Cenozoic paleogeography of the western United States." Society of Economic Paleontologists and Mineralogists: 219–38.

Hansen, W. R. 1964. "Curecanti Pluton—an unusual intrusive body in the Black Canyon of the Gunnison." *U.S. Geol. Survey Bull.* 1181–D.

———. 1965. "The Black Canyon of the Gunnison." *U.S. Geol. Survey Bull.* 1191.

———. 1969. "The geologic story of the Uinta Mountains." *U.S. Geol. Survey Bull.* 1291.

Harris, A. G., and E. Tuttle. 1983. *Geology of National Parks.* Dubuque, Iowa: Kendall/Hunt Publ. Co.

Harris, D. V. 1963. "Geomorphology of Larimer County, Colorado." In *Geology of the Northern Denver Basin Guidebook.* Rocky Mountain Assoc. of Geologists.

———. 1969. "Observations of Glacier Variations, Glacier Bay, Alaska, 1969." Open file report, National Park Service.

———. 1980. *The Geologic Story of the National Parks and Monuments,* 3d ed. New York: John Wiley & Sons.

Harris, Stephen L. 1980. *Fire and Ice, the Cascade Volcanoes.* Seattle: Pacific Search Press.

Harris, L. D., and K. C. Bayer. 1979. "Sequential development of the Appalachian orogen above a master decollement—a hypothesis." *Geology* 7:568–72.

Harrison, A. E. 1961. "Fluctuations of the Coleman Glacier, Mt. Baker, Wash." *Jour. Geophys. Res.* 66:649–50.

Harrison, W. D. 1975. "Yakutat: The turbulent crescent." *Alaska Geographic* 2, no. 4.

Hay, Edwards. 1974. "New Cave: a new look." *National Parks and Conservation Magazine* 48, no. 6:21–24.

Hayden, F. V. 1871. Preliminary Report of U.S. Geol. Survey of Wyoming and Contiguous Territories.

Hayes, P. T. 1964. "Geology of the Guadalupe Mountains, New Mexico." U.S. Geol. Survey Prof. Paper 446.

Haynes, D. D. 1964. *Geology of the Mammoth Cave quadrangle, Kentucky.* U.S. Geol. Survey, Map GQ-351.

Haynes, J. E. 1961. *A handbook of Yellowstone National Park (Hayne's Guide).* Bozeman, Mont.: Haynes Studios.

Hazen, R. M. 1974. "The founding of geology in America. 1771–1818." *Geol. Soc. America Bull.* 85:1827–33.

Heiken, G. 1981. "Holocene plinian tephra deposits of the Medicine Lake Highland, California." In "Guides to some volcanic terranes in Washington, Idaho, Oregon, and northern California," D. A. Johnston and J. Donnelly-Nolan. U.S. Geol. Survey Circular 838.

Herrick, C. L. 1900. "The geology of the White Sands of New Mexico." *Jour. Geol.* 8:112–28.

Hess, H. H., ed. 1966. "Caribbean geological investigations." Geol. Soc. America Memoir 98.

Hewett, D. F. 1931. "Geology and ore deposits of the Goodsprings quadrangle." U.S. Geol. Survey Prof. Paper 162.

Hill, C. A. 1981. "Origin of cave saltpeter." *Jour. Geology* 89:252–259.

Hill, M. L., and T. W. Dibblee, Jr. 1953. "San Andreas, Garlock, and Big Pine faults, California." *Geol. Soc. America Bull.* 64:443–58.

Hill, M. R. 1970. "Mt. Lassen is in eruption." Mineral Information Service, Calif. Div. Mines and Geol. 23, no. 11 (Nov. 1970):211.

Hillhouse, J. W. 1977. "Paleomagnetism of the Triassic Nikolai greenstone, McCarthy Quadrangle, Alaska." *Can. Jour. Earth Science* 14:2578–92.

Hinds, N. E. A. 1952. "Evolution of the California landscape." *Calif. Div. Mines Bull.* 158.

Hite, R. J. 1972. "Pennsylvanian rocks." In *Geologic Atlas of the Rocky Mountain Region,* 135–37. Denver, Colorado: Rocky Mountain Association of Geologists.

Hoffmeister, J. E. 1974. *The Geologic Story of South Florida.* Coral Gables, Florida: Univ. of Miami Press.

Holden, Constance. 1983. "Scientists describe 'nuclear winter.'" *Science* 222:822–23.

Hooper, R. Peter. 1982. "The Columbia River basalts." *Science* 215:1463–68.

Hopkins, D. M. 1967. *The Bering Land Bridge.* Stanford, Calif.: Stanford Univ. Press.

Hopson, C. A., A. C. Waters, V. A. Bender, and Meyer Rubin. 1962. "The latest eruptions from Mount Rainier volcano." *Jour. Geol.* 70:635–47.

Hough, J. L. 1958. *Geology of the Great Lakes.* Chicago: Univ. Illinois Press.

Howard, A. D. 1937. "History of the Grand Canyon of the Yellowstone." Geol. Soc. America Spec. Paper 6.

Howarth, Peter C. 1982. *Channel Islands, the story behind the scenery.* Las Vegas, Nevada: KC Publications.

Hoyt, John H. 1967. "Barrier island formation." *Geol. Soc. America Bull.* 78, no. 9:1125–35.

———. 1970. "Development and migration of barrier islands, northern Gulf of Mexico." *Geol. Soc. America Bull.* 81, no. 12:3779–82.

Huber, N. K. 1975. "The geologic story of Isle Royale National Park." *U.S. Geol. Survey Bull.* 1309.

———. 1977. *Devils Postpile National Monument.* (Ed. by W. Eckhardt) Sequoia Nat. Hist. Assn.

Huber, N. K., and C. D. Rinehart. 1965. "The Devils Postpile National Monument." Calif. Div. Mines and Geology Mineral Information Service: 18, no. 6:109–118.

Hubley, R. C. 1956. Glaciers of the Washington Cascades and Olympic Mountains; their activity and its relation to local climate trends. *Jour. Glac.* 2:669–74.

Hudson, T., and G. Plafker. 1980. "Emplacement age of the Crillon-La Perouse pluton, Fairweather Range, Alaska." U.S. Geol. Survey Circ. 823-B:90–94.

Hunt, C. B. 1956. "Cenozoic geology of the Colorado Plateau." U.S. Geol. Survey Prof. Paper 279.

———. 1975. *Death Valley: Geology, Ecology and Archeology.* Berkeley and Los Angeles, California: Univ. of Calif. Press.

Hunter, R. E., R. L. Watson, K. A. Dickinson, and E. W. Hill. 1981. *Padre Island National Seashore Field Guide.* Corpus Christi Geol. Society.

Hurlbut, C. S., Jr. 1959. *Dana's Manual of Mineralogy,* 17th ed. New York: John Wiley and Sons.

Hyde, J. H. 1973. *Late Quaternary volcano stratigraphy, south flank of Mount St. Helens,*

Washington (Ph.D. thesis, Univ. of Washington, Seattle.)

Hymans, L. 1978. "The flowing wilderness." *Not Man Apart,* mid-March issue, Friends of the Earth.

Irving, W. 1868. *The adventures of Captain Bonneville, U.S.A.* New York: Putnam and Son.

Jackson, D. D. 1982. *Underground Worlds.* Alexandria: Time-Life Books.

Jackson, E. 1970. *The Natural History Story of Chiricahua National Monument, Arizona.* Globe, Arizona: Southwest Parks and Monuments Assoc.

Jackson, V. L. 1978. *Discover Zion* (full color guide with maps). Zion Nat. Hist. Assn.

Jaggar, T.A., and E. Howe. 1900. "The laccoliths of the Black Hills." U.S. Geol. Survey, 21st Report, pt. 2:163–303.

Jeffery, D. 1975. "Preserving America's last great wilderness." *National Geographic* 147, no. 6:769–91.

Jenkins, H. O. 1958. *Glaciation in Big Bend National Park, Texas.* Sacramento, Calif.: Sacramento State College Foundation.

Jennings, J. N. 1971. *Karst.* Cambridge, Mass.: MIT Press.

Jenson, P. 1966. "Arches National Monument." In *America's Wonderlands: The National Parks.* Nat. Geog. Soc.

Johnson, Arthur. 1980. "Grinnell and Sperry glaciers, Glacier National Park, Montana—a record of vanishing ice." U.S. Geological Survey Professional Paper 1180.

Johnson, A. S. 1972. "McKinley: new freedom to enjoy." *National Parks and Conservation Magazine* 46, no. 12:19–26.

Johnson, Ross B. 1971. "The Great Sand Dunes of Southern Colorado." In *San Luis Basin Guidebook.* New Mexico Geol. Soc.

Jones, D. L., W. P. Irwin, and A. T. Ovenshine.

1972. "Southeastern Alaska—a displaced continental fragment?" U.S. Geol. Survey Prof. Paper 800-B:211–17.

Jones, D. L., and J. G. Moore. 1973. "Lower Jurassic Ammonite from the south-central Sierra Nevada, Calif." *Jour. of Research, U.S. Geol. Survey* 1, no. 4:453–58.

Jones, D. L., and N. J. Silberling. 1979. "Mesozoic stratigraphy—the key to tectonic analysis of southern and central Alaska." U.S. Geol. Survey Open-file Rept. 79-1200.

Jones, D. L., N. J. Silberling, and J. W. Hillhouse. 1977. "Wrangellia—A Displaced Terrane in Northwestern North America." *Can. Jour. Earth Science* 14:2565–77.

————. 1978. "Microplate tectonics of Alaska—significance for the Mesozoic history of the Pacific coast of North America." *In* D. G. Howell and K. A. McDougal (eds.), "Mesozoic Paleogeography of the Western United States, Pacific Coast Section." *Soc. Econ. Paleontologists and Minerologists, Pacific Coast Paleogeography Symposium,* vol. 2.

Jones, S. B. 1938. "Geomorphology of the Hawaiian Islands: A review." *Jour. Geomorph.* 1:55–61.

Karlstrom, T. N. V. 1964. "Quaternary geology of the Kenai Lowland and glacial history of the Cook Inlet region, Alaska." U.S. Geol. Survey Prof. Paper 443:1–69.

Kaufman, D. 1983. "Acid Rain costs U.S. $5 billion annually." *National Parks and Conservation Magazine* 57, no. 7:42.

Kaye, Glen. 1980. *Cape Cod, the story behind the scenery.* Las Vegas, Nevada: KC Publications.

Kaye, J. M. 1974. "Compositional sorting of topographically high Tennessee River gravels: A glacial hypothesis." *Geology* 2, no. 1:45–47.

Keefer, W. R. 1971. "The geologic story of Yellowstone National Park." *U.S. Geol. Survey Bull.* 1347.

Keith, A. 1896a. "Some stages of Appalachian erosion." *Geol. Soc. America Bull.* 7:519–25.

———. 1896b. "Crystalline groups of the southern Appalachians." *Science* 3:215–16.

———. 1927. "The Great Smoky overthrust." *Geol. Soc. America Bull.* 38:154–55 (abs.).

Kenney, N. T., and J. P. Blair. 1968. "The spectacular North Cascades." *Nat. Geog.* 133, no. 5:642–67.

Kerr, R. A. 1983. "Suspect Terranes and Continental Growth." *Science* 222:36–38.

———. 1983. "Volcanoes to keep an eye on." *Science* 221:634–35.

King, P. B. 1935. "Outline of structural development of Trans-Pecos, Texas." *Amer. Assoc. Petrol. Geol. Bull.* 19:221–61.

———. 1949. "The floor of the Shenandoah Valley." *Am. Jour. Sci.* 247:73–93.

———. 1959. *Evolution of North America.* Princeton, N.J.: Princeton Univ. Press.

———. 1964. "Geology of the Central Great Smoky Mountains, Tenn." U.S. Geol. Survey Prof. Paper 349C.

King, P. B., R. B. Neuman, and J. B. Hadley. 1968. "Geology of the Great Smoky Mountains National Park, Tennessee and North Carolina." U.S. Geol. Survey Prof. Paper 587.

King, P. B., and S. A. Schumm. 1980. "The Physical Geography (Geomorphology) of William Morris Davis." *Geo Abstracts,* Norwich, England.

King, P. B., and A. Stupka. 1950. "The Great Smoky Mountains. *Sci. Monthly*: 71:31–43.

———. 1961. "The great Smoky Mountains, their geology and natural history." Explanatory text with U.S. Geological Survey Great Smoky Mountains National Park and vicinity map (scale = 1 : 125,000).

Kittleman, L. R. 1979. "Tephra." *Sci. Amer.* 241, no. 6:160–177 (includes LANDSAT picture of Crater Lake Nat. Park).

Kiver, E. P. 1968. "Geomorphology and glacial geology of the southern Medicine Bow Mountains, Colorado and Wyoming." Ph.D. thesis, University of Wyoming.

———. 1978. "Mount Baker's changing fumaroles." *The Ore Bin* 40, no. 8:133–45.

———. 1982. "The Cascades volcanoes—comparison of geologic and historic records." *Mount St. Helens, one year later,* proceedings volume, Eastern Washington University: 3–12.

Kiver, E. P., and M. D. Mumma. 1971. "Summit Firn Caves, Mt. Rainier, Washington." *Science* 173:320–22.

Kiver, E. P., and W. K. Steele. 1975. "Firn caves in the volcanic craters of Mount Rainier, Washington." *National Speleological Society Bull.* 37:34–55.

Kiver, Eugene P., and Dale F. Stradling. 1982. "Quaternary geology of the Spokane area." Tobacco Root Geological Society 1980 field conference guidebook: 26–44.

Knappen, R. S. 1929. "Geology and mineral resources of the Aniakchak district, Alaska." *U.S. Geol. Survey Bull.* 797:161–223.

Knopf, Adolph. 1957. "The Boulder batholith of Montana." *Am. Jour. Sci.* 255:81–103.

LaChapelle, E. 1965. "The mass budget of Blue Glacier, Washington." *Jour. Glac.* 5:609–23.

Lanphere, M. A. 1978. "Displacement history of the Denali Fault system, Alaska and Canada." *Can. Jour. Earth Sci.* 15:817–22.

Lawrence, D. B. 1941. "The 'Floating Island' lava flow of Mt. St. Helens." *Mazama* 23, no. 12:55–60.

Lee, W. T. 1917. *The Geologic Story of Rocky Mountain National Park, Colo.* Nat. Park Service.

———. 1923. "Peneplains of the Front Range and Rocky Mountain National Park." *U.S. Geol. Survey Bull.* 730.

Leopold, Estella. 1974. "Florissant, a photograph in rock." *Nat. Park Service Newsletter* 9, no. 1:1–2.

Lessentine, R. H. 1969. "Kaiparowits and Black Mesa Basins: Stratigraphic Synthesis." In *Geol. and Nat. Hist. of the Grand Canyon Region.* Four Corners Geol. Soc.

Lidstrom, J. W. 1971. "A new model for the formation of Crater Lake caldera, Oregon." Ph.D. thesis, Oregon State University, Corvallis.

Lipman, P. W., and D. R. Mullineaux, eds. 1981. "The 1980 eruptions of Mt. St. Helens, Washington." U.S. Geol. Survey Prof. Paper 1250.

Lisle, Lorance Dix, and Robert Dolan. 1983. "Coastal erosion at Cape Hatteras: a lighthouse in danger." National Park Service Research/Resources Management Report SER-65.

Livesay, A. 1953. (Rev. by Preston McGrain. 1962.) *Geology of the Mammoth Cave National Park area.* Univ. Kentucky Spec. Publ. 7.

Lobeck, A. K. 1928. "The geology and physiography of the Mammoth Cave National Park." Ky. Geol. Survey Series 6, no. 21.

Lohman, S. W. 1963. "The geology and artesian water supply of the Grand Junction Area, Colorado." U.S. Geol. Survey Prof. Paper 451.

———. 1974. "The geologic story of Canyonlands National Park." *U.S. Geol. Survey Bull.* 1327.

———. 1975. "The geologic story of Arches National Park." *U.S. Geol. Survey Bull.* 1393.

———. 1981. "The Geologic Story of Colorado National Monument." *U.S. Geol. Survey Bull.* 1508.

Longwell, C. R. 1928. "Geology of the Muddy Mountains, Nevada, with a section through the Virgin Range to the Grand Wash Cliff, Arizona." *U.S. Geol. Survey Bull.* 798.

———. 1946. "How old is the Colorado River?" *Am. Jour. Sci.* 244, no. 12:817–35.

Lonsdale, J. T., R. A. Maxwell, J. A. Wilson, and R. T. Hazzard. 1955. *Geology of Big Bend National Park.* West Texas Geol. Soc. Guidebook, 1955 Spring Field Trip, Midland, Texas.

Look, Al. 1951. *In my back yard.* Denver, Colorado: Univ. of Denver Press.

Loomis, B. F. 1948. *Pictorial history of the Lassen Volcano.* San Francisco: Calif. Press.

———. 1966. *Eruptions of Lassen Peak,* 3rd ed. Mineral, Calif: Loomis Museum Assoc., Lassen Volcanic National Park.

Louderback, G. D. 1951. "Geologic history of San Francisco Bay." *Calif. Div. Mines Bull.* 154.

Love, J. D., and J. de la Montagne. 1956. *Pleistocene and Recent tilting of Jackson Hole, Teton County, Wyoming.* Wyo. Geol. Assoc. Guidebook, Field Conf. for 1956: 169–78.

Love, J. D., and J. C. Reed. 1971. *Creation of the Teton landscape.* Moose, Wyo: Grand Teton Natural History Assoc.

Lovering, T. S., and E. N. Goddard. 1950. "Geology and ore deposits of the Front Range, Colorado." U.S. Geol. Survey Prof. Paper 223.

Lukert, M. T. 1982. "Uranium-lead isotope age of the Old Rag Granite, northern Virginia." *Am. Jour. Sci.* 282:391–98.

Luyendyk, Bruce P., Marc J. Kamerling, and Richard Terres. 1980. "Geometric model for Neogene crustal rotations in southern California." *Geol. Soc. America Bulletin* 91:211–17.

Lyell, Chas. 1833. *Principles of Geology* (3 vol.). London: J. Murray.

————. 1873. *The geological evidences of the antiquity of man,* 4th ed. London: J. Murray.

Macdonald, G. A. 1966. "Geology of the Cascade Range and Modoc Plateau." In *Geology of northern California.* Calif. Div. Mines and Geology Bull. 190:65–96.

Macdonald, G. A., and A. T. Abbott. 1970. *Volcanoes in the sea.* Univ. of Hawaii Press.

Macdonald, G. A., and D. H. Hubbard. 1961. *Volcanoes of the National Parks in Hawaii.* Hawaii Nat. Hist. Assoc.

Macdonald, G. A., and T. Katsura. 1965. "Eruption of Lassen Peak, Cascade Range, California, in 1915 — Example of mixed magmas." *Geol. Soc. America Bull.* 76, no. 5:475–82.

MacGinitie, H. D. 1953. *Fossil plants of the Florissant Beds, Colo.* Carnegie Inst. of Wash., Publ. 599.

MacKevett, E. M., Jr. 1971. "Stratigraphy and general geology of the McCarthy C-5 Quadrangle, Alaska." *U.S. Geol. Survey Bull.* 1323.

Mackin, J. H. 1960. "Structural significance of Tertiary volcanic rocks in southwestern Utah." *Am. Jour. Sci.* 258:81–131.

Marshall, Robert. 1970. *Alaska Wilderness.* Berkeley, Calif.: Univ. of Calif. Press.

Martin, A. J. 1970. "Structure and tectonic history of the western Brooks Range, DeLong Mountains and Lisburne Hills, northern Alaska." *Geol. Soc. Am. Bull.* 81:3605–22.

Martin, G. C., B. L. Johnson, and U.S. Grant. 1915. "Geology and mineral resources of Kenai Peninsula, Alaska." *U.S. Geol. Survey Bull.* 587.

Mather, K. F., and S. A. Wengerd. 1962. "Pleistocene age of the 'Eocene' Ridgeway till, Colorado." Geol. Soc. America Spec. Paper 73 (abs.): 203.

Matthes, F. E. 1930. "Geologic history of the Yosemite Valley." U.S. Geol. Survey Prof. Paper 160.

————. 1937. "The geologic history of Mt. Whitney." *Sierra Club Bull.* 22.

————. 1942. "Glaciers." In *Hydrology,* edited by O. E. Meinzer. New York: Dover Publications, 149–210.

————. 1950a. *The Incomparable Valley, a geologic interpretation of the Yosemite.* Berkeley, Calif.: Univ. Calif. Press.

————. 1950b. *Sequoia National Park, A Geological Album* (F. Fryxell, ed.). Berkeley, Calif.: Univ. of Calif. Press.

————. 1956a. "Crater Lake." (Text with USGS map of Crater Lake National Park and Vicinity, Oregon.)

————. 1956b. Edited by F. Fryxell. *The Incomparable Valley: A geological interpretation of the Yosemite.* Berkeley, Calif.: Univ. Calif. Press.

————. 1965. "Glacial reconnaissance of Sequoia National Park, Calif." U.S. Geol. Survey Prof. Paper 504-A:A1-A58.

Matthews, S. W. 1973. "This changing earth." *National Geographic* (Jan.).

Matthews, Vincent. 1972. "Test of new global tectonics." *Amer. Assoc. Petrol. Geol. Bull.* 56, no. 2:371–74.

————. 1973. "Pinnacles-Neenach correlation: a restriction for models of the origin of the Transverse Ranges and the big bend in the San Andreas Fault." *Geol. Soc. America Bull.* 84:683–88.

————. 1976. "Correlation of Pinnacles and Neenach volcanic formations and their bearing on San Andreas Fault problem." *Amer. Assoc. Petrol. Geol. Bull.* 60, no. 12:2128–141.

Matthews, W. H., III. 1960. *Texas fossils: An amateur collector's handbook.* Univ. Texas, Bur. Econ. Geology Guidebook no. 2.

————. 1964. *National Parks*. New York: Barnes & Noble.

————. 1968. *A Guide to the National Parks: Their landscape and geology*. New York: The Natural History Press.

Maxson, J. H. 1950. "Lava flows in the Grand Canyon of the Colorado, Arizona." *Geol. Soc. America Bull.* 61:9–16.

Maxwell, J. C. 1974. "Anatomy of an orogen." *Geol. Soc. Amer. Bull.* 85:1195–1204.

Maxwell, R. A. 1968. *The Big Bend of the Rio Grande*. Bur. of Econ. Geol., Univ. of Texas Guidebook 7.

McBirney, A. R. 1978. "Volcanic evolution of the Cascade Range." *Ann. Rev. Planet Earth, Science* 6:437–456.

McDougall, I., and D. A. Swanson. 1972. "Potassium-argon ages of lavas from the Hawi and Pololu Volcanic Series, Kohala Volcano, Hawaii." *Geol. Soc. America Bull.* 83:3731–738.

McGee, W. J. 1897. "Sheetflood erosion." *Geol. Soc. America Bull.* 8:87–112.

McGowen, J. H., and J. L. Brewton. 1975. *Historical changes and related coastal processes, gulf and mainland shorelines, Matagoras Bay area, Texas*. Austin, Texas: Bureau of Economic Geol., Univ. of Texas.

McGrew, P. O., and M. Casilliano. 1975. "The geological history of Fossil Butte National Monument and Fossil Basin." National Park Service Occasional Paper no. 3.

McGuinness, C. L. 1963. "Role of ground water in the national water situation." U.S. Geol. Survey Water-Supply Paper 1800.

McKee, Bates. 1972. *Cascadia, the geologic evolution of the Pacific northwest*. New York: McGraw-Hill Book Co.

McKee, E. D. 1966. "Structures of dunes at White Sands National Monument, New Mexico." *Sedimentology* 7, no. 1.

McKee, E. D., J. R. Douglass, and G. Rittenhouse. 1971. "Growth and movement of dunes at White Sands National Monument." U.S. Geol. Survey Prof. Paper 750-D:D108-D114.

McKee E. D., W. K. Hamblin, and P. E. Damon. 1968. "K-Ar age of Lava Dam in Grand Canyon." *Geol. Soc. America Bull.* 79:133.

McKee, E. D., and E. H. McKee. 1972. "Pliocene uplift of the Grand Canyon region—time of drainage adjustment." *Geol. Soc. America Bull.* 83:1923–32.

McKee, E. D., R. F. Wilson, W. J. Breed, and C. S. Breed, eds. 1967. "Evolution of the Colorado River in Arizona." Museum Northern Arizona, Flagstaff, Bull. 44.

McLean, J. S. 1971. "The microclimate in Carlsbad Caverns, New Mexico." U.S. Geol. Survey Open-file Report (May).

McPhee, John. 1971. *Encounters with the Archdruid*. New York: Farrar, Straus and Giroux.

Means, T. H. 1932. *Death Valley*. Sierra Club Bull., 17:67–76.

Meier, M. 1965. "Glaciers and climate." In *INQUA VII Congress*: 745–805.

Melham, Tom. 1975. "Cape Cod's circle of seasons." *National Geog.* 148, no. 1:40–65.

Mendenhall, W. C. 1902. "Reconnaissance from Fort Hamlin to Kotzebue Sound, Alaska." U.S. Geol. Survey Prof. Paper 10.

Miller, D. J. 1960. "Giant waves in Lituya Bay, Alaska." U.S. Geol. Survey Prof. Paper 354-C.

Miller, M. 1967. "Alaska's mighty rivers of ice." *Nat. Geog.* 131, no. 2:194–217.

Miller, Russell. 1983. *Continents in Collision*. Alexandria, Virginia: Time-Life Books.

Miller, T. P., and R. L. Smith. 1975. "'New' volcanoes in the Aleutian volcanic arc." U.S. Geol. Survey Circ. 733:11.

————. 1977. "Spectacular mobility of ash

flows around Aniakchak and Fisher calderas, Alaska." *Geology* 5, no. 3:173–76.

Milling, M. E., and E. W. Behrens. 1966. "Sedimentary structures of beach and dune deposits, Mustang Island, Texas." Publications Institute of Marine Science, Texas, v. 11:135–48.

Molenaar, P., and P. Tapponnier. 1977. "The collision between India and Eurasia." *Sci. American* 236, no. 4:30–41.

Moore, J. G., R. L. Phillips, R. W. Grigg, D. W. Peterson, and D. A. Swanson. 1973. "Flow of lava into the sea, 1968–1971, Kilauea Volcano, Hawaii." *Geol. Soc. America Bull.* 84:537–46.

Moxham, R. M., D. R. Crandell, and W. E. Marlatt. 1965. "Thermal features at Mount Rainier, Washington, as revealed by infrared surveys." In *Geological Survey Research 1965*. U.S. Geol. Survey Prof. Paper 525-D:D93–D100.

Mudge, M. R. 1970. "Origin of the disturbed belt in northwestern Montana." *Geol. Soc. America Bull.* 81, no. 2:377–92.

Mudge, M. R., and R. L. Earhart. 1980. "The Lewis Thrust Fault and related structures in the disturbed belt, northwestern Montana." U.S. Geological Survey Professional Paper 1174.

Muir, John. 1912. *The Yosemite*. The Century Co. (Reprinted 1962, Doubleday and Co., Inc.)

————. 1915. *Travels in Alaska*. Boston: Houghton Mifflin Co.

————. 1916. *Our National Parks*. Boston: Houghton Mifflin Co.

Mull, C. G. 1977. "The Brooks Range." *Alaska Geographic* 4, no. 2:12–29.

————. 1977. "Apparent south vergent folding and possible nappes in the Schwatka Mountains, Alaska." U.S. Geol. Survey Circ. 751-B:29–30.

————. 1979. "Nanushuk Group deposition

and the Late Mesozoic structural evolution of the central and western Brooks Range and Arctic Slope." U.S. Geol. Survey Circ. 794:5–13.

————. 1981. "The Wrangell, St. Elias and Chugach Mountains." *Alaska Geographic* 8, no. 1:10–21.

Murie, Louise. 1978. "Denali wilderness." *National Parks and Conservation Magazine* 52, no. 1:3–9.

Murie, Margaret E. 1962. *Two in the Far North*. New York: Alfred A. Knopf, Inc.

Murphy, W. D. 1982. "Agate Fossil Beds." *National Parks Magazine* (July/August):15–17.

National Geographic, 1979. Special July issue: "Our National Parks" (includes a visitor's guide to all 320 park areas).

————. 1980. *The New America's Wonderlands*. Washington, D.C.: Nat. Geog. Soc.

National Parks and Conservation Association. *National Parks and Conservation Magazine.* 1701 18th St., N.W., Washington, D.C.

National Parks of New Zealand. 1965. R. E. Owen, Government Printer, Wellington, New Zealand.

Nelson, S. W., and D. Grybeck. 1978. "The Arrigetch Peaks and Mount Igikpak plutons, Survey Pass quadrangle, Alaska." U.S. Geol. Survey Circ. 772-B:7–9.

Neuman, Robert B., and Willis H. Helsen. 1965. "Geology of the western Great Smoky Mountains, Tenn." U.S. Geol. Survey Prof. Paper 349D.

Newell, N. D. 1953. *The Permian Reef Complex of the Guadalupe Mountains, Texas and New Mexico: A study in paleo-ecology*. San Francisco: W. H. Freeman and Co.

Oldale, Robert N. 1980. *Geologic history of Cape Cod, Massachusetts*. U.S. Geological Survey, Popular Publications Series.

Olson, V. J. 1972. *Capitol Reef, the story be-*

hind the scenery. Las Vegas, Nevada: KC Publications.

Orth, D. J. 1967. "Dictionary of Alaska Place Names." U.S. Geol. Survey Prof. Paper 567.

Ostrom, John H. 1978. "A new look at dinosaurs." *National Geographic* 154, no. 1:154–85.

Otvos, E. G., Jr. 1970. "Development and migration of barrier islands, northern Gulf of Mexico." *Geol. Soc. America Bull.* 81, no. 1:241–46.

Page, B. M. 1966. "Geology of the Coast Ranges of California." In *Geology of Northern California.* Cal. Div. of Mines and Geology Bull. 190, edited by E. H. Bailey.

Palmer, A. N. 1981. *Geological Guide to Mammoth Cave National Park.* Teaneck, N.J.: Zephyrus Press.

Pardee, J. T. 1942. "Unusual currents in Glacial Lake Missoula." *Geol. Soc. Am. Bull.* 53:1569–1600.

Parrish, Jay B., and Peter M. Lavin. 1982. "Tectonic model for kimberlite emplacement in the Appalachian Plateau of Pennsylvania." *Geology* 10:344–47.

———. 1983. "Reply to comment on 'Tectonic model for kimberlite emplacement in the Appalachian Plateau of Pennsylvania.'" *Geology* 11:235–54.

Parsons, W. H. 1973. "Kilauea volcano, 1972." *Ward's Bull.* 12, no. 90, Ward's Nat. Sci. Estab., Rochester, New York.

Patton, W. W., Jr., and I. L. Tailleur. 1964. "Geology of the Killik-Itkillik region, Alaska." U.S. Geol. Survey Prof. Paper 303-G:409.

Peattie, R. 1943. *The Great Smokies and the Blue Ridge.* New York: The Vanguard Press.

Peck, D. L. 1960. "Cenozoic volcanism in the Oregon Cascades." U.S. Geol. Survey Prof. Paper 400-B:B308–B310.

Perry, J. H. 1904. "Geol. of Monadnock Mountain, New Hampshire." *Jour. Geol.* 12:1–14.

Pewe, T. L. 1975. "Quaternary geology of Alaska." U.S. Geol. Survey Prof. Paper 835.

Pewe, T. L., D. M. Hopkins, and J. L. Giddings. 1965. "The Quaternary geology and archeology of Alaska." In *Quaternary of the U.S.:* 355–74.

Pewe, T. L., and R. G. Updike. 1970. *Guidebook to the geology of the San Francisco Peaks, Arizona.* XV Field Conference of the Rocky Mountain Section of the Friends of the Pleistocene.

Pierce, J. W., and D. J. Colquhoun. 1970. "Holocene evolution of a portion of the North Carolina coast." *Geol. Soc. America Bull.* 81:3697–3714.

Pierce, K. L. 1979. "History and Dynamics of Glaciation in the Northern Yellowstone Park Area." U.S. Geol. Survey Prof. Paper 729-F.

Plafker, G., and R. B. Campbell. 1979. "The Border Ranges Fault in the Saint Elias Mountains." U.S. Geol. Survey Circ. 804-B:102–4.

Plafker, G., R. L. Detterman, and T. Hudson. 1975. "New data on the displacement history of the Lake Clark Fault." In *Alaska Programs 1975,* edited by M. E. Yount. U.S. Geol. Survey Circ. 722:44–5.

Plafker, G., T. Hudson, T. Bruns, and M. Rubin. 1978. "Late Quaternary offsets along the Fairweather Fault and crustal plate interactions in southern Alaska." *Can. Jour. Earth Sci.* 15:805–16.

Pollack, J. B., O. B. Toon, T. P. Ackerman, and C. P. McKay. 1983. "Environmental effects of an impact-generated dust cloud; implications for the Cretaceous-Tertiary extinctions." *Science* 219:287–89.

Porter, Eliot. 1969. *Down the Colorado, illustrating the Diary of the first trip of John Wesley Powell through the Grand Canyon.* New York: E. P. Dutton and Co., Inc.

Porter S. C. 1964. "Late Pleistocene glacial chronology of the north-central Brooks Range, Alaska." *Am. Jour. Sci.* 262:446–60.

———. 1972. "Distribution, morphology, and size frequency of cinder cones on Mauna Kea Volcano." *Geol. Soc. America Bull.* 83:3607–12.

Post, A. 1969. "Distribution of surging glaciers in western North America." *Jour. Glaciol.* 8, no. 53:229–40.

Post, A., and E. R. LaChapelle. 1971. *Glacier Ice*. Seattle: Univ. of Washington Press.

Post, A., and M. F. Meier. 1979. "Glacier inventories, Alaska." U.S. Geol. Survey Circ. 804-B:7.

Post, A., D. Richardson, W. V. Tangborn, and F. L. Rosselot. 1971. "Inventory of glaciers in the North Cascades, Washington." U.S. Geol. Survey Prof. Paper 705-A.

Powell, J. W. 1873. *Exploration of the Colorado River of the west and its tributaries. Explored in 1869–1872*. Smithsonian Inst., p. 94–196.

———. 1875. "Exploration of the Colorado River of the west and its tributaries." Smithsonian Inst. Ann. Report 291.

Powell, W. J. 1958. "Groundwater resources of the San Luis Valley, Colorado." U.S. Geol. Survey Water-Supply Paper 1379:1–24.

Powers, H. A. 1958. "Alaska Peninsula-Aleutian Islands." In *Landscapes of Alaska,* edited by H. Williams. Berkeley, Calif.: Univ. Calif. Press: 61–75.

Price, R. A. 1971. "Gravitational sliding and the foreland thrust and fold belt of the North American Cordillera." *Geol. Soc. America Bull.* 82, no. 6:1133–38.

Prinz, M. 1970. "Idaho rift system, Snake River plain, Idaho." *Geol. Soc. America Bull.* 81:941–47.

Proceedings of the First Conference on Scientific Research in the National Parks, Vol. II, 1979. Includes the following:

Anderson, A. B., and C. V. Haynes, Jr. "How old is Capulin Mountain, New Mexico?"

Barnett, J. A., and R. Herrman. "Geology, parks and oil: energy with preservation."

Field, W. O. "Observations of glacier variations in Glacier Bay National Monument."

Goldthwait, R. P., and G. D. McKenzie. "Research by the Institute of Polar Studies in Glacier Bay, Alaska."

Hamilton, W. L. "Holocene and Pleistocene lakes in Zion National Park, Utah."

Lindquist R. C. "Genesis of erosional forms of Bryce Canyon National Park."

Prostka, H. J. 1968. "Facies of Eocene volcanics in northeastern Yellowstone National Park and their relations to eruptive centers." (Abs.) Geol. Soc. America, Rocky Mountain Section, Program of Meetings.

Quinlan, J. F., and R. O. Ewers. 1981. "Preliminary speculations on the evolution of groundwater basins in the Mammoth Cave Region, Kentucky." GSA Cincinnati '81 Field Trip Guidebooks, vol. 3, pp. 496–501. American Geological Institute, Washington, D.C.

Quinlan, J. F., R. O. Ewers, J. A. Ray, R. L. Powell, and N. C. Krothe. 1983. "Groundwater hydrology and geomorphology of the Mammoth Cave Region, Kentucky, and of the Mitchell Plain, Indiana." In *Field Trips in Midwestern Geology,* edited by R. H. Shaver and J. A. Sunderman. Bloomington, IN.: Geological Society of America and Indiana Geological Survey, v. 2:1–85.

Quinlan, J. F., and J. A. Ray. 1981. "Groundwater basins in the Mammoth Cave Region, Kentucky." Friends of the Karst, Occas. Pub. No. 1.

Rakestraw, L. 1965. *Historic mining on Isle Royale*. Isle Royale Historic Assoc.

Rau, W. W. 1980. "Washington coastal geology between the Hoh and Quillayute

Rivers." State of Wash. Div. of Geology and Earth Res. Bull. 72.

Raup, Omer B., Robert L. Earhart, James W. Whipple, and Paul E. Carrara. 1983. *Geology along Going-to-the Sun road, Glacier National Park, Montana*. Glacier Natural History Association.

Reed, B. L., and M. A. Lanphere. 1969. "Age and Chemistry of Mesozoic and Tertiary plutonic rocks in south-central Alaska." *Geol. Soc. America Bull.* 80:23–44.

———. 1973. "Alaska-Aleutian Range Batholith: Geochronology, chemistry and relation to circum-Pacific plutonism." *Geol. Soc. America Bull.* 84:2583–2610.

———. 1974. "Offset plutons and history of movement along the McKinley segment of the Denali fault system, Alaska." *Geol. Soc. America Bull.* 85:1883–93.

Reed, J. C. 1961. "Geology of the Mt. McKinley Quadrangle, Alaska." *U.S. Geol. Survey Bull.* 1108-A:1–36.

Reed, J. C., and R. E. Zartman. 1973. "Geochronology of Precambrian rocks of the Teton Range, Wyoming." *Geol. Soc. America Bull.* 84:561–82.

Reid, J. B. 1970. "Late Wisconsin and Neoglacial history of the Martin River Glacier, Alaska." *Geol. Soc. Am. Bull.* 81:3593–3604.

Retzer, J. L. 1954. "Glacial advances and soil development, Grand Mesa, Colorado." *Am. Jour. Sci.* 252:26–37.

Rich, J. L. 1938. "Recognition and significance of multiple erosion surface." *Geol. Soc. America Bull.* 49:1695–1722.

Richmond, G. M. 1957. "Three pre-Wisconsin glacial stages in the Rocky Mountain region." *Geol. Soc. America Bull.* 68:239–62.

———. 1960a. "Glaciation of the east slope of Rocky Mountain National Park, Colorado." *Geol. Soc. America Bull.* 71:1371–82.

———. 1960b. "Correlation of alpine and continental glacial deposits of Glacier National Park and adjacent High Plains." U.S. Geol. Survey Prof. Paper 400-B:B223–B224.

———. 1965. "Glaciation of the Rocky Mountains." In *Quaternary of United States:* 217–30.

———. 1969. *Development and stagnation of the last Pleistocene ice cap in the Yellowstone Lake Basin, Yellowstone National Park, U.S.A.* Eiszeitalter u. Gegenwart, Band 20: 196–203, Ohringen/Wurtt.

———. 1974. *Raising the roof of the Rockies, a geologic history of the mountains and of the ice age in Rocky Mountain National Park.* Rocky Mountain Nature Assoc.

Richmond, G. M., R. Fryxell, G. E. Neff, and P. L. Weis. 1965. "The Cordilleran Ice Sheet of the Northern Rocky Mountains, and related Quaternary history of the Columbia Plateau." In *Quaternary of the United States:* 231–42.

Richter, D. H., and D. L. Jones. 1973. "Structure and Stratigraphy of the Eastern Alaska Range, Alaska." Am. Assoc. of Petrol. Geol. Memoir 19:408–420.

Robinson, C. S. 1956. "Geology of Devils Tower, Wyoming." *U.S. Geol. Survey Bull.* 1021-I.

Rocky Mountain Association of Geologists. 1972. *Geologic atlas of the Rocky Mountain Region.*

Rodgers, John. 1953. "Geologic map of East Tennessee with explanatory text." *Tennessee Div. Geology Bull.* 58.

———. 1972. "Latest Precambrian Rocks of the Appalachian Region." *Am. Jour. Sci.* 272:507–520.

Roeder, D., and C. G. Mull. 1978. "Tectonics of Brooks Range ophiolites, Alaska." *Am. Assoc. Petrol. Geol Bull.* 62, no. 9:1696–1702.

Ross, C. P. 1959. "Geology of Glacier Park and

the Flathead region, northwestern Montana." U.S. Geol. Survey Prof. Paper 296.

Rowe, R. C. 1974. *Geology of our Western National Parks and Monuments*. Portland, Ore.: Binfords and Mort, Publishers.

Ruhle, G. C. 1957. *Guide to Glacier National Park*. Minneapolis, Minn: J. W. Forney, Foshay Tower.

Ruppel, E. T. 1972. "Geology of Pre-Tertiary rocks in the northern part of Yellowstone National Park, Wyoming." U.S. Geol. Survey Prof. Paper 729-A.

Russell, Dale A. 1982. "The mass extinctions of the late Mesozoic." *Scientific American* 246, no. 1:58 – 65.

Russell, F. 1975. *The Mountains of America*. New York, N.Y.: Bonanza Books, Crown Publ. Co.

Ryall, A., and Floriana Ryall. 1983. "Spasmodic tremor and possible magma ejection in Long Valley Caldera, eastern California." *Science* 219:1432 – 33.

Sackett, Russell. 1983. *Edge of the Sea*. Alexandria, Virginia: Time-Life Books.

Schenk, P. E. 1971. "Southeastern Atlantic Canada, northwestern Africa, and continental drift." *Can. Jour. Earth Sci.* 8:1218 – 51.

Schoch, Henry A. 1974. *Theodore Roosevelt, the story behind the scenery*. Las Vegas, Nevada: KC Publications.

Schulz, P. E. 1959. *Geology of Lassen's Landscape*. Mineral, Calif.: Loomis Museum Assoc., Lassen Volcanic Nat. Park.

———. 1966. *Road guide to Lassen Volcanic National Park*. Loomis Museum Assoc.

Schumm, S. A. 1956. "The role of creep and rainwash on the retreat of badland slopes." *Am. Jour. Sci.* 254:693 – 706.

Schumm, S. A., and R. J. Chorley. 1964. "The fall of Threatening Rock." *Am. Jour. of Sci.* 262:1041 – 54.

———. 1966. "Talus weathering and scarp recession in the Colorado Plateaus." *Geomorph. Ann.* 10:11 – 36.

Scopel, L. J. 1956. *The volcanic rocks of the Gros Ventre Buttes, Jackson Hole, Wyoming*. Wyo. Geol. Assoc. Guidebook for the 1956 Field Conf.: 126 – 28.

Scott, W. R. 1951. *Mammoth Cave National Park, Kentucky*. Mammoth Cave, Kentucky: National Park Concessions, Inc.

Sharp, R. P. 1951. "Glacial history of Wolf Creek, St. Elias Range, Canada." *Jour. Geol.* 59:97 – 115.

———. 1957. "Geomorphology of Cima Dome, Mohave Desert, California." *Geol. Soc. America Bull.* 68:273 – 90.

———. 1958. "Malaspina Glacier, Alaska." *Geol. Soc. of Am. Bull.* 69:617 – 46.

Sharp, R. P., and J. H. Birman. 1963. "Additions to classical sequence of Pleistocene glaciations, Sierra Nevada, California." *Geol. Soc. America Bull.* 74:635 – 59.

Shepherd, R. G., and S. A. Schumm. 1974. "Experimental study of river incision." *Geol. Soc. of Am. Bull.* 85:257 – 68.

Shimer, J. A. 1959. *This Sculptured Earth: Landscapes of America*. New York: Columbia Univ. Press.

Sims, P. K., and G. S. Morey. 1972. *Geology of Minnesota: A centennial volume*. Minnesota Geol. Survey.

Sinkankas, John. 1959. *Gemstones of North America*. New York: Van Nostrand Reinhold Co.

Smith, R. A. 1968. *The Frontier States, Alaska and Hawaii*. New York: Time-Life Books.

Smith, R. B., and Robert L. Christiansen. 1980. "Yellowstone Park as a window on the earth's interior." *Scientific American* 242, no. 2:104 – 117.

Smith, R. B., and M. L. Sbar. 1974. "Contemporary tectonics and seismicity of the western United States, with emphasis on the Intermountain Seismic Belt." *Geol. Soc. Amer. Bull.* 85:1205–13.

Smith, R. L., R. A. Bailey, and C. S. Ross. 1961. "Structural evolution of the Valles caldera, New Mexico." U.S. Geol. Survey Prof. Paper 424D:D145–D149.

Smith, W. D., and C. R. Swartzlow. 1936. "Mount Mazama; explosion versus collapse." *Geol. Soc. America Bull.*, v. 47:1809–30.

Sperry, O. E. 1938. "A check list of fern, gymnosperms, and flowering plants in the proposed Big Bend National Park of Texas." *Sul Ross State Teachers College Bull.* XIX, no. 4:9–98, Alpine, Texas.

Spring, R. and I. 1969. *The North Cascades National Park.* North Seattle, Wash.: Superior Publ. Co.

Spurr, J. E. 1900. "A reconnaissance in southwestern Alaska in 1894." U.S. Geol. Survey 20th Ann. Rept., pt. 7:31–264.

Stagner, Howard. 1952. *Behind the scenery of Mount Rainier National Park.* Longmire, Washington: Mount Rainier Nat. History Assoc.

Stanley, G. M. 1932. "Abandoned strands of Isle Royale and northeastern Lake Superior." Ph.D. thesis, Univ. of Michigan.

Starck, W. A., II. 1966. "Marvels of a Coral Reef." *National Geographic* 130, no. 5:710–738.

Stearns, H. T. 1945. "Glaciation of Mauna Kea, Hawaii." *Geol. Soc. America Bull.* 56:267–74.

———. 1959. *A Guide to the Craters of the Moon National Monument, Idaho.* (Reprinted from *Idaho Bur. of Mines and Geology Bull.,* v. 13, 1928.) Caldwell, Idaho: The Caxton Printers, Ltd.

———. 1966. *Geology of the State of Hawaii.* Palo Alto and Santa Clara, Calif.: Pacific Books.

Stokes, W. L. 1948. *Geology of the Utah-Colo-rado Salt Dome Region with emphasis on Gypsum Valley, Colorado.* Guidebook to the Geology of Utah, no. 3. Salt Lake City, Utah: Utah Geol. Soc.

Stokes, W. L. 1964. "Incised, wind-aligned stream patterns of the Colorado Plateau." *Am. Jour. Sci.* 262:808–16.

Stone, David. 1982. "Putting the pieces together." *Alaska Geographic* 9, no. 4:4–8.

Stout, J. H., J. B. Brady, F. Weber, and R. A. Page. 1973. "Evidence for Quaternary movement on the McKinley strand of the Denali fault in the Delta River area, Alaska." *Geol. Soc. America Bull.* 84:939–48.

Strahler, Arthur N. 1966. *A geologist's view of Cape Cod.* Garden City, New York: Natural History Press.

Sugden, D. E., and B. S. John. 1976. *Glaciers and Landscape.* New York: John Wiley & Sons.

Sunset Magazine Editors. 1965. *National Parks of the West.* Menlo Park, Calif.: Lane Magazine and Book Co.

Swanson, D. A. 1973. "Pahoehoe flows from the 1969–1971 Mauna Ulu eruption, Kilauea Volcano, Hawaii." *Geol. Soc. America Bull.* 84:615–626.

Swanson, D. A., and R. L. Christiansen. 1973. "Tragic base surge in 1790 at Kilauea Volcano." *Geology* 1, no. 2:83–96.

Swanson, D. A., T. L. Wright, P. R. Hooper, and R. D. Bentley. 1979. "Revisions in stratigraphic nomenclature of the Columbia River Basalt Group." U.S. Geol. Survey Bull. 1475-G.

Swift, Donald J. P. 1975. "Barrier-island genesis: evidence from the central Atlantic shelf, eastern U.S.A." *Sed. Geol.* 14:1–43.

Tabor, R. W. 1975. *Guide to the geology of Olympic National Park.* Seattle, Washington: Univ. of Washington Press.

Tailleur, I. L., and W. P. Brosge. 1970. Tectonic history of northern Alaska. In *Geological seminar on the North Slope of Alaska, Palo Alto, Calif., 1970: Pacific Section Meeting Proceedings.* Edited by W. L. Adkison and M. M. Brosge. Los Angeles, Calif., American Association of Petroleum Geologists: E1-E19.

Taylor, Steven R., and M. Jafi Toksoz. 1982. "Crust and upper-mantle velocity structure in the Appalachian orogenic belt: implications for tectonic evolution." *Geol. Soc. of America Bull.* 93:315–29.

Teale, E. W., ed. 1954. *The wilderness world of John Muir.* Boston: Houghton Mifflin Co.

Thayer, T. P. 1974. "The geologic setting of the John Day Country, Grant County, Oregon." U.S. Geological Survey Information Circular 69-10 (R, 4), 23

Thompson, G. A., and D. B. Burke. 1973. "Rate and direction of spreading in Dixie Valley, Basin and Range Province, Nevada." *Geol. Soc. America Bull.* 84:627–32.

Thornbury, W. D. 1965. *Regional Geomorphology of the United States.* New York: John Wiley & Sons.

Trabant, D. C. 1976. "Alaska glaciology studies." U.S. Geol. Survey Circ. 733:45–47.

Turner, D. L., T. E. Smith, and R. B. Forbes. 1974. "Geochronology of offset along the Denali fault system in Alaska." Geol. Soc. America, Cordilleran Section, abstract with programs, v. 6, no. 3:268–69.

Twenhofel, W. S., and C. L. Sainsbury. 1958. "Fault patterns in southeastern Alaska." *Geol. Soc. America Bull.* 69:1431–42.

Tyers, J. A. 1965. *The Significance of Wind Cave.* Hot Springs, South Dakota: Wind Cave Natural History Assoc.

Udall, S. L. 1966. *National Parks of America.* New York: Putnam and Sons.

———. 1971. *America's Natural Treasures: National Monuments and Seashores.* Waukesha, Wisconsin: Country Beautiful Corp.

Unger, D., and K. F. Mills. 1973. "Earthquakes near Mt. St. Helens, Washington." *U.S. Geol. Survey Bull.* 84:1065–68.

Untermann, G. E., and B. R. Untermann. 1969. *A popular guide to the geology of Dinosaur National Monument.* Dinosaur Nature Assn.

Van Tuyl, F. M., and T. S. Lovering. 1935. "Physiographic development of the Front Range, Colorado." *Geol. Soc. America Bull.* 46:1291–1350.

Veatch, F. M. 1969. "Analysis of a 24-year photographic record of Nisqually Glacier, Mount Rainier National Park, Washington." U.S. Geol. Survey Prof. Paper 631.

Vine, F. J. 1969. "Sea-floor spreading: New evidence." *Jour. Geol. Education* XVII, no. 1.

———. 1970. "Sea-floor spreading and continental drift." *Jour. Geol. Education* XVIII, no. 2.

Von Engeln, O. D. 1932. "The Ubehebe craters and explosion breccias in Death Valley." *Jour. Geol.* 40:726–34.

Wahlstrom, E. E. 1944. "Structure and petrology of Specimen Mountain, Colorado." *Geol. Soc. America Bull.* 55:77–90.

———. 1947. "Cenozoic physiographic history of the Front Range, Colorado." *Geol. Soc. America Bull.* 58:551–72.

Wahrhaftig, C. 1965. "Physiographic Divisions of Alaska." U.S. Geol. Survey Prof. Paper 482.

Wahrhaftig, C., and J. H. Birman. 1965. "The Quaternary of the Pacific Mountain system in California." In *The Quaternary of the United States,* edited by H. E. Wright, Jr. and D. G. Frey. A review volume for the VII Congress of the International Assoc. for Quaternary Research, Princeton Univ. Press, Princeton, New Jersey: 299–340.

Wahrhaftig, C., and A. Cox. 1959. "Rock glaciers in the Alaska Range." *Geol. Soc. America Bull.* 70:383–436.

Waitt, Richard B., Jr. 1980. "About forty last-Glacial Lake Missoula jokulhaups through southern Washington." *Journal of Geology* 88:653–79.

Walcott, C. D. 1895. "Algonkian rocks of the Grand Canyon of the Colorado." *Jour. Geol.* 3:312–30.

Wallace, R. 1973. *Hawaii.* New York: Time-Life Wilderness Series.

Waters, A. C. 1981. "Captain Jack's Stronghold (the geologic events that created a natural fortress)." U.S. Geol. Survey Circular 838.

Watson, P. J. (ed.) 1974. Archeology of the Mammoth Cave Area. New York: Academic Press.

Wayburn, E. 1982. "The Alaska Lands Act." *Sierra Club Bull.* 67, no. 5.

Wegemen, C. H. 1961. *A guide to the geology of Rocky Mountain National Park.* U.S. Govt. Printing Office, Washington, D.C.

Weise, B. R., and W. A. White. 1980. "Padre Island National Seashore: A Guide to the Geology, Natural Environment and History of a Texas Barrier Island." Bur. of Econ. Geol., Univ. of Texas, Austin.

Weiss, Paul L. 1976. "The channeled scablands of eastern Washington." U.S. Geological Survey: INN-72-2.

Wells, S. G., and D. J. Des Marais. 1973. "The Flint-Mammoth connection." *NSS News* 31, no. 2. National Speleological Society.

Wentworth, C. K. 1927. "Estimates of marine and fluvial erosion in Hawaii." *Jour. Geol.* 35:117–53.

———. 1936. "Geomorphic divisions of the Island of Hawaii." Univ. of Hawaii Occasional Paper No. 29, Honolulu.

———. 1938. "Marine bench-forming processes: Water-level weathering." *Jour. Geomorph.* 1:6–32.

Wentworth, C. K., and G. A. Macdonald.

1953. "Structures and forms of basaltic rocks in Hawaii." *U.S. Geol. Survey Bull.* 994.

White, J. R., and S. Pusateri. 1965. *Illustrated Guide—Sequoia and Kings Canyon National Parks.* Palo Alto, Calif.: Stanford Univ. Press.

Williams, C. 1980. *Mount St. Helens: A Changing Landscape.* Portland, Oregon: Graphic Arts Center Publ. Co.

Williams, Howel. 1932a. "Mount Shasta, a Cascade volcano." *Jour. Geol.* 40:417–29.

———. 1932b. "Geology of the Lassen Volcanic National Park, California." *Univ. Calif. Pub. Geol. Sci.* 21:195–386.

———. 1941. "Calderas and their origin." Univ. Calif. Pubs., Dept. Geol. Sci. Bull., 25:239–346.

———. 1942. *The geology of Crater Lake National Park, Oregon.* Publ. No. 540, Carnegie Inst. of Washington, Washington, D.C.

———. ed. 1958. *Landscapes of Alaska.* Berkeley, Calif.: Univ. of Calif. Press.

Williams, H., and G. Goles. 1968. "Volume of the Mazama ash-fall and the origin of Crater Lake caldera." In Andesite Conference Guidebook, Oregon Department of Geology and Mineral Industries Bulletin 62, edited by H. M. Dole: 37–41.

Williams, H., and A. McBirney. 1979. *Volcanology:* San Francisco: Freeman, Cooper & Co.

Williams, M. W. 1973. "Wild Wrangells: Jewels of Alaska." *National Parks Magazine* (July).

Willis, Bailey. 1903. "Physiography and deformation of the Wenatchee-Chelan district, Cascade Range." U.S. Geol. Survey Prof. Paper 19:49–97.

Wilson, J. T. 1963. "Continental drift." *Sci. Amer.* 208, no. 4:86–100.

Witkind, I. J. 1960. *The Hebgen Lake, Mon-*

tana, earthquake of August 17, 1959. Billings Geol. Soc. Guidebook, 11th Annual Field Conf.

Yeend, W. E. 1969."Quaternary geology of the Grand and Battlement Mesas area, Colorado." U.S. Geol. Survey Prof. Paper 617.

Yetman, R. P. 1974."The geology of Chiricahua National Monument." Unpublished report, National Park Service.

Ziegler, J. M. 1959. "Origin of the sea islands of the southeastern United States." *Geog. Rev.* XLIX, no. 2:222–37.

GLOSSARY-INDEX

Terms defined or explained in the text are shown in boldface page numbers; photographs and other illustrations are indicated by asterisks, except for color photographs on pages C-1, C-2, etc., in the color section.

Author - DAVID V. HARRIS

Dave Harris received degrees in geology at DePauw, Northwestern and Colorado Universities. His introduction to preservation and conservation began at Turkey Run State Park near the Indiana farm where he grew up. During his university years a summer field trip to Yellowstone kindled his interest in National Park geology.

For several years he was engaged in conservation work in New Mexico, Colorado and California. From 1946 until his retirement in 1975 he taught geology at Colorado State University; as department head, he directed the development of the new geology department.

His travels have taken him to almost all of the National Parks and to many of the National Monuments. Glaciers are a part-time obsession, and he has traveled to New Zealand, Alaska, Canada, Iceland, Norway and the Swiss Alps. He developed one of the first courses in "Geology of the National Parks" in the country. The earlier editions of this book are the culmination of that teaching experience.

Co-Author - EUGENE P. KIVER

An early fascination with limestone caverns in the Appalachian region led Eugene Kiver to a career in geology. After graduating from Case Western Reserve, he discovered the rugged beauty of the alpine regions and the glaciers in the Rocky Mountains while completing his doctorate at the University of Wyoming. He began his teaching career in 1968 at Eastern Washington University, where his interest in glaciers led him onto the slopes and eventually to the eruptive history of the Cascade Mountains.

He has undertaken numerous geological studies, including detailed investigations of the summit ice caves on Mt. Rainier and, most recently, the Ice Age history of Grand Coulee National Recreation Area. His long-time enthusiasm for the national parks and his incurable compulsion to photograph scenery led to his development in 1972 of a course on the geology of the national parks, in order to share with his students the remarkable geology of these national treasures. His contributions to this book are a continuation of all these interests.